LONDON MATHEMATICAL SOCIETY LECTURE NOTE SERIES

Managing Editor: Professor M. Reid, Mathematics Institute, University of Warwick, Coventry CV4 7AL, United Kingdom

The titles below are available from booksellers, or from Cambridge University Press at www.cambridge.org/mathematics

London Mathematical Society Lecture Note Series: 376

Permutation Patterns

St Andrews 2007

Edited by

STEVE LINTON
University of St Andrews

NIK RUŠKUC
University of St Andrews

VINCENT VATTER
Dartmouth College

CAMBRIDGE
UNIVERSITY PRESS

CAMBRIDGE
UNIVERSITY PRESS

University Printing House, Cambridge CB2 8BS, United Kingdom

One Liberty Plaza, 20th Floor, New York, NY 10006, USA

477 Williamstown Road, Port Melbourne, VIC 3207, Australia

314-321, 3rd Floor, Plot 3, Splendor Forum, Jasola District Centre, New Delhi - 110025, India

103 Penang Road, #05-06/07, Visioncrest Commercial, Singapore 238467

Cambridge University Press is part of the University of Cambridge.

It furthers the University's mission by disseminating knowledge in the pursuit of education, learning and research at the highest international levels of excellence.

www.cambridge.org
Information on this title: www.cambridge.org/9780521728348

First published 2010

A catalogue record for this publication is available from the British Library

ISBN 978-0-521-72834-8 Paperback

Contents

Preface

The Permutation Patterns 2007 conference was held 11–15 June 2007 at the University of St Andrews. This was the fifth Permutation Patterns conference; the previous conferences were held at Otago University (Dundein, New Zealand), Malaspina College (Vancouver Island, British Columbia), the University of Florida (Gainesville, Florida), and Reykjavík University (Reykjavík, Iceland). The organizing committee was comprised of Miklós Bóna, Lynn Hynd, Steve Linton, Nik Ruškuc, Einar Steingrímsson, Vincent Vatter, and Julian West. A half-day excursion was taken to Falls of Bruar, Blair Athol Castle and The Queen's View on Loch Tummel.

There were two invited talks:

- Mike Atkinson (Otago University, Dunedin, New Zealand), "Simple permutations and wreath-closed pattern classes".
- Martin Klazar (Charles University, Prague, Czech Republic), "Polynomial counting".

There were 35 participants, 23 talks, and a problem session (the problems from which are included at the end of these proceedings). All the main strands of research in permutation patterns were represented, and we hope this is reflected by the articles of these proceedings, especially the eight surveys at the beginning. The conference was supported by the EPSRC and Edinburgh Mathematical Society.

Some general results in combinatorial enumeration

Martin Klazar

Department of Applied Mathematics
Charles University
118 00 Praha Czech Republic

Abstract

This survey article is devoted to general results in combinatorial enumeration. The first part surveys results on growth of hereditary properties of combinatorial structures. These include permutations, ordered and unordered graphs and hypergraphs, relational structures, and others. The second part advertises four topics in general enumeration: 1. counting lattice points in lattice polytopes, 2. growth of context-free languages, 3. holonomicity (i.e., P-recursiveness) of numbers of labeled regular graphs and 4. ultimate modular periodicity of numbers of MSOL-definable structures.

1 Introduction

We survey some general results in combinatorial enumeration. A *problem* in enumeration is (associated with) an infinite sequence $P = (S_1, S_2, \dots)$ of finite sets S_i. Its *counting function* f_P is given by $f_P(n) = |S_n|$, the cardinality of the set S_n. We are interested in results of the following kind on *general* classes of problems and their counting functions.

Scheme of general results in combinatorial enumeration. *The counting function f_P of every problem P in the class $\mathcal{C}$ belongs to the class of functions $\mathcal{F}$. Formally, $\{f_P \mid P \in \mathcal{C}\} \subset \mathcal{F}$.*

The larger $\mathcal{C}$ is, and the more specific the functions in $\mathcal{F}$ are, the stronger the result. The present overview is a collection of many examples of this scheme.

3

One can distinguish general results of two types. In *exact results*, $\mathcal{F}$ is a class of explicitly defined functions, for example polynomials or functions defined by recurrence relations of certain type or functions computable in polynomial time. In *asymptotic results*, $\mathcal{F}$ consists of functions defined by asymptotic equivalences or asymptotic inequalities, for example functions growing at most exponentially or functions asymptotic to $n^{(1-1/k)n+o(n)}$ as $n \to \infty$, with the constant $k \geq 2$ being an integer.

The sets S_n in P usually constitute sections of a fixed infinite set. Generally speaking, we take an infinite universe U of combinatorial structures and introduce problems and classes of problems as subsets of U and families of subsets of U, by means of *size functions* $s : U \to \mathbb{N}_0 = \{0, 1, 2, \dots\}$ and/or (mostly binary) relations between structures in U. More specifically, we will mention many results falling within the framework of growth of downsets in partially order sets, or posets.

Downsets in posets of combinatorial structures. We consider a nonstrict partial ordering $(U, \prec)$, where $\prec$ is a containment or a substructure relation on a set U of combinatorial structures, and a size function $s : U \to \mathbb{N}_0$. Problems P are *downsets* in $(U, \prec)$, meaning that $P \subset U$ and $A \prec B \in P$ implies $A \in P$, and the counting function of P is

$$f_P(n) = \#\{A \in P \mid s(A) = n\}.$$

(More formally, the problem is the sequence of sections $(P \cap U_1, P \cap U_2, \dots)$ where $U_n = \{A \in U \mid s(A) = n\}$.) Downsets are exactly the sets of the form

$$\mathrm{Av}(F) := \{A \in U \mid A \not\succ B \text{ for every } B \text{ in } F\}, \ F \subset U.$$

There is a one-to-one correspondence $P \mapsto F = \min(U \backslash P)$ and $F \mapsto P = \mathrm{Av}(F)$ between the family of downsets P and the family of *antichains* F, which are sets of mutually incomparable structures under $\prec$. We call the antichain $F = \min(U \backslash P)$ corresponding to a downset P the *base of* P.

We illustrate the scheme by three examples, all for downsets in posets.

1.1 Three examples

Example 1. Downsets of partitions. Here U is the family of partitions of $[n] = \{1, 2, \dots, n\}$ for n ranging in $\mathbb{N}$, so U consists of finite sets $S = \{B_1, B_2, \dots, B_k\}$ of disjoint and nonempty finite subsets B_i of

N, called *blocks*, whose union $B_1 \cup B_2 \cup \cdots \cup B_k = [n]$ for some n in
N. Two natural size functions on U are order and size, where the *order*,
$\|S\|$, of S is the cardinality, n, of the underlying set and the *size*, $|S|$, of
S is the number, k, of blocks. The formula for the number of partitions
of $[n]$ with k blocks

$$S(n,k) := \#\{S \in U \mid \|S\| = n, |S| = k\} = \sum_{i=0}^{k} \frac{(-1)^i (k-i)^n}{i!(k-i)!}$$

is a classical result (see [111]); $S(n,k)$ are called *Stirling numbers*. It is
already a simple example of the above scheme but we shall go further.

For fixed k, the function $S(n,k)$ is a linear combination with rational
coefficients of the exponentials $1^n, 2^n, \ldots, k^n$. So is the sum $S(n,1) +
S(n,2) + \cdots + S(n,k)$ counting partitions with order n and size at most k.
We denote the set of such partitions $\{S \in U \mid |S| \le k\}$ as $U_{\le k}$. Consider
the poset $(U, \prec)$ with $S \prec T$ meaning that there is an *increasing* injection
$f : \bigcup S \to \bigcup T$ such that every two elements x, y in $\bigcup S$ lie in the same
block of S if and only if $f(x), f(y)$ lie in the same block of T. In other
words, $S \prec T$ means that $\bigcup T$ has a subset X of size $\|S\|$ such that
T induces on X a partition order-isomorphic to S. Note that $U_{\le k}$ is
a downset in $(U, \prec)$. We know that the counting function of $U_{\le k}$ with
respect to order n equals $a_1 1^n + \cdots + a_k k^n$ with a_i in $\mathcal{Q}$. What are
the counting functions of other downsets? If the size is bounded, as
for $U_{\le k}$, they have similar form as shown in the next theorem, proved
by Klazar [77]. It is our first example of an exact general enumerative
result.

Theorem 1.1 (Klazar). *If P is a downset in the poset of partitions
such that $\max_{S \in P} |S| = k$, then there exist a natural number n_0 and
polynomials $p_1(x), p_2(x), \ldots, p_k(x)$ with rational coefficients such that
for every $n > n_0$,*

$$f_P(n) = \#\{S \in P \mid \|S\| = n\} = p_1(n)1^n + p_2(n)2^n + \cdots + p_k(n)k^n.$$

If $\max_{S \in P} |S| = +\infty$, the situation is much more intricate and we are far
from having a complete description but the growths of $f_P(n)$ below 2^{n-1}
have been determined (see Theorem 2.17 and the following comments).
We briefly mention three subexamples of downsets with unbounded size,
none of which has $f_P(n)$ in the form of Theorem 1.1. If P consists of all
partitions of $[n]$ into intervals of length at most 2, then $f_P(n) = F_n$, the
n^{th} Fibonacci number, and so $f_P(n) = b_1 \alpha^n + b_2 \beta^n$ where $\alpha = \frac{\sqrt{5}-1}{2}$,

$\beta = \frac{\sqrt{5}+1}{2}$ and $b_1 = \frac{\alpha}{\sqrt{5}}, b_2 = \frac{\beta}{\sqrt{5}}$. If P is given as $P = \text{Av}(\{C\})$ where $C = \{\{1,3\}, \{2,4\}\}$ (the partitions in P are so called *noncrossing partition*, see the survey of Simion [106]) then $f_P(n) = \frac{1}{n+1}\binom{2n}{n}$, the nth Catalan number which is asymptotically $cn^{-3/2}4^n$. Finally, if $P = U$, so P consists of all partitions, then $f_P(n) = B_n$, the nth Bell number which grows superexponentially.

Example 2. Hereditary graph properties. Here U is the universe of finite simple graphs $G = ([n], E)$ with vertex sets $[n]$, n ranging over $\mathbb{N}$, and $\prec$ is the induced subgraph relation; $G_1 = ([n_1], E_1) \prec G_2 = ([n_2], E_2)$ means that there is an injection from $[n_1]$ to $[n_2]$ (not necessarily increasing) that sends edges to edges and nonedges to nonedges. The size, $|G|$, of a graph G is the number of vertices. Problems are downsets in $(U, \prec)$ and are called *hereditary graph properties*. The next theorem, proved by Balogh, Bollobás and Weinreich [18], describes counting functions of hereditary graph properties that grow no faster than exponentially.

Theorem 1.2 (Balogh, Bollobás and Weinreich). *If P is a hereditary graph property such that for some constant $c > 1$, $f_P(n) = \#\{G \in P \mid |G| = n\} < c^n$ for every n in $\mathbb{N}$, then there exists a natural numbers k and n_0 and polynomials $p_1(x), p_2(x), \ldots, p_k(x)$ with rational coefficients such that for every $n > n_0$,*

$$f_P(n) = p_1(n)1^n + p_2(n)2^n + \cdots + p_k(n)k^n.$$

The case of superexponential growth of $f_P(n)$ is discussed below in Theorem 2.11.

In both examples we have the same class of functions $\mathcal{F}$, linear combinations $p_1(n)1^n + p_2(n)2^n + \cdots + p_k(n)k^n$ with $p_i \in \mathcal{Q}[x]$. It would be nice to find a common extension of Theorems 1.1 and 1.2. It would be also of interest to determine if the two classes of functions realizable as counting functions in both theorems coincide and how they differ from $\mathcal{Q}[x, 2^x, 3^x, \ldots]$.

Example 3. Downsets of words. Here U is the set of finite words over a finite alphabet A, so $U = \{u = a_1 a_2 \ldots a_k \mid a_i \in A\}$. The size, $|u|$, of such a word is its length k. The subword relation (also called the *factor relation*) $u = a_1 a_2 \ldots a_k \prec v = b_1 b_2 \ldots b_l$ means that $b_{i+1} = a_1, b_{i+2} = a_2, \ldots, b_{i+k} = a_k$ for some i. We associate with an infinite word $v = b_1 b_2 \ldots$ over A the set $P = P_v$ of all its finite subwords, thus $P_v = \{b_r b_{r+1} \ldots b_s \mid 1 \le r \le s\}$. Note that P_v is a downset in

$(U, \prec)$. The next theorem was proved by Morse and Hedlund [92], see also Allouche and Shallit [7, Theorem 10.2.6].

Theorem 1.3 (Morse and Hedlund). *Let P be the set of all finite subwords of an infinite word v over a finite alphabet A. Then $f_P(n) = \#\{u \in P \mid |u| = n\}$ is either larger than n for every n in $\mathbb{N}$ or is eventually constant. In the latter case the word v is eventually periodic.*

The case when P is a general downset in $(U, \prec)$, not necessarily coming from an infinite word (cf. Subsection 2.4), is discussed below in Theorem 2.19.

Examples 1 and 2 are exact results and example 3 combines a tight form of an asymptotic inequality with an exact result. Examples 1 and 2 involve only countably many counting functions $f_P(n)$ and, as follows from the proofs, even only countably many downsets P. In example 3 we have uncountably many distinct counting functions. To see this, take $A = \{0, 1\}$ and consider infinite words v of the form $v = 10^{n_1} 10^{n_2} 10^{n_3} 1 \ldots$ where $1 \leq n_1 < n_2 < n_2 < \ldots$ is a sequence of integers and $0^m = 00 \ldots 0$ with m zeros. It follows that for distinct words v the counting functions f_{P_v} are distinct; Proposition 2.1 presents similar arguments in more general settings.

1.2 Content of the overview

The previous three examples illuminated to some extent general enumerative results we are interested in but they are not fully representative because we shall cover a larger area than the growth of downsets. We do not attempt to set forth any more formalized definition of a general enumerative result than the initial scheme but in Subsections 2.4 and 3.4 we will discuss some general approaches of finite model theory based on relational structures. Not every result or problem mentioned here fits naturally the scheme; Proposition 2.1 and Theorem 2.6 are rather results to the effect that $\{f_P \mid P \in \mathcal{C}\}$ is too big to be contained in a small class $\mathcal{F}$. This collection of general enumerative results is naturally limited by the author's research area and his taste but we do hope that it will be of interest to others and that it will inspire a quest for further generalizations, strengthenings, refinements, common links, unifications etc.

For the lack of space, time and expertise we do not mention results on growth in algebraic structures, especially the continent of growth in groups; we refer the reader for information to de la Harpe [52] (and also

to Cameron [43]). Also, this is not a survey on the class of problems
#P in computational complexity theory (see Papadimitriou [94, Chap-
ter 18]). There are other areas of general enumeration not mentioned
properly here, for example 0-1 laws (see Burris [42] and Spencer [109]).

In the next subsection we review some notions and definitions from
combinatorial enumeration, in particular we recall the notion of Wilfian
formula (polynomial-time counting algorithm). In Section 2 we review
results on growth of downsets in posets of combinatorial structures. Subsection 2.1 is devoted to pattern avoiding permutations, Subsections 2.2
and 2.3 to graphs and related structures, and Subsection 2.4 to relational
structures. Most of the results in Subsections 2.2 and 2.3 were found by
Balogh and Bollobás and their coauthors [11, 13, 12, 15, 14, 16, 17, 18,
19, 20, 21]. We recommend the comprehensive survey of Bollobás [30]
on this topic. In Section 3 we advertise four topics in general enumeration together with some related results. 1. The Ehrhart–Macdonald
theorem on numbers of lattice points in lattice polytopes. 2. Growth
of context-free languages. 3. The theorem of Gessel on numbers of labeled regular graphs. 4. The Specker–Blatter theorem on numbers of
MSOL-definable structures.

1.3 Notation and some specific counting functions

As above, we write $\mathbb{N}$ for the set $\{1, 2, 3, \dots\}$, $\mathbb{N}_0$ for $\{0, 1, 2, \dots\}$ and $[n]$
for $\{1, 2, \dots, n\}$. We use $\#X$ and $|X|$ to denote the cardinality of a set.
By the phrase "for every n" we mean "for every n in $\mathbb{N}$" and by "for large
n" we mean "for every n in $\mathbb{N}$ with possibly finitely many exceptions".
Asymptotic relations are always based on $n \to \infty$. The *growth constant*
$c = c(P)$ of a problem P is $c = \limsup f_P(n)^{1/n}$; the reciprocal $1/c$ is
then the radius of convergence of the power series $\sum_{n \geq 0} f_P(n) x^n$.

We review several counting sequences appearing in the mentioned results. *Fibonacci numbers* $(F_n) = (1, 2, 3, 5, 8, 13, \dots)$ are given by the
recurrence $F_0 = F_1 = 1$ and $F_n = F_{n-1} + F_{n-2}$ for $n \geq 2$. They
are a particular case $F_n = F_{n,2}$ of the *generalized Fibonacci numbers*
$F_{n,k}$, given by the recurrence $F_{n,k} = 0$ for $n < 0$, $F_{0,k} = 1$ and
$F_{n,k} = F_{n-1,k} + F_{n-2,k} + \cdots + F_{n-k,k}$ for $n > 0$. Using the notation $[x^n]G(x)$ for the coefficient of x^n in the power series expansion of
the expression $G(x)$, we have

$$F_{n,k} = [x^n] \frac{1}{1 - x - x^2 - \cdots - x^k}.$$

Standard methods provide asymptotic relations $F_{n,2} \sim c_2(1.618\ldots)^n$, $F_{n,3} \sim c_3(1.839\ldots)^n$, $F_{n,4} \sim c_4(1.927\ldots)^n$ and generally $F_{n,k} \sim c_k\alpha_k^n$ for constants $c_k > 0$ and $1 < \alpha_k < 2$; $1/\alpha_k$ is the least positive root of the denominator $1 - x - x^2 - \cdots - x^k$ and $\alpha_2, \alpha_3, \ldots$ monotonically increase to 2. The unlabeled exponential growth of tournaments (Theorem 2.21) is governed by the *quasi-Fibonacci numbers* F_n^* defined by the recurrence $F_0^* = F_1^* = F_2^* = 1$ and $F_n^* = F_{n-1}^* + F_{n-3}^*$ for $n \geq 3$; so

$$F_n^* = [x^n]\frac{1}{1 - x - x^3}$$

and $F_n^* \sim c(1.466\ldots)^n$.

We introduced *Stirling numbers* $S(n, k)$ in Example 1. The *Bell numbers* $B_n = \sum_{k=1}^{n} S(n, k)$ count all partitions of an n-elements set and follow the recurrence $B_0 = 1$ and $B_n = \sum_{k=0}^{n-1} \binom{n-1}{k} B_k$ for $n \geq 1$. Equivalently,

$$B_n = [x^n]\sum_{k=0}^{\infty} \frac{x^k}{(1 - x)(1 - 2x)\ldots(1 - kx)}.$$

The asymptotic form of the Bell numbers is

$$B_n = n^{n(1 - \log\log n/\log n + O(1/\log n))}.$$

The numbers p_n of *integer partitions* of n count the ways to express n as a sum of possibly repeated summands from $\mathbb{N}$, with the order of summands being irrelevant. Equivalently,

$$p_n = [x^n]\prod_{k=1}^{\infty} \frac{1}{1 - x^k}.$$

The asymptotic form of p_n is $p_n \sim cn^{-1}\exp(d\sqrt{n})$ for some constants $c, d > 0$. See Andrews [8] for more information on these asymptotics and for recurrences satisfied by p_n.

A sequence $f : \mathbb{N} \to \mathbb{C}$ is a *quasipolynomial* if for every n we have $f(n) = a_k(n)n^k + \cdots + a_1(n)n + a_0(n)$ where $a_i : \mathbb{N} \to \mathbb{C}$ are periodic functions. Equivalently,

$$f(n) = [x^n]\frac{p(x)}{(1 - x)(1 - x^2)\ldots(1 - x^l)}$$

for some l in $\mathbb{N}$ and a polynomial $p \in \mathbb{C}[x]$. We say that the sequence f is *holonomic* (other terms are *P-recursive* and *D-finite*) if it satisfies for every n (equivalently, for large n) a recurrence

$$p_k(n)f(n + k) + p_{k-1}(n)f(n + k - 1) + \cdots + p_0(n)f(n) = 0$$

with polynomial coefficients $p_i \in \mathbb{C}[x]$, not all zero. Equivalently, the power series $\sum_{n \geq 0} f(n)x^n$ satisfies a linear differential equation with polynomial coefficients. Holonomic sequences generalize sequences satisfying linear recurrences with constant coefficients. The sequences $S(n,k)$, $F_{n,k}$, and F_n^* for each fixed k satisfy a linear recurrence with constant coefficients and are holonomic. The sequences of Catalan numbers $\frac{1}{n+1}\binom{2n}{n}$ and of factorial numbers $n!$ are holonomic as well. The sequences B_n and p_n are not holonomic [112]. It is not hard to show that if (a_n) is holonomic and every a_n is in $\mathcal{Q}$, then the polynomials $p_i(x)$ in the recurrence can be taken with integer coefficients. In particular, there are only countably many holonomic rational sequences.

Recall that a power series $F = \sum_{n \geq 0} a_n x^n$ with a_n in $\mathbb{C}$ is *algebraic* if there exists a nonzero polynomial $Q(x,y)$ in $\mathbb{C}[x,y]$ such that $Q(x, F(x)) = 0$. F is *rational* if Q has degree 1 in y, that is, $F(x) = R(x)/S(x)$ for two polynomials in $\mathbb{C}[x]$ where $S(0) \neq 0$. It is well known (Comtet [50], Stanley [112]) that algebraic power series have holonomic coefficients and that the coefficients of rational power series satisfy (for large n) linear recurrence with constant coefficients.

Wilfian formulas. A counting function $f_P(n)$ has a *Wilfian formula* (Wilf [117]) if there exists an algorithm that calculates $f_P(n)$ for every input n effectively, that is to say, in polynomial time. More precisely, we require (extending the definition in [117]) that the algorithm calculates $f_P(n)$ in the number of steps polynomial in the quantity

$$t = \max(\log n, \log f_P(n)).$$

This is (roughly) the minimum time needed for reading the input and writing down the answer. In the most common situations when $\exp(n^c) < f_P(n) < \exp(n^d)$ for large n and some constants $d > c > 0$, this amounts to requiring a number of steps polynomial in n. But if $f_P(n)$ is small (say $\log n$) or big (say doubly exponential in n), then one has to work with t in place of n. The class of counting functions with Wilfian formulas includes holonomic sequences but is much more comprehensive than that.

2 Growth of downsets of combinatorial structures

We survey results in the already introduced setting of downsets in posets of combinatorial structures $(U, \prec)$. The function $f_P(n)$ counts structures of size n in the downset P and P can also be defined in terms of forbidden substructures as $P = \text{Av}(F)$. Besides the containment relation

$\prec$ we employ also isomorphism equivalence relation $\sim$ on U and will count *unlabeled* (i.e., nonisomorphic) structures in P. We denote the corresponding counting function $g_P(n)$, so

$$g_P(n) = \#(\{A \in P \mid s(A) = n\}/\sim)$$

is the number of isomorphism classes of structures with size n in P.

Restrictions on $f_P(n)$ and $g_P(n)$ defining the classes of functions $\mathcal{F}$ often have the form of *jumps in growth*. A jump is a region of growth prohibited for counting functions—every counting function resides either below it or above it. There are many kinds of jumps but the most spectacular is perhaps the *polynomial–exponential jump* from polynomial to exponential growth, which prohibits counting functions satisfying $n^k < f_P(n) < c^n$ for large n for any constants $k > 0$ and $c > 1$. For groups, Grigorchuk constructed a finitely generated group having such intermediate growth (Grigorchuk [70], Grigorchuk and Pak [69], [52]), which excludes the polynomial–exponential jump for general finitely generated groups, but a conjecture says that this jump occurs for every finitely presented group. We have seen this jump in Theorems 1.1 and 1.2 (from polynomial growth to growth at least 2^n) and will meet new examples in Theorems 2.4, 2.17, 2.18, 2.21, and 3.3.

If $(U, \prec)$ has an infinite antichain A, then under natural conditions we get uncountably many functions $f_P(n)$. This was observed several times in the context of permutation containment and for completeness we give the argument here again. These natural conditions, which will always be satisfied in our examples, are *finiteness*, for every n there are finitely many structures with size n in U, and *monotonicity*, $s(G) \geq s(H)$ & $G \prec H$ implies $G = H$ for every G, H in U. (Recall that $G \prec G$ for every G.)

Proposition 2.1. *If $(U, \prec)$ and the size function $s(\cdot)$ satisfy the monotonicity and finiteness conditions and $(U, \prec)$ has an infinite antichain A, then the set of counting functions $f_P(n)$ is uncountable.*

Proof. By the assumption on U we can assume that the members of A have distinct sizes. We show that all the counting functions $f_{\mathrm{Av}(F)}$ for $F \subset A$ are distinct and so this set of functions is uncountable. We write simply f_F instead of $f_{\mathrm{Av}(F)}$. If X, Y are two distinct subsets of A, we express them as $X = T \cup \{G\} \cup U$ and $Y = T \cup \{H\} \cup V$ so that, without loss of generality, $m = s(G) < s(H)$, and $G_1 \in T, G_2 \in U$ implies $s(G_1) < s(G) < s(G_2)$ and similarly for Y (the sets T, U, V may be empty). Then, by the assumption on $\prec$ and $s(\cdot)$,

$$f_X(m) = f_{T \cup \{G\}}(m) = f_T(m) - 1 = f_{T \cup \{H\} \cup V}(m) - 1 = f_Y(m) - 1$$

and $f_X \neq f_Y$. $\qquad\qquad\qquad\qquad\qquad\qquad\qquad\qquad\qquad\qquad\qquad\square$

An infinite antichain thus gives not only uncountably many downsets but in fact uncountably many counting functions. Then, in particular, almost all counting functions are not computable because we have only countably many algorithms. Recently, Albert and Linton [4] significantly refined this argument by showing how certain infinite antichains of permutations produce even uncountably many growth constants, see Theorem 2.6.

On the other hand, if every antichain is finite then there are only countably many functions $f_P(n)$. Posets with no infinite antichain are called *well quasiorderings* or shortly *wqo*. (The second part of the wqo property, nonexistence of infinite strictly descending chains, is satisfied automatically by the monotonicity condition.) But even if $(U, \prec)$ has infinite antichains, there still may be only countably many downsets P with slow growth functions $f_P(n)$. For example, this is the case in Theorems 1.1 and 1.2. It is then of interest to determine for which growth uncountably many downsets appear (cf. Theorem 2.5). The posets $(U, \prec)$ considered here usually have infinite antichains, with two notable wqo exceptions consisting of the minor ordering on graphs and the subsequence ordering on words over a finite alphabet.

2.1 Permutations

Let U denote the universe of permutations represented by finite sequences $b_1 b_2 \ldots b_n$ such that $\{b_1, b_2, \ldots, b_n\} = [n]$. The size of a permutation $\pi = a_1 a_2 \ldots a_m$ is its length $|\pi| = m$. The containment relation on U is defined by $\pi = a_1 a_2 \ldots a_m \prec \rho = b_1 b_2 \ldots b_n$ if and only if for some increasing injection $f : [m] \to [n]$ one has $a_r < a_s \iff b_{f(r)} < b_{f(s)}$ for every r, s in $[m]$. Problems P are downsets in $(U, \prec)$ and their counting functions are $f_P(n) = \#\{\pi \in P \mid |\pi| = n\}$. The poset of permutations $(U, \prec)$ has infinite antichains (see Aktinson, Murphy, and Ruškuc [10]). For further information and background on the enumeration of downsets of permutations see Bóna [34].

Recall that $c(P) = \limsup f_P(n)^{1/n}$. We define

$$E = \{c(P) \in [0, +\infty] \mid P \text{ is a downset of permutations}\}$$

to be the set of growth constants of downsets of permutations. E contains elements $0, 1$ and $+\infty$ because of the downsets $\emptyset$, $\{(1, 2, \ldots, n) \mid n \in \mathbb{N}\}$ and U (all permutations), respectively. How much does $f_P(n)$ drop from $f_U(n) = n!$ if $P \neq U$? The *Stanley–Wilf conjecture* (Bóna [33, 34]) asserted that it drops to exponential growth. The conjecture was proved in 2004 by Marcus and Tardos [87].

Theorem 2.2 (Marcus and Tardos). *If P is a downset of permutations that is not equal to the set of all permutations, then, for some constant c, $f_P(n) < c^n$ for every n.*

Thus, with the sole exception of U, every P has a finite growth constant. Arratia [9] showed that if F consists of a single permutation then $c(P) = c(\mathrm{Av}(F))$ is attained as a limit $\lim f_P(n)^{1/n}$. It would be nice to extend this result.

Problem 2.3. Does $\lim f_P(n)^{1/n}$ always exist when F in $P = \mathrm{Av}(F)$ has more than one forbidden permutation?

For infinite F there conceivably might be oscillations between two different exponential growths (similar oscillations occur for hereditary graph properties and for downsets of words). It would be surprising if oscillations occurred for finite F.

Kaiser and Klazar [76] determined growths of downsets of permutations in the range up to 2^{n-1}.

Theorem 2.4 (Kaiser and Klazar). *If P is a downset of permutations, then exactly one of the following four cases occurs.*

(i) *For large n, $f_P(n)$ is constant.*

(ii) *There are integers $a_0, \ldots, a_k$, $k \geq 1$ and $a_k > 0$, such that $f_P(n) = a_0 \binom{n}{0} + \cdots + a_k \binom{n}{k}$ for large n. Moreover, $f_P(n) \geq n$ for every n.*

(iii) *There are constants c, k in $\mathbb{N}$, $k \geq 2$, such that $F_{n,k} \leq f_P(n) \leq n^c F_{n,k}$ for every n, where $F_{n,k}$ are the generalized Fibonacci numbers.*

(iv) *One has $f_P(n) \geq 2^{n-1}$ for every n.*

The lower bounds in cases 2, 3, and 4 are best possible.

This implies that

$$E \cap [0, 2] = \{0, 1, 2, \alpha_2, \alpha_3, \alpha_4, \ldots\},$$

α_k being the growth constants of $F_{n,k}$, and that $\lim f_P(n)^{1/n}$ exists and equals to $0, 1$ or to some α_k whenever $f_P(n) < 2^{n-1}$ for one n. Note that 2 is the single accumulation point of $E \cap [0, 2]$. We shall see that Theorem 2.4 is subsumed in Theorem 2.17 on ordered graphs. Case 2 and case 3 with $k = 2$ give the *polynomial–Fibonacci jump*: If P is a downset of permutations, then either $f_P(n)$ grows at most polynomially (and in fact equals to a polynomial for large n) or at least Fibonaccially. Huczynska and Vatter [72] gave a simpler proof for this jump. Theorem 2.18 extends it to edge-colored cliques. Theorem 2.4 combines an exact result in case 1 and 2 with an asymptotic result in case 3. It would be nice to have an exact result in case 3 as well and to determine precise forms of the corresponding functions $f_P(n)$ (it is known that in cases 1–3 the generating function $\sum_{n \geq 0} f_P(n) x^n$ is rational, see the remarks at the end of this subsection). Klazar [80] proved that cases 1–3 comprise only countably many downsets, more precisely: if $f_P(n) < 2^{n-1}$ for one n, then $P = \mathrm{Av}(F)$ has finite base F. In the other direction he showed [80] that there are uncountably many downsets P with $f_P(n) < (2.336\dots)^n$ for large n. Recently, Vatter [115] determined the uncountability threshold precisely and extended the description of E above 2.

Theorem 2.5 (Vatter). *Let* $\kappa = 2.205\dots$ *be the real root of* $x^3 - 2x^2 - 1$. *There are uncountably many downsets of permutations* P *with* $c(P) \leq \kappa$ *but only countably many of them have* $c(P) < \kappa$ *and for each of these* $\lim f_P(n)^{1/n}$ *exists. Moreover, the countable intersection*

$$E \cap (2, \kappa)$$

consists exactly of the largest positive roots of the polynomials

> *(i)* $3 + 2x + x^2 + x^3 - x^4$,
> *(ii)* $1 + 3x + 2x^2 + x^3 + x^4 - x^5$,
> *(iii)* $1 + 3x + 2x^2 + x^3 + x^4 - x^5$,
> *(iv)* $1 + 3x + x^2 + x^3 - x^4$

and the two families $(k, l$ *range over* $\mathbb{N})$

> *(v)* $1 + x^l - x^{k+l} - 2x^{k+l+2} + x^{k+l+3}$, *and*
> *(vi)* $1 - x^k - 2x^{k+2} + x^{k+3}$.

The set $E \cap (2, \kappa)$ has no accumulation point from above but it has infinitely many accumulation points from below: κ is the smallest element of E which is an accumulation point of accumulation points. The smallest element of $E \cap (2, \kappa)$ is $2.065\dots$ $(k = l = 1$ in the family (v)).

In [12] it was conjectured that all elements of E (even in the more general situation of ordered graphs) are algebraic numbers and that E has no accumulation point from above. These conjectures were refuted by Albert and Linton [4]. Recall that a subset of $\mathbb{R}$ is *perfect* if it is closed and has no isolated point. Due to the completeness of $\mathbb{R}$ such a set is inevitably uncountable.

Theorem 2.6 (Albert and Linton). *The set E of growth constants of downsets of permutations contains a perfect subset and therefore is uncountable. Also, E contains accumulation points from above.*

The perfect subset constructed by Albert and Linton has smallest element $2.476\ldots$ and they conjectured that E contains some real interval $(\lambda, +\infty)$, which has been recently established by Vatter [114]. However, a typical downset produced by their construction has infinite base. It seems that the refuted conjectures should have been phrased for finitely based downsets.

Problem 2.7. Let E^* be the countable subset of E consisting of the growth constants of finitely based downsets of permutations. Show that every α in E^* is an algebraic number and that for every α in E^* there is a $\delta > 0$ such that $(\alpha, \alpha + \delta) \cap E^* = \emptyset$.

We know from [115] that $E^* \cap [0, \kappa) = E \cap [0, \kappa)$.

We turn to the questions of exact counting. In view of Proposition 2.1 and Theorem 2.6, we restrict to downsets of permutations with finite bases. The next problem goes back to Gessel [67, the final section].

Problem 2.8. Is it true that for every finite set of permutations F the counting function $f_{\mathrm{Av}(F)}(n)$ is holonomic?

All explicit $f_P(n)$ found so far are holonomic. Zeilberger conjectures (see [57]) that $P = \mathrm{Av}(1324)$ has nonholonomic counting function (see Marinov and Radoičić [88] and Albert et al. [3] for the approaches to counting $\mathrm{Av}(1324)$). We remarked earlier that almost all infinitely based P have nonholonomic $f_P(n)$.

More generally, one may pose (Vatter [116]) the following question.

Problem 2.9. Is it true that for every finite set of permutations F the counting function $f_{\mathrm{Av}(F)}(n)$ has a Wilfian formula, that is, can be evaluated by an algorithm in number of steps polynomial in n?

Wilfian formulas were shown to exist for several classes of finitely based downsets of permutations. We refer the reader to Vatter [116] for further

information and mention here only one such result due to Albert and
Atkinson [1]. Recall that $\pi = a_1 a_2 \ldots a_n$ is a *simple permutation* if
$\{a_i, a_{i+1}, \ldots, a_j\}$ is not an interval in $[n]$ for every $1 \leq i \leq j \leq n$,
$0 < j - i < n - 1$.

Theorem 2.10 (Albert and Atkinson). *If P is a downset of permutations containing only finitely many simple permutations, then P is
finitely based and the generating function $\sum_{n \geq 0} f_P(n) x^n$ is algebraic
and thus $f_P(n)$ has a Wilfian formula.*

Brignall, Ruškuc and Vatter [41] show that it is decidable whether a
downset given by its finite basis contains finitely many simple permutations and Brignall, Huczynska and Vatter [40] extend Theorem 2.10 by
showing that many subsets of downsets with finitely many simple permutations have algebraic generating functions as well. See also Brignall's
survey [39].

We conclude this subsection by looking back at Theorems 2.4 and
2.5 from the standpoint of effectivity. Let a downset of permutations
$P = \mathrm{Av}(F)$ be given by its finite base F. Then it is decidable whether
$c(P) < 2$ and (as noted in [115]) for $c(P) < 2$ the results of Albert, Linton
and Ruškuc [5] provide effectively a Wilfian formula for $f_P(n)$, in fact,
the generating function is rational. Also, it is decidable whether $f_P(n)$
is a polynomial for large n (see [72], Albert, Atkinson and Brignall [2]).
By [115], it is decidable whether $c(P) < \kappa$ and Vatter conjectures that
even for $c(P) < \kappa$ the generating function of P is rational.

2.2 Unordered graphs

U is the universe of finite simple graphs with normalized vertex sets
$[n]$ and $\prec$ is the induced subgraph relation. Problems P are hereditary graph properties, that is, downsets in $(U, \prec)$, and $f_P(n)$ counts the
graphs in P with n vertices. *Monotone properties*, which are hereditary
properties that are closed under taking any subgraph, constitute a more
restricted family. An even more restricted family consists of *minor-closed
classes*, which are monotone properties that are closed under contract-
ing edges. By Proposition 2.1 there are uncountably many counting
functions of monotone properties (and hence of hereditary properties as
well) because, for example, the set of all cycles is an infinite antichain
under the subgraph ordering. On the other hand, by the monumental
theorem of Robertson and Seymour [100] there are no infinite antichains

in the minor ordering and so there are only countably many minor-closed classes. The following remarkable theorem describes growths of hereditary properties.

Theorem 2.11 (Balogh, Bollobás, Weinreich, Alekseev, Thomason). *If P is a proper hereditary graph property then exactly one of the four cases occurs.*

(i) *There exist rational polynomials $p_1(x), p_2(x), \ldots, p_k(x)$ such that $f_P(n) = p_1(n) + p_2(n)2^n + \cdots + p_k(n)k^n$ for large n.*

(ii) *There is a constant k in $\mathbb{N}$, $k \geq 2$, such that $f_P(n) = n^{(1-1/k)n+o(n)}$ for every n.*

(iii) *One has $n^{n+o(n)} < f_P(n) < 2^{o(n^2)}$ for every n.*

(iv) *There is a constant k in $\mathbb{N}$, $k \geq 2$, such that for every n we have $f_P(n) = 2^{(1/2-1/2k)n^2+o(n^2)}$.*

We mentioned case 1 as Theorem 1.2. The first three cases were proved by Balogh, Bollobás and Weinreich in [18]. The fourth case is due to Alekseev [6] and independently Bollobás and Thomason [31].

Now we will discuss further strengthenings and refinements of Theorem 2.11. Scheinerman and Zito in a pioneering work [103] obtained its weaker version. They showed that for a hereditary graph property P either (i) $f_P(n)$ is constantly 0, 1 or 2 for large n or (ii) $an^k < f_P(n) < bn^k$ for every n and some constants k in $\mathbb{N}$ and $0 < a < b$ or (iii) $n^{-c}k^n \leq f_P(n) \leq n^c k^n$ for every n and constants c, k in $\mathbb{N}$, $k \geq 2$, or (iv) $n^{cn} \leq f_P(n) \leq n^{dn}$ for every n and some constants $0 < c < d$ or (v) $f_P(n) > n^{cn}$ for large n for every constant $c > 0$.

In cases 1, 2, and 4 growths of $f_P(n)$ settle to specific asymptotic values and these can be characterized by certain minimal hereditary properties, as shown in [18]. Case 3, the penultimate rate of growth (see [19]), is very different. Balogh, Bollobás and Weinreich proved in [19] that for every $c > 1$ and $\varepsilon > 1/c$ there is a monotone property P such that

$$f_P(n) \in [n^{cn+o(n)}, \ 2^{(1+o(1))n^{2-\varepsilon}}]$$

for every n and $f_P(n)$ attains either extremity of the interval infinitely often. Thus in case 3 the growth may oscillate (infinitely often) between the bottom and top parts of the range. The paper [19] contains further examples of oscillations (we stated here just one simplified version) and a conjecture that for finite F the functions $f_{\mathrm{Av}(F)}(n)$ do not oscillate. As for the upper boundary of the range, in [19] it is proven that for every monotone property P,

$$f_P(n) = 2^{o(n^2)} \Rightarrow f_P(n) < 2^{n^{2-1/t+o(1)}} \quad \text{for some } t \text{ in } \mathbb{N}.$$

For hereditary properties this jump is only conjectured. What about the lower boundary? The paper [21] is devoted to the proof of the following theorem.

Theorem 2.12 (Balogh, Bollobás and Weinreich). *If P is a hereditary graph property, then exactly one of the two cases occurs.*

   (i) *There is a constant k in $\mathbb{N}$ such that $f_P(n) < n^{(1-1/k)n+o(n)}$ for every n.*
  (ii) *For large n, one has $f_P(n) \geq B_n$ where B_n are Bell numbers. This lower bound is the best possible.*

By this theorem, the growth of Bell numbers is the lower boundary of the penultimate growth in case 3 of Theorem 2.11.

   Monotone properties of graphs are hereditary and therefore their counting functions follow Theorem 2.11. Their more restricted nature allows for simpler proofs and simple characterizations of minimal monotone properties, which is done in the paper [20]. Certain growths of hereditary properties do not occur for monotone properties, for example if P is monotone and $f_P(n)$ is unbounded, then $f_P(n) \geq \binom{n}{2} + 1$ for every n (see [20]) but, P consisting of complete graphs with possibly an additional isolated vertex is a hereditary property with $f_P(n) = n + 1$ for $n \geq 3$. More generally, Balogh, Bollobás and Weinreich show in [20] that if P is monotone and $f_P(n)$ grows polynomially, then

$$f_P(n) = a_0 \binom{n}{0} + a_1 \binom{n}{1} + \cdots + a_k \binom{n}{k} \quad \text{for large } n$$

and some integer constants $0 \leq a_j \leq 2^{j(j-1)/2}$. In fact, [20] deals mostly with general results on the extremal functions $e_P(n) := \max\{|E| \mid G = ([n], E) \in P\}$ for monotone properties P.

   For the top growths in case 4 of Theorem 2.11, Alekseev [6] and Bollobás and Thomason [32] proved that for $P = \mathrm{Av}(F)$ with $f_P(n) = 2^{(1/2-1/2k)n^2+o(n^2)}$ the parameter k is equal to the maximum r such that there is an s, $0 \leq s \leq r$, with the property that no graph in F can have its vertex set partitioned into r (possibly empty) blocks inducing s complete graphs and $r - s$ empty graphs. For monotone properties this reduces to $k = \min\{\chi(G) - 1 \mid G \in F\}$ where χ is the usual chromatic number of graphs. Balogh, Bollobás and Simonovits [17] replaced for monotone properties the error term $o(n^2)$ by $O(n^\gamma)$, $\gamma = \gamma(F) < 2$. Ishigami [74] recently extended case 4 to k-uniform hypergraphs.

Minor-closed classes of graphs were recently looked at from the point of view of counting functions as well. They again follow Theorem 2.11, with possible simplifications due to their more restricted nature. One is that there are only countably many minor-closed classes. Another simplification is that, with the trivial exception of the class of all graphs, case 4 does not occur as proved by Norine et al. [93].

Theorem 2.13 (Norine, Seymour, Thomas and Wollan). *If P is a proper minor-closed class of graphs then $f_P(n) < c^n n!$ for every n in $\mathbb{N}$ for a constant $c > 1$.*

Bernardi, Noy and Welsh [26] obtained the following theorem; we shorten its statement by omitting characterizations of classes P with the given growth rates.

Theorem 2.14 (Bernardi, Noy and Welsh). *If P is a proper minor-closed class of graphs then exactly one of the six cases occurs.*

(i) *The counting function $f_P(n)$ is constantly 0 or 1 for large n.*

(ii) *For large n, $f_P(n) = p(n)$ for a rational polynomial p of degree at least 2.*

(iii) *For every n, $2^{n-1} \le f_P(n) < c^n$ for a constant $c > 2$.*

(iv) *There exist constants k in $\mathbb{N}$, $k \ge 2$, and $0 < a < b$ such that $a^n n^{(1-1/k)n} < f_P(n) < b^n n^{(1-1/k)n}$ for every n.*

(v) *For every n, $B_n \le f_P(n) = o(1)^n n!$ where B_n is the nth Bell number.*

(vi) *For every n, $n! \le f_P(n) < c^n n!$ for a constant $c > 1$.*

The lower bounds in cases 3, 5 and 6 are best possible.

In fact, in case 3 the formulas of Theorem 1.2 apply. Using the strongly restricted nature of minor-closed classes, one could perhaps obtain in case 3 an even more specific exact result. The work in [26] gives further results on the growth constants $\lim (f_P(n)/n!)^{1/n}$ in case 6 and states several open problems, of which we mention the following analogue of Theorem 2.2 for unlabeled graphs. A similar conjecture was also made by McDiarmid, Steger and Welsh [89].

Problem 2.15. Does every proper minor-closed class of graphs contain at most c^n nonisomorphic graphs on n vertices, for a constant $c > 1$?

This brings us to the unlabeled count of hereditary properties. The following theorem was obtained by Balogh et al. [16].

Theorem 2.16 (Balogh, Bollobás, Saks and Sós). *If P is a hereditary graph property and $g_P(n)$ counts nonisomorphic graphs in P by the number of vertices, then exactly one of the three cases occurs.*

(i) *For large n, $g_P(n)$ is constantly $0, 1$ or 2.*

(ii) *For every n, $g_P(n) = cn^k + O(n^{k-1})$ for some constants k in $\mathbb{N}$ and c in $\mathcal{Q}$, $c > 0$.*

(iii) *For large n, $g_P(n) \geq p_n$ where p_n is the number of integer partitions of n. This lower bound is best possible.*

(We have shortened the statement by omitting the characterizations of P with given growth rates.) The authors of [16] remark that with more effort case 2 can be strengthened, for large n, to an exact result with the error term $O(n^{k-1})$ replaced by a quasipolynomial $p(n)$ of degree at most $k - 1$. It turns out that a weaker form of the jump from case 2 to case 3 was proved already by Macpherson [86, 85]: If $G = (\mathbb{N}, E)$ is an infinite graph and $g_G(n)$ is the number of its unlabeled n-vertex induced subgraphs then either $g_G(n) \leq n^c$ for every n and a constant $c > 0$ or $g_G(n) > \exp(n^{1/2-\varepsilon})$ for large n for every constant $\varepsilon > 0$. Pouzet [96] showed that in the former case $c_1 n^d < g_G(n) < c_2 n^d$ for every n and some constants $0 < c_1 < c_2$ and d in $\mathbb{N}$.

2.3 Ordered graphs and hypergraphs, edge-colored cliques, words, posets, tournaments, and tuples

Ordered graphs.   As previously, U is the universe of finite simple graphs with vertex sets $[n]$ but $\prec$ is now the *ordered* induced subgraph relation, which means that $G_1 = ([m], E_1) \prec G_2 = ([n], E_2)$ if and only if there is an increasing injection $f : [m] \rightarrow [n]$ such that $\{u, v\} \in E_1 \iff \{f(u), f(v)\} \in E_2$. Problems are downsets P in $(U, \prec)$, are called *hereditary properties of ordered graphs*, and $f_P(n)$ is the number of graphs in P with vertex set $[n]$. The next theorem, proved by Balogh, Bollobás and Morris [12], vastly generalizes Theorem 2.4.

Theorem 2.17 (Balogh, Bollobás and Morris). *If P is a hereditary property of ordered graphs, then exactly one of the four cases occurs.*

(i) *For large n, $f_P(n)$ is constant.*

(ii) *There are integers $a_0, \ldots, a_k$, $k \geq 1$ and $a_k > 0$, such that $f_P(n) = a_0 \binom{n}{0} + \cdots + a_k \binom{n}{k}$ for large n. Moreover, $f_P(n) \geq n$ for every n.*

(iii) *There are constants c, k in $\mathbb{N}$, $k \geq 2$, such that $F_{n,k} \leq f_P(n) \leq n^c F_{n,k}$ for every n, where $F_{n,k}$ are the generalized Fibonacci numbers.*

(iv) *One has $f_P(n) \geq 2^{n-1}$ for every n.*

The lower bounds in cases 2, 3, and 4 are best possible.

This is an extension of Theorem 2.4 because the poset of permutations is embedded in the poset of ordered graphs via representing a permutation $\pi = a_1 a_2 \ldots a_n$ by the graph $G_\pi = ([n], \{\{i, j\} \mid i < j \text{ \& } a_i < a_j\})$. One can check that $\pi \prec \rho \iff G_\pi \prec G_\rho$ and that the graphs G_π form a downset in the poset of ordered graphs, so Theorem 2.17 implies Theorem 2.4. Similarly, the poset of set partitions of Example 1 in Introduction is embedded in the poset of ordered graphs, via representing partitions by graphs whose components are cliques. Thus the growths of downsets of set partitions in the range up to 2^{n-1} are described by Theorem 2.17. As for permutations, it would be nice to have in case 3 an exact result. Balogh, Bollobás and Morris conjecture that $2.031\ldots$ (the largest real root of $x^5 - x^4 - x^3 - x^2 - 2x - 1$) is the smallest growth constant for ordered graphs above 2 and Vatter [115] notes that this is not an element of E and thus here the growth constants for permutations and for ordered graphs part ways.

Edge-colored cliques. Klazar [81] considered the universe U of pairs (n, χ) where n ranges over $\mathbb{N}$ and χ is a mapping from the set $\binom{[n]}{2}$ of two-element subsets of $[n]$ to a finite set of colors C. The containment $\prec$ is defined by $(m, \phi) \prec (n, \chi)$ if and only if there is an increasing injection $f : [m] \to [n]$ such that $\chi(\{f(x), f(y)\}) = \phi(\{x, y\})$ for every x, y in $[m]$. For two colors we recover ordered graphs with induced ordered subgraph relation. In [81] the following theorem was proved.

Theorem 2.18 (Klazar). *If P is a downset of edge-colored cliques, then exactly one of the three cases occurs.*

(i) *The function $f_P(n)$ is constant for large n.*

(ii) *There is a constant c in $\mathbb{N}$ such that $n \leq f_P(n) \leq n^c$ for every n.*

(iii) *One has $f_P(n) \geq F_n$ for every n, where F_n are the Fibonacci numbers.*

The lower bounds in cases 2 and 3 are best possible.

This extends the bounded-linear jump and the polynomial-Fibonacci jump of Theorem 2.17. It would be interesting to have full Theorem 2.17 in this more general setting. As explained in [81], many posets of structures can be embedded in the poset of edge-colored cliques (as we have just seen for permutations) and thus Theorem 2.18 applies to them. With more effort, case 2 can be strengthened to the exact result $f_P(n) = p(n)$ with rational polynomial $p(x)$.

Words over finite alphabet. We revisit Example 3 from Introduction. Recall that $U = A^*$ consists of all finite words over finite alphabet A and $\prec$ is the subword ordering. This ordering has infinite antichains, for example $11, 101, 1001, \ldots$ for $A = \{0,1\}$. Balogh and Bollobás [11] investigated general downsets in $(A^*, \prec)$ and proved the following extension of Theorem 1.3.

Theorem 2.19 (Balogh and Bollobás). *If P is a downset of finite words over a finite alphabet A in the subword ordering, then $f_P(n)$ is either bounded or $f_P(n) \geq n + 1$ for every n.*

In contrast with Theorem 1.3, for general downsets, a bounded function $f_P(n)$ need not be eventually constant. Balogh and Bollobás [11] showed that for fixed s in $\mathbb{N}$ function $f_P(n)$ may oscillate infinitely often between the maximum and minimum values s^2 and $2s - 1$, and $s^2 + s$ and $2s$. These are, however, the wildest bounded oscillations possible since they proved, as their main result, that if $f_P(n) = m \leq n$ for some n then $f_P(N) \leq (m+1)^2/4$ for every N, $N \geq n + m$. They also gave examples of unbounded oscillations of $f_P(n)$ between $n + g(n)$ and $2^{n/g(n)}$ for any increasing and unbounded function $g(n) = o(\log n)$, with the downset P coming from an infinite word over two-letter alphabet.

Another natural ordering on A^* is the *subsequence ordering* where $a_1 a_2 \ldots a_k \prec b_1 b_2 \ldots b_l$ if and only if $b_{i_1} = a_1, b_{i_2} = a_2, \ldots, b_{i_k} = a_k$ for some indices $1 \leq i_1 < i_2 < \cdots < i_k \leq l$. Downsets in this ordering remain downsets in the subword ordering and thus their counting functions are governed by Theorem 2.19. But they can be also embedded in the poset of edge-colored complete graphs (associate with $a_1 a_2 \ldots a_n$ the pair (n, χ) where $\chi(\{i,j\}) = \{a_i, a_j\}$ for $i < j$) and Theorem 2.18 applies. In particular, if $P \subset A^*$ is a downset in the subsequence ordering, then $f_P(n)$ is constant for large n or $f_P(n) \geq n + 1$ for every n (by Theorems 2.18 and 2.19). The subsequence ordering on A^* is a wqo by Higman's theorem [71] and therefore has only countably many downsets.

A variation on the subsequence ordering is the ordering on A^* given by $u = a_1 a_2 \ldots a_k \prec v = b_1 b_2 \ldots b_l$ if and only if there is a permutation π of the alphabet A such that $a_1 a_2 \ldots a_k \prec \pi(b_1)\pi(b_2)\ldots\pi(b_l)$ in the subsequence ordering, that is, u becomes a subsequence of v after the letters in v are injectively renamed. This ordering on A^* gives Example 1 in Introduction and leads to Theorem 1.1. It is a wqo as well.

Posets and tournaments. U is the set of all pairs $S = ([n], \leq_S)$ where $\leq_S$ is a non-strict partial ordering on $[n]$. We set $R = ([m], \leq_R) \prec S = ([n], \leq_S)$ if and only if there is an injection $f : [m] \to [n]$ such that $x \leq_R y \iff f(x) \leq_S f(y)$ for every x, y in $[m]$. Thus $R \prec S$ means that the poset R is an induced subposet of S. Downsets in $(U, \prec)$, *hereditary properties of posets*, and their growths were investigated by Balogh, Bollobás and Morris in [14]. For the unlabeled count they obtained the following result.

Theorem 2.20 (Balogh, Bollobás and Morris). *If P is a hereditary property of posets and $g_P(n)$ counts nonisomorphic posets in P by the number of vertices, then exactly one of the three cases occurs.*

   (i) *Function $g_P(n)$ is bounded.*

   (ii) *There is a constant $c > 0$ such that, for every n, $\lceil \frac{n+1}{2} \rceil \leq g_P(n) \leq \lceil \frac{n+1}{2} \rceil + c$.*

   (iii) *For every n, $g_P(n) \geq n$.*

The lower bounds in cases 2 and 3 are best possible.

As for the labeled count $f_P(n)$, using case 1 of Theorem 2.11 they proved in [14, Theorem 2] that if P is a hereditary property of posets then either (i) $f_P(n)$ is constantly 1 for large n or (ii) there are k integers $a_1, \ldots, a_k$, $a_k \neq 0$, such that $f_P(n) = a_0 \binom{n}{0} + \cdots + a_k \binom{n}{k}$ for large n or (iii) $f_P(n) \geq 2^n - 1$ for every n, $n \geq 6$. Moreover, the lower bound in case (iii) is best possible and in case (ii) one has $f_P(n) \geq \binom{n}{0} + \cdots + \binom{n}{k}$ for every n, $n \geq 2k + 1$ and this bound is also best possible.

A *tournament* is a pair $T = ([n], T)$ where T is a binary relation on $[n]$ such that xTx for no x in $[n]$ and for every two distinct elements x, y in $[n]$ exactly one of xTy and yTx holds. U consists of all tournaments for n ranging in $\mathbb{N}$ and $\prec$ is the induced subtournament relation. Balogh, Bollobás and Morris considered in [15, 14] unlabeled counting functions of hereditary properties of tournaments. We merge their results in a single theorem.

Theorem 2.21 (Balogh, Bollobás and Morris). *If P is a hereditary* property of tournaments and $g_P(n)$ counts nonisomorphic tournaments in P by the number of vertices, then exactly one of the three cases occurs.

(i) For large n, function $g_P(n)$ is constant.

(ii) There are constants k in $\mathbb{N}$ and $0 < c < d$ such that $cn^k < g_P(n) < dn^k$ for every n. Moreover, $g_P(n) \geq n - 2$ for every n, $n \geq 4$.

(iii) For every n, $n \neq 4$, one has $g_P(n) \geq F_n^*$ where F_n^* are the quasi-Fibonacci numbers.

The lower bounds in cases 2 and 3 are best possible.

Case 1 and the second part of case 2 were proved in [14] and the rest of the theorem in [15]. A closely related and in one direction stronger theorem was independently obtained by Boudabbous and Pouzet [35] (see also [97, Theorem 22]): If $g_T(n)$ counts unlabeled n-vertex subtournaments of an infinite tournament T, then either $g_T(n)$ is a quasipolynomial for large n or $g_T(n) > c^n$ for large n for a constant $c > 1$.

Ordered hypergraphs. U consists of all hypergraphs, which are the pairs $H = ([n], H)$ with n in $\mathbb{N}$ and H being a set of nonempty and non-singleton subsets of $[n]$, called edges. Note that U extends both the universe of finite simple graphs and the universe of set partitions. The containment $\prec$ is ordered but non-induced and is defined by $([m], G) \prec ([n], H)$ if and only if there is an increasing injection $f : [m] \to [n]$ and an injection $g : G \to H$ such that for every edge E in G we have $f(E) \subset g(E)$. Equivalently, one can omit some vertices from $[n]$, some edges from H and delete some vertices from the remaining edges in H so that the resulting hypergraph is order-isomorphic to G. Downsets in $(U, \prec)$ are called *strongly monotone properties of ordered hypergraphs*. Again, $f_P(n)$ counts hypergraphs in P with the vertex set $[n]$. We define a special downset Π: we associate with every permutation $\pi = a_1 a_2 \ldots a_n$ the (hyper)graph $G_\pi = ([2n], \{\{i, n+a_i\} \mid i \in [n]\})$ and let Π denote the set of all hypergraphs in U contained in some graph G_π; the graphs in Π differ from G_π's only in adding in all ways isolated vertices. Note that $\pi \prec \rho$ for two permutations if and only if $G_\pi \prec G_\rho$ for the corresponding (hyper)graphs. The next theorem was conjectured by Klazar in [77] for set partitions and in [78] for ordered hypergraphs.

Theorem 2.22 (Balogh, Bollobás, Morris, Klazar, Marcus). *If P is a strongly monotone property of ordered hypergraphs, then exactly one of the two cases occurs.*

(i) *There is a constant $c > 1$ such that $f_P(n) \leq c^n$ for every n.*

(ii) *One has $P \supset \Pi$, which implies that*

$$f_P(n) \geq \sum_{k=0}^{\lfloor n/2 \rfloor} \binom{n}{2k} k! = n^{n+O(n/\log n)}$$

for every n and that the lower bound is best possible.

The theorem was proved by Balogh, Bollobás and Morris [13] and independently by Klazar and Marcus [81] (by means of results from [79, 78, 87]). It follows that Theorem 2.22 implies Theorem 2.2. Theorem 2.22 was motivated by efforts to extend the Stanley–Wilf conjecture, now the Marcus–Tardos theorem, from permutations to more general structures. A further extension would be to have it for the wider class of hereditary properties of ordered hypergraphs. These correspond to the containment $\prec$ defined by $([m], G) \prec ([n], H)$ if and only if there is an increasing injection $f : [m] \to [n]$ such that $\{f(E) \mid E \in G\} = \{f([m]) \cap E \mid E \in H\}$. The following conjecture was proposed in [13].

Problem 2.23. If P is a hereditary property of ordered hypergraphs, then either $f_P(n) \leq c^n$ for every n for some constant $c > 1$ or one has $f_P(n) \geq \sum_{k=0}^{\lfloor n/2 \rfloor} \binom{n}{2k} k!$ for every n. Moreover, the lower bound is best possible.

As noted in [13], now it is no longer true that in the latter case P must contain Π.

Tuples of nonnegative integers. For a fixed k in $\mathbb{N}$, we set $U = \mathbb{N}_0^k$, so U contains all k-tuples of nonnegative integers. We define the containment by $a = (a_1, a_2, \ldots, a_k) \prec b = (b_1, b_2, \ldots, b_k)$ if and only if $a_i \leq b_i$ for every i. By Higman's theorem, $(U, \prec)$ is wqo. So there are only countably many downsets. The size function $\| \cdot \|$ on U is given by $\|a\| = a_1 + a_2 + \cdots + a_k$. For a downset P in $(U, \prec)$, $f_P(n)$ counts all tuples in P whose entries sum up to n. Stanley [110] (see also [111, Exercise 6 in Chapter 4]) obtained the following result.

Theorem 2.24 (Stanley). *If P is a downset of k-tuples of nonnegative integers, then there is a rational polynomial $p(x)$ such that $f_P(n) = p(n)$ for large n.*

In fact, the theorem holds for upsets as well because they are complements of downsets and $\#\{a \in \mathbb{N}_0^k \mid \|a\| = n\} = \binom{n+k-1}{k-1}$ is a rational polynomial in n. Jelínek and Klazar [75] noted that the theorem holds for the larger class of sets $P \subset \mathbb{N}_0^k$ that are finite unions of the generalized orthants $\{a \in \mathbb{N}_0^k \mid a_i \geq_i b_i, i \in [k]\}$, here $(b_1, \ldots, b_k)$ is in $\mathbb{N}_0^k$ and each $\geq_i$ is either $\geq$ or equality $=$; we call such P *simple sets*. It appears that this generalization of Theorem 2.24 provides a unified explanation of the exact polynomial results in Theorems 1.1, 2.4, 2.11, and 2.17, by mapping downsets of structures in a size-preserving manner onto simple sets in $\mathbb{N}_0^k$ for some k.

2.4 Growths of profiles of relational structures

In this subsection we mostly follow the survey article of Pouzet [97], see also Cameron [44]. This approach of relational structures was pioneered by Fraïssé [64, 65, 66]. A *relational structure* $R = (X, (R_i \mid i \in I))$ on X is formed by the underlying set X and relations $R_i \subset X^{m_i}$ on X; the sets X and I may be infinite. The *size* of R is the cardinality of X and R is called finite (infinite) if X is finite (infinite). The *signature* of R is the list $(m_i \mid i \in I)$ of arities $m_i \in \mathbb{N}_0$ of the relations R_i. It is *bounded* if the numbers m_i are bounded and is *finite* if I is finite.

Consider two relational structures $R = (X, (R_i \mid i \in I))$ and $S = (Y, (S_i \mid i \in I))$ with the same signature and an injection $f : X \to Y$ satisfying, for every i in I and every m_i-tuple $(a_1, a_2, \ldots, a_{m_i})$ in X^{m_i},

$$(a_1, a_2, \ldots, a_{m_i}) \in R_i \iff (f(a_1), f(a_2), \ldots, f(a_{m_i})) \in S_i.$$

If such an injection f exists, we say that R is *embeddable* in S and write $R \prec S$. If in addition f is an identity (in particular, $X \subset Y$), R is a *substructure* of S and we write $R \prec^* S$. If the injection f is onto Y, we say that R and S are *isomorphic*.

The *age* of a (typically infinite) relational structure R on X is the set P of all finite substructures of R. Note that the age forms a downset in the poset $(U, \prec^*)$ of all finite relational structures with the signature of R whose underlying sets are subsets of X. The *kernel* of R is the set of elements x in X such that the deletion of x changes the age. The *profile* of R is the unlabeled counting function $g_R(n)$ that counts nonisomorphic structures with size n in the age of R. We get the same function if we replace the age of R by the set P of all finite substructures embeddable in R whose underlying sets are $[n]$:

$$g_R(n) = \#(\{S = ([n], (S_i \mid i \in I)) \mid S \prec R\}/\sim)$$

where $\sim$ is the isomorphism relation.

The next general result on growth of profiles was obtained by Pouzet [95], see also [97].

Theorem 2.25 (Pouzet). *If $g_R(n)$ is the profile of an infinite relational structure R with bounded signature or with finite kernel, then exactly one of the three cases occurs.*

   (i) *The function $g_R(n)$ is constant for large n.*
   (ii) *There are constants k in $\mathbb{N}$ and $0 < c < d$ such that $cn^k < g_R(n) < dn^k$ for every n.*
   (iii) *One has $g_R(n) > n^k$ for large n for every constant k in $\mathbb{N}$.*

Case 1 follows from the interesting fact that every infinite R has a nondecreasing profile (Pouzet [96], see [97, Theorem 4] for further discussion) and cases 2 and 3 were proved in [95] (see [97, Theorems 7 and 42]). It is easy to see ([97, Theorem 10]) that for unbounded signature one can get arbitrarily slowly growing unbounded profiles. Also, it turns out ([97, Fact 2]) that for bounded signature and finite-valued profile, one may assume without loss of generality that the signature is finite. An infinite graph $G = (\mathbb{N}, E)$ whose components are cliques and that for every n has infinitely many components (cliques) of size n shows that for the signature (2) the numbers of integer partitions p_n appear as a profile (cf. Theorem 2.16)—for relational structures in general (and unlabeled count) there is no polynomial-exponential jump (but cf. Theorem 2.21).

The survey [97] contains, besides further results and problems on profiles of relational structures, the following attractive conjecture which was partially resolved by Pouzet and Thiéry [98].

Problem 2.26. In the cases 1 and 2 of Theorem 2.25 function $g_R(n)$ is a quasipolynomial for large n.

As remarked in [97], since $g_P(n)$ is nondecreasing, if the conjecture holds then the leading coefficient in the quasipolynomial must be constant and, in cases 1 and 2, $g_R(n) = an^k + O(n^{k-1})$ for some constants $a > 0$ and k in $\mathbb{N}_0$ (cf. Theorem 2.16 and the following comment).

Relational structures are quite general in allowing arbitrarily many relations with arbitrary arities and therefore they can accommodate many previously discussed combinatorial structures and many more. On the other hand, ages of relational structures are less general than downsets of structures, every age is a downset in the substructure ordering but not vice versa—many downsets of finite structures do not come from a single infinite structure (a theorem due to Fraïssé [97, Lemma 7], [44]

characterizes downsets that are ages). An interesting research direction
may be to join the general sides of both approaches.

3 Four topics in general enumeration

In this section we review four topics in general enumeration. As we shall
see, there are connections to the results on growth of downsets presented
in the previous section.

3.1 Counting lattice points in polytopes

A *polytope* P in $\mathbb{R}^k$ is a convex hull of a finite set of points. If these points
have rational, respectively integral, coordinates, we speak of *rational*,
respectively *lattice*, polytope. For a polytope P and n in $\mathbb{N}$ we consider
the dilation $nP = \{nx \mid x \in P\}$ of P and the number of lattice points
in it,

$$f_P(n) = \#(nP \cap \mathbb{Z}^k).$$

The following useful result was derived by Ehrhart [56] and Macdonald [83, 84].

Theorem 3.1 (Ehrhart, Macdonald). *If P is a lattice polytope, respectively rational polytope, and $f_P(n)$ counts lattice points in the dilation
nP, then there is a rational polynomial, respectively rational quasipolynomial, $p(x)$ such that $f_P(n) = p(n)$ for every n.*

For further refinements and ramifications of this result and its applications see Beck and Robins [25] (also Stanley [111]). Barvinok [23]
and Barvinok and Woods [24] developed a beautiful and powerful theory producing polynomial-time algorithms for counting lattice points in
rational polytopes. In way of specializations one obtains from it many
Wilfian formulas. We will not say more on it because in its generality it
is out of scope of this overview (as we said, this not a survey on #P).

3.2 Context-free languages

A *language* P is a subset of A^*, the infinite set of finite words over a
finite alphabet A. The natural size function $|\cdot|$ measures length of words
and

$$f_P(n) = \#\{u \in P \mid |u| = n\}$$

is the number of words in P with length n. In this subsection the alphabet A is always finite, thus $f_P(n) \le |A|^n$.

We review the definition of context-free languages, for further information on (formal) languages see Salomaa [101]. A *context-free grammar* is a quadruple $G = (A, B, c, D)$ where A, B are finite disjoint sets, $c \in B$ (starting variable) and D (production rules) is a finite set of pairs (d, u) where $d \in B$ and $u \in (A \cup B)^*$. A *rightmost derivation* of a word $v \in (A \cup B)^*$ in G is a sequence of words $v_1 = c, v_2, \ldots, v_r = v$ in $(A \cup B)^*$ such that v_i is obtained from v_{i-1} by replacing the rightmost occurrence of a letter d from B in v_{i-1} by the word u, according to some production rule $(d, u) \in D$. (Note that no v_i with $i < r$ is in A^*.) We let $L(G)$ denote the set of words in A^* that have rightmost derivation in G. If in addition every v in $L(G)$ has a unique rightmost derivation in G, then G is an *unambiguous context-free grammar*. A language $P \subset A^*$ is *context-free* if $P = L(G)$ for a context-free grammar $G = (A, B, c, D)$. P is, in addition, *unambiguous* if it can be generated by an unambiguous context-free grammar. If P is context-free but not unambiguous, we say that P is *inherently ambiguous*. We associate with a context-free grammar $G = (A, B, c, D)$ a digraph $H(G)$ on the vertex set B by putting an arrow $d_1 \to d_2$, $d_i \in B$, if and only if there is a production rule $(d_1, u) \in D$ such that d_2 appears in u. We call a context-free language *ergodic* if it can be generated by a context-free grammar G such that the digraph $H(G)$ is strongly connected.

Chomsky and Schützenberger [49] obtained the following important result.

Theorem 3.2 (Chomsky and Schützenberger). *If P is an unambiguous context-free language and $f_P(n)$ counts words of length n in P, then the generating function*

$$F(x) = \sum_{n \ge 0} f_P(n) x^n$$

of P is algebraic over $\mathcal{Q}(x)$.

The algebraicity of a power series $F(x) = \sum_{n \ge 0} a_n x^n$ with a_n in $\mathbb{N}_0$ has two important practical corollaries for the counting sequence $(a_n)_{n \ge 1}$. First, as we already mentioned, it is holonomic. Second, it has a nice asymptotics. More precisely, $F(x)$ determines a function analytic in a neighborhood of 0 and if $F(x)$ is not a polynomial, it has a finite radius of convergence ρ, $0 < \rho \le 1$, and finitely many (dominating) singularities on the circle of convergence $|x| = \rho$. In the case of single dominating singularity we have

$$a_n \sim cn^\alpha r^n$$

where $c > 0$ is in $\mathbb{R}$, $r = 1/\rho \geq 1$ is an algebraic number, and the exponent α is in $\mathcal{Q} \setminus \{-1, -2, -3, \ldots\}$ (if $F(x)$ is rational, then α is in $\mathbb{N}_0$). For example, for Catalan numbers $C_n = \frac{1}{n+1}\binom{2n}{n}$ and their generating function $C(x) = \sum_{n \geq 0} C_n x^n$ we have

$$xC^2 - C + 1 = 0 \quad \text{and} \quad C_n \sim \pi^{-1/2} n^{-3/2} 4^n.$$

For more general results on asymptotics of coefficients of algebraic power series see Flajolet and Sedgewick [63, Chapter VII].

Flajolet [62] used Theorem 3.2 to prove the inherent ambiguity of certain context-free languages. For further information on rational and algebraic power series in enumeration and their relation to formal languages see Barcucci et al. [22], Bousquet-Mélou [36, 37], Flajolet and Sedgewick [63] and Salomaa and Soittola [102].

How fast do context-free languages grow? Trofimov [113] proved for them a polynomial to exponential jump.

Theorem 3.3 (Trofimov). *If $P \subset A^*$ is a context-free language over the alphabet A, then either $f_P(n) \leq |A|n^k$ for every n or $f_P(n) > c^n$ for large n, where $k > 0$ and $c > 1$ are constants.*

Trofimov proved that in the former case in fact $P \subset w_1^* w_2^* \ldots w_k^*$ for some k words w_i in A^*. Later this theorem was independently rediscovered by Incitti [73] and Bridson and Gilman [38]. D'Alessandro, Intrigila and Varricchio [51] show that in the former case the function $f_P(n)$ is in fact a quasipolynomial $p(n)$ for large n (and that $p(n)$ and the bound on n can be effectively determined from P).

Recall that for a language $P \subset A^*$ the growth constant is defined as $c(P) = \limsup f_P(n)^{1/n}$. P is *growth-sensitive* if $c(P) > 1$ and $c(P \cap Q) < c(P)$ for every downset Q in $(A^*, \prec)$, where $\prec$ is the subword ordering, such that $P \cap Q \neq P$. In other words, forbidding any word u such that $u \prec v \in P$ for some v as a subword results in a significant decrease in growth. Yet in other words, in growth-sensitive languages an analogue of Marcus–Tardos theorem (Theorem 2.2) holds. In a series of papers Ceccherini-Silberstein, Machì and Scarabotti [46], Ceccherini-Silberstein and Woess [47, 48], and Ceccherini-Silberstein [45] the following theorem on growth-sensitivity was proved.

Theorem 3.4 (Ceccherini-Silberstein, Machì, Scarabotti, Woess). *Ev-*

ery unambiguous context-free language P that is ergodic and has $c(P) > 1$ is growth-sensitive.

See [47] for extensions of the theorem to the ambiguous case and [46] for the more elementary case of regular languages.

3.3 Exact counting of regular and other graphs

We consider finite simple graphs and, for a given set $P \subset \mathbb{N}_0$, the counting function ($\deg(v) = \deg_G(v)$ is the degree of a vertex v in G, the number of incident edges)

$$f_P(n) = \#\{G = ([n], E) \mid \deg_G(v) \in P \text{ for every } v \text{ in } [n]\}.$$

For example, for $P = \{k\}$ we count labeled k-regular graphs on $[n]$. The next general theorem was proved by Gessel [67, Corollary 11], by means of symmetric functions in infinitely many variables.

Theorem 3.5 (Gessel). *If P is a finite subset of $\mathbb{N}_0$ and $f_P(n)$ counts labeled graphs on $[n]$ with all degrees in P, then the sequence $(f_P(n))_{n \geq 1}$ is holonomic.*

This theorem was conjectured and partially proved for the k-regular case, $k \leq 4$, by Goulden and Jackson [68]. As remarked in [67], the theorem holds also for graphs with multiple edges and/or loops. Domocoş [54] extended it to 3-regular and 3-partite hypergraphs (and remarked that Gessel's method works also for general k-regular and k-partite hypergraphs). For more information see also Mishna [91, 90].

Consequently, the numbers of labeled graphs with degrees in fixed finite set have Wilfian formula. Now we demonstrate this directly by a more generally applicable argument. For d in $\mathbb{N}_0^{k+1}$, we say that a graph G is a d-graph if $|V(G)| = d_0 + d_1 + \cdots + d_k$, $\deg(v) \leq k$ for every v in $V(G)$ and exactly d_i vertices in $V(G)$ have degree i. Let

$$p(d) = p(d_0, d_1, \ldots, d_k) = \#\{G = ([n], E) \mid G \text{ is a } d\text{-graph}\}$$

be the number of labeled d-graphs with vertices $1, 2, \ldots, n = d_0 + \cdots + d_k$.

Proposition 3.6. *For fixed k, the list of numbers*

$$(p(d) \mid d \in \mathbb{N}_0^{k+1}, d_0 + d_1 + \cdots + d_k = m \leq n)$$

can be generated in time polynomial in n.

Proof. A natural idea is to construct graphs G with d_i vertices of degree i by adding vertices $1, 2, \ldots, n$ one by one, keeping track of the numbers d_i. In the first phase of the algorithm we construct an auxiliary $(n+1)$-partite graph

$$H = (V_0 \cup V_1 \cup \cdots \cup V_n, E)$$

where we start with V_m consisting of all $(k+1)$-tuples $d = (d_0, d_1, \ldots, d_k)$ in $\mathbb{N}_0^{k+1}$ satisfying $d_0 + \cdots + d_k = m$ and the edges will go only between V_m and V_{m+1}. An edge joins $d \in V_m$ with $e \in V_{m+1}$ if and only if there exist numbers $\Delta_0, \Delta_1, \ldots, \Delta_{k-1}$ in $\mathbb{N}_0$ such that: $0 \le \Delta_i \le d_i$ for $0 \le i \le k-1$, $r := \Delta_0 + \cdots + \Delta_{k-1} \le k$, and

$$e_i = d_i + \Delta_{i-1} - \Delta_i \text{ for } i \in \{0, 1, \ldots, k\} \setminus \{r\} \text{ but } e_r = d_r + \Delta_{r-1} - \Delta_r + 1$$

where we set $\Delta_{-1} = \Delta_k = 0$. We omit from H (or better, do not construct at all) the vertices d in V_m not reachable from $V_0 = \{(0, 0, \ldots, 0)\}$ by a path $v_0, v_1, \ldots, v_m = d$ with v_i in V_i. For example, the k vertices in V_1 with $d_0 = 0$ are omitted and only $(1, 0, \ldots, 0)$ remains. Also, we label the edge $\{e, d\}$ with the k-tuple $\Delta = (\Delta_0, \ldots, \Delta_{k-1})$. (It follows that Δ and r are uniquely determined by d, e.) The graph H together with its labels can be constructed in time polynomial in n. It records the changes of the numbers d_i of vertices with degree i caused by adding to $G = ([m], E)$ new vertex $m + 1$; Δ_i are the numbers of neighbors of $m + 1$ with degree i in G and r is the degree of $m + 1$.

In the second phase we evaluate a function $p : V_0 \cup \cdots \cup V_n \to \mathbb{N}$ defined on the vertices of H by this inductive rule: $p(0, 0, \ldots, 0) = 1$ on V_0 and, for e in V_m with $m > 0$,

$$p(e) = \sum_d p(d) \prod_{i=0}^{k-1} \binom{d_i}{\Delta_i}$$

where we sum over all d in V_{m-1} such that $\{d, e\} \in E(H)$ and Δ is the label of the edge $\{d, e\}$. It is easy to see that all values of p can be obtained in time polynomial in n and that $p(d)$ for d in V_m is the number of labeled d-graphs on $[m]$. $\square$

Now we can in time polynomial in n easily calculate

$$f_P(n) = \sum_d p(d)$$

as a sum over all $d = (d_0, \ldots, d_k)$ in V_n, $k = \max P$, satisfying $d_i = 0$ when $i \notin P$. Of course, this algorithm is much less effective than the holonomic recurrence ensured by (and effectively obtainable by the proof of) Theorem 3.5. But by this approach we can get Wilfian formula also for some infinite sets of degrees P, for example when P is an arithmetical progression. (We leave to the reader as a nice exercise to count labeled graphs with even degrees.) On the other hand, it seems to fail for many classes of graphs, for example, for triangle-free graphs.

Problem 3.7. Is there a Wilfian formula for the number of labeled triangle-free graphs on $[n]$? Can this number be calculated in time polynomial in n?

The problem of enumeration of labeled triangle-free graphs was mentioned by Read in [99, Chapter 2.10]. A quarter century ago, Wilf [117] posed the following similar problem.

Problem 3.8. Can one calculate in time polynomial in n the number of unlabeled graphs on $[n]$?

3.4 Ultimate modular periodicity

One general aspect of counting functions $f_P(n)$ not touched so far is their modular behavior. For given modulus m in $\mathbb{N}$, what can be said about the sequence of residues $(f_P(n) \bmod m)_{n \geq 1}$. Before presenting a rather general result in this area, we motivate it by two examples.

First, Bell numbers B_n counting partitions of $[n]$. Recall that

$$\sum_{n \geq 0} B_n x^n = \sum_{k=0}^{\infty} \frac{x^k}{(1-x)(1-2x)\dots(1-kx)}.$$

Reducing modulo m we get, denoting $v(x) = (1-x)(1-2x)\dots(1-(m-1)x)$,

$$\sum_{n \geq 0} B_n x^n \equiv_m \sum_{j=0}^{\infty} \frac{x^{mj}}{v(x)^j} \sum_{i=0}^{m-1} \frac{x^i}{(1-x)(1-2x)\dots(1-ix)}$$

$$= \frac{1}{1-x^m/v(x)} \sum_{i=0}^{m-1} \frac{x^i}{(1-x)(1-2x)\dots(1-ix)}$$

$$= \frac{a(x)}{v(x)-x^m} = \frac{a(x)}{1+b_1x+\dots+b_m x^m}$$

$$= \sum_{n \geq 0} c_n x^n$$

where $a(x) \in \mathbb{Z}[x]$ has degree at most $m-1$ and b_i are integers, $b_m = -1$. Thus the sequence of integers $(c_n)_{n \geq 1}$ satisfies for $n > m$ the linear recurrence $c_n = -b_1 c_{n-1} - \cdots - b_m c_{n-m}$ of order m. By the pigeonhole principle, the sequence $(c_n \bmod m)_{n \geq 1}$ is periodic for large n. Since $B_n \equiv c_n \bmod m$, the sequence $(B_n \bmod m)_{n \geq 1}$ is periodic for large n as well. (For modular periods of Bell numbers see Lunnon, Pleasants and Stephens [82]).

Second, Catalan numbers $C_n = \frac{1}{n+1}\binom{2n}{n}$ counting, for example, non-crossing partitions of $[n]$. The shifted version $D_n = C_{n-1} = \frac{1}{n}\binom{2n-2}{n-1}$ satisfies the recurrence $D_1 = 1$ and, for $n > 1$,

$$D_n = \sum_{i=1}^{n-1} D_i D_{n-i} = 2 \sum_{i=1}^{\lfloor n/2 \rfloor - 1} D_i D_{n-i} + \sum_{i=\lfloor n/2 \rfloor}^{\lceil n/2 \rceil} D_i D_{n-i}.$$

Thus, modulo 2, $D_1 \equiv 1$, $D_n \equiv 0$ for odd $n > 1$ and $D_n \equiv D_{n/2}^2 \equiv D_{n/2}$ for even n. It follows that $D_n \equiv 1$ if and only if $n = 2^m$ and that the sequence $(C_n \bmod 2)_{n \geq 1}$ has 1's for $n = 2^m - 1$ and 0's elsewhere. In particular, it is not periodic for large n. (For modular behavior of Catalan numbers see Deutsch and Sagan [53] and Eu, Liu and Yeh [58]).

Bell numbers come out as a special case of a general setting. Consider a *relational system* which is a set P of relational structures R with the same finite signature and underlying sets $[n]$ for n ranging in $\mathbb{N}$. We say that P is *definable in MSOL*, monadic second-order logic, if P coincides with the set of finite models (on sets $[n]$) of a closed formula ϕ in MSOL. (MSOL has, in addition to the language of the first-order logic, variables S for sets of elements, which can be quantified by $\forall, \exists$, and atomic formulas of the type $x \in S$; see Ebbinghaus and Flum [55].) Let $f_P(n)$ be the number of relational structures in P on the set $[n]$, that is, the number of models of ϕ on $[n]$ when P is defined by ϕ. For example, the (first-order) formula ϕ given by (a, b, c are variables for elements, $\sim$ is a binary predicate)

$$\forall a, b, c : (a \sim a) \;\&\; (a \sim b \Rightarrow b \sim a) \;\&\; ((a \sim b \;\&\; b \sim c) \Rightarrow a \sim c)$$

has as its models equivalence relations and $f_P(n) = f_\phi(n) = B_n$, the Bell numbers.

Let us call a sequence $(s_1, s_2, \dots)$ *ultimately periodic* if it is periodic for large n: there are constants p, q in $\mathbb{N}$ such that $s_{n+p} = s_n$ whenever $n \geq q$. Specker and Blatter [27, 28, 29] (see also Specker [107]) proved the following remarkable general theorem.

Theorem 3.9 (Specker and Blatter). *If a relational system P defin-able in MSOL uses only unary and binary relations and $f_P(n)$ counts* *its members on the set $[n]$, then the sequence $(f_P(n) \bmod m)_{n \geq 1}$ is ultimately periodic for every $m \in \mathbb{N}$.*

The general reason for ultimate modular periodicity in Theorem 3.9 is the same as in our example with B_n, residues satisfy a linear recurrence with constant coefficients. Fischer [59] constructed counterexamples to the theorem for quaternary relations, see also Specker [108]. Fischer and Makowski [60] extended Theorem 3.9 to CMSOL (monadic second-order logic with modular counting) and to relations with higher arities when vertices have bounded degrees. Note that regular graphs and triangle-free graphs are first-order definable. Thus the counting sequences mentioned in Theorem 3.5 and in Problem 3.7 are ultimately periodic to any modulus. More generally, any hereditary graph property $P = \mathrm{Av}(F)$ with finite base F is first-order definable and similarly for other structures. Many hereditary properties with infinite bases (and also many sets of graphs or structures which are not hereditary) are MSOL-definable; this is the case, for example, for forests ($P = \mathrm{Av}(F)$ where F is the set of cycles) and for planar graphs (use Kuratowski's theorem). To all of them Theorem 3.9 applies. On the other hand, as the example with Catalan numbers shows, counting of ordered structures is in general out of reach of Theorem 3.9.

A closely related circle of problems is the determination of spectra of relational systems P and more generally of finite models; the *spectrum of P* is the set

$$\{n \in \mathbb{N} \mid f_P(n) > 0\}$$

—the set of sizes of members of P. In several situation it was proved that the spectrum is an ultimately periodic subset of $\mathbb{N}$. See Fischer and Makowski [61] (and the references therein), Shelah [104] and Shelah and Doron [105]. We conclude with a problem posed in [59].

Problem 3.10. Does Theorem 3.9 hold for relational systems with ternary relations?

Acknowledgments My thanks go to the organizers of the conference Permutations Patterns 2007 in St Andrews—Nik Ruškuc, Lynn Hynd, Steve Linton, Vince Vatter, Miklós Bóna, Einar Steingrímsson and Julian West—for a very nice conference which gave me opportunity and incentive to write this overview article. I thank also Vít Jelínek for reading the manuscript and an anonymous referee for making many improvements and corrections in my style and grammar.

References

[1] M. H. Albert and M. D. Atkinson. Simple permutations and pattern restricted permutations. *Discrete Math.*, 300(1-3):1–15, 2005.

[2] M. H. Albert, M. D. Atkinson, and R. Brignall. Permutation classes of polynomial growth. *Ann. Comb.*, 11(3–4):249–264, 2007.

[3] M. H. Albert, M. Elder, A. Rechnitzer, P. Westcott, and M. Zabrocki. On the Wilf-Stanley limit of 4231-avoiding permutations and a conjecture of Arratia. *Adv. in Appl. Math.*, 36(2):95–105, 2006.

[4] M. H. Albert and S. Linton. Growing at a perfect speed. *Combin. Probab. Comput.*, 18:301–308, 2009.

[5] M. H. Albert, S. Linton, and N. Ruškuc. The insertion encoding of permutations. *Electron. J. Combin.*, 12(1):Research paper 47, 31 pp., 2005.

[6] V. E. Alekseev. Range of values of entropy of hereditary classes of graphs. *Diskret. Mat.*, 4(2):148–157, 1992.

[7] J.-P. Allouche and J. Shallit. *Automatic sequences.* Cambridge University Press, Cambridge, 2003.

[8] G. E. Andrews. *The theory of partitions.* Addison-Wesley Publishing Co., Reading, Mass.Mass.-London-Amsterdam, 1976.

[9] R. Arratia. On the Stanley-Wilf conjecture for the number of permutations avoiding a given pattern. *Electron. J. Combin.*, 6:Note, N1, 4 pp., 1999.

[10] M. D. Atkinson, M. M. Murphy, and N. Ruškuc. Partially well-ordered closed sets of permutations. *Order*, 19(2):101–113, 2002.

[11] J. Balogh and B. Bollobás. Hereditary properties of words. *Theor. Inform. Appl.*, 39(1):49–65, 2005.

[12] J. Balogh, B. Bollobás, and R. Morris. Hereditary properties of ordered graphs. In M. Klazar, J. Kratochvíl, M. Loebl, J. Matoušek, R. Thomas, and P. Valtr, editors, *Topics in discrete mathematics*, volume 26 of *Algorithms Combin.*, pages 179–213. Springer, Berlin, 2006.

[13] J. Balogh, B. Bollobás, and R. Morris. Hereditary properties of partitions, ordered graphs and ordered hypergraphs. *European J. Combin.*, 27(8):1263–1281, 2006.

[14] J. Balogh, B. Bollobás, and R. Morris. Hereditary properties of combinatorial structures: posets and oriented graphs. *J. Graph Theory*, 56(4):311–332, 2007.

[15] J. Balogh, B. Bollobás, and R. Morris. Hereditary properties of tournaments. *Electron. J. Combin.*, 14(1):Research Paper 60, 25 pp., 2007.

[16] J. Balogh, B. Bollobás, M. Saks, and V. T. Sós. The unlabeled speed of a hereditary graph property. Preprint.

[17] J. Balogh, B. Bollobás, and M. Simonovits. The number of graphs without forbidden subgraphs. *J. Combin. Theory Ser. B*, 91(1):1–24, 2004.

[18] J. Balogh, B. Bollobás, and D. Weinreich. The speed of hereditary properties of graphs. *J. Combin. Theory Ser. B*, 79(2):131–156, 2000.

[19] J. Balogh, B. Bollobás, and D. Weinreich. The penultimate rate of growth for graph properties. *European J. Combin.*, 22(3):277–289, 2001.

[20] J. Balogh, B. Bollobás, and D. Weinreich. Measures on monotone properties of graphs. *Discrete Appl. Math.*, 116(1-2):17–36, 2002.

[21] J. Balogh, B. Bollobás, and D. Weinreich. A jump to the Bell number for hereditary graph properties. *J. Combin. Theory Ser. B*, 95(1):29–48, 2005.

[22] E. Barcucci, A. Del Lungo, A. Frosini, and S. Rinaldi. From rational functions to regular languages. In *Formal power series and algebraic combinatorics (Moscow, 2000)*, pages 633–644. Springer, Berlin, 2000.

[23] A. Barvinok. The complexity of generating functions for integer points in polyhedra and beyond. In *International Congress of Mathematicians. Vol. III*, pages 763–787. Eur. Math. Soc., Zürich, 2006.

[24] A. Barvinok and K. Woods. Short rational generating functions for lattice point problems. *J. Amer. Math. Soc.*, 16(4):957–979, 2003.

[25] M. Beck and S. Robins. *Computing the continuous discretely.* Undergraduate Texts in Mathematics. Springer, New York, 2007.

[26] O. Bernardi, M. Noy, and D. Welsh. On the growth rate of minor-closed classes of graphs. arXiv:06710.2995 [math.CO].

[27] C. Blatter and E. Specker. Le nombre de structures finies d'une théorie à caractère fini. *Sci. Math. Fonds Nat. Rec. Sci. Bruxelles*, pages 41–44, 1981.

[28] C. Blatter and E. Specker. Modular periodicity of combinatorial sequences. *Abstract Amer. Math. Soc.*, 4:313, 1983.

[29] C. Blatter and E. Specker. Recurrence relations for the number of labeled structures on a finite set. In E. Börger, G. Hasenjaeger, and D. Rödding, editors, *Logic and Machines*, pages 43–61, 1983.

[30] B. Bollobás. Hereditary and monotone properties of combinatorial structures. In A. Hilton and J. Talbot, editors, *Surveys in Combinatorics 2007*, number 346 in London Mathematical Society Lecture Note Series, pages 1–39. Cambridge University Press, 2007.

[31] B. Bollobás and A. Thomason. Projections of bodies and hereditary properties of hypergraphs. *Bull. London Math. Soc.*, 27(5):417–424, 1995.

[32] B. Bollobás and A. Thomason. Hereditary and monotone properties of graphs. In *The mathematics of Paul Erdős, II*, volume 14 of *Algorithms Combin.*, pages 70–78. Springer, Berlin, 1997.

[33] M. Bóna. Permutations avoiding certain patterns: the case of length 4 and some generalizations. *Discrete Math.*, 175(1-3):55–67, 1997.

[34] M. Bóna. *Combinatorics of permutations*. Discrete Mathematics and its Applications (Boca Raton). Chapman & Hall/CRC, Boca Raton, FL, 2004.

[35] Y. Boudabbous and M. Pouzet. The morphology of infinite tournaments. application to the growth of their profile. arXiv:0801.4069 [math.CO].

[36] M. Bousquet-Mélou. Algebraic generating functions in enumerative combinatorics and context-free languages. In *STACS 2005*, volume 3404 of *Lecture Notes in Comput. Sci.*, pages 18–35. Springer, Berlin, 2005.

[37] M. Bousquet-Mélou. Rational and algebraic series in combinatorial enumeration. In *International Congress of Mathematicians. Vol. III*, pages 789–826. Eur. Math. Soc., Zürich, 2006.

[38] M. R. Bridson and R. H. Gilman. Context-free languages of sub-exponential growth. *J. Comput. System Sci.*, 64(2):308–310, 2002.

[39] R. Brignall. A survey of simple permutations. In *this volume*, 41–65.

[40] R. Brignall, S. Huczynska, and V. Vatter. Simple permutations and algebraic generating functions. *J. Combin. Theory Ser. A*, 115(3):423–441, 2008.

[41] R. Brignall, N. Ruškuc, and V. Vatter. Simple permutations: decidability and unavoidable substructures. *Theoret. Comput. Sci.*, 391(1-2):150–163, 2008.

[42] S. N. Burris. *Number theoretic density and logical limit laws*, volume 86 of *Mathematical Surveys and Monographs*. American Mathematical Society, Providence, RI, 2001.

[43] P. J. Cameron. *Oligomorphic permutation groups*, volume 152 of *London Mathematical Society Lecture Note Series*. Cambridge University Press, Cambridge, 1990.

[44] P. J. Cameron. Some counting problems related to permutation groups. *Discrete Math.*, 225(1-3):77–92, 2000.

[45] T. Ceccherini-Silberstein. Growth and ergodicity of context-free languages. II. The linear case. *Trans. Amer. Math. Soc.*, 359(2):605–618, 2007.

[46] T. Ceccherini-Silberstein, A. Machi, and F. Scarabotti. On the entropy of regular languages. *Theoret. Comput. Sci.*, 307(1):93–102, 2003.

[47] T. Ceccherini-Silberstein and W. Woess. Growth and ergodicity of context-free languages. *Trans. Amer. Math. Soc.*, 354(11):4597–4625, 2002.

[48] T. Ceccherini-Silberstein and W. Woess. Growth-sensitivity of context-free languages. *Theoret. Comput. Sci.*, 307(1):103–116, 2003.

[49] N. Chomsky and M. P. Schützenberger. The algebraic theory of context-free languages. In *Computer programming and formal systems*, pages 118–161. North-Holland, Amsterdam, 1963.

[50] L. Comtet. *Advanced combinatorics*. D. Reidel Publishing Co., Dordrecht,

1974.

[51] F. D'Alessandro, B. Intrigila, and S. Varricchio. On the structure of the counting function of sparse context-free languages. *Theoret. Comput. Sci.*, 356(1-2):104–117, 2006.

[52] P. de la Harpe. *Topics in geometric group theory*. Chicago Lectures in Mathematics. University of Chicago Press, Chicago, IL, 2000.

[53] E. Deutsch and B. E. Sagan. Congruences for Catalan and Motzkin numbers and related sequences. *J. Number Theory*, 117(1):191–215, 2006.

[54] V. Domocoş. Minimal coverings of uniform hypergraphs and *P*-recursiveness. *Discrete Math.*, 159(1-3):265–271, 1996.

[55] H.-D. Ebbinghaus and J. Flum. *Finite model theory*. Springer Monographs in Mathematics. Springer-Verlag, Berlin, enlarged edition, 2006.

[56] E. Ehrhart. Sur les polyèdres rationnels homothétiques à *n* dimensions. *C. R. Acad. Sci. Paris*, 254:616–618, 1962.

[57] M. Elder and V. Vatter. Problems and conjectures presented at the Third International Conference on Permutation Patterns, University of Florida, March 7–11, 2005. arXiv:0505504 [math.CO].

[58] S.-P. Eu, S.-C. Liu, and Y.-N. Yeh. Catalan and Motzkin numbers modulo 4 and 8. *European J. Combin.*, 29(6):1449–1466, 2008.

[59] E. Fischer. The Specker-Blatter theorem does not hold for quaternary relations. *J. Combin. Theory Ser. A*, 103(1):121–136, 2003.

[60] E. Fischer and J. A. Makowsky. The Specker-Blatter theorem revisited. In *Computing and combinatorics*, volume 2697 of *Lecture Notes in Comput. Sci.*, pages 90–101. Springer, Berlin, 2003.

[61] E. Fischer and J. A. Makowsky. On spectra of sentences of monadic second order logic with counting. *J. Symbolic Logic*, 69(3):617–640, 2004.

[62] P. Flajolet. Analytic models and ambiguity of context-free languages. *Theoret. Comput. Sci.*, 49(2-3):283–309, 1987.

[63] P. Flajolet and R. Sedgewick. *Analytic combinatorics*. Cambridge University Press, Cambridge, 2009.

[64] R. Fraïssé. *Sur quelques classifications des systèmes de relations*. PhD thesis, Université de Paris, 1953.

[65] R. Fraïssé. Sur l'extension aux relations de quelques propriétés des ordres. *Ann. Sci. Ecole Norm. Sup. (3)*, 71:363–388, 1954.

[66] R. Fraïssé. *Theory of relations*, volume 145 of *Studies in Logic and the Foundations of Mathematics*. North-Holland Publishing Co., Amsterdam, revised edition, 2000.

[67] I. M. Gessel. Symmetric functions and *P*-recursiveness. *J. Combin. Theory Ser. A*, 53(2):257–285, 1990.

[68] I. P. Goulden and D. M. Jackson. Labelled graphs with small vertex degrees and *P*-recursiveness. *SIAM J. Algebraic Discrete Methods*, 7(1):60–66, 1986.

[69] R. Grigorchuk and I. Pak. Groups of intermediate growth: an introduction. *Enseign. Math. (2)*, 54(3-4):251–272, 2008.

[70] R. I. Grigorchuk. On the Milnor problem of group growth. *Dokl. Akad. Nauk SSSR*, 271(1):30–33, 1983.

[71] G. Higman. Ordering by divisibility in abstract algebras. *Proc. London Math. Soc. (3)*, 2:326–336, 1952.

[72] S. Huczynska and V. Vatter. Grid classes and the Fibonacci dichotomy for restricted permutations. *Electron. J. Combin.*, 13:Research paper 54, 14 pp., 2006.

[73] R. Incitti. The growth function of context-free languages. *Theoret. Comput. Sci.*, 255(1-2):601–605, 2001.

[74] Y. Ishigami. The number of hypergraphs and colored hypergraphs with hereditary properties. arXiv:0712.0425v1 [math.CO].

[75] V. Jelínek and M. Klazar. Generalizations of Khovanskii's theorems on the growth of sumsets in abelian semigroups. *Adv. in Appl. Math.*, 41(1):115–132, 2008.

[76] T. Kaiser and M. Klazar. On growth rates of closed permutation classes. *Electron. J. Combin.*, 9(2):Research paper 10, 20 pp., 2003.

[77] M. Klazar. Counting pattern-free set partitions. I. A generalization of Stirling numbers of the second kind. *European J. Combin.*, 21(3):367–378, 2000.

[78] M. Klazar. Counting pattern-free set partitions. II. Noncrossing and other hypergraphs. *Electron. J. Combin.*, 7:Research Paper 34, 25 pp., 2000.

[79] M. Klazar. Extremal problems for ordered (hyper) graphs: applications of Davenport-Schinzel sequences. *European J. Combin.*, 25(1):125–140, 2004.

[80] M. Klazar. On the least exponential growth admitting uncountably many closed permutation classes. *Theoret. Comput. Sci.*, 321(2-3):271–281, 2004.

[81] M. Klazar. On growth rates of permutations, set partitions, ordered graphs and other objects. *Electron. J. Combin.*, 15(1):Research paper 75, 22 pp., 2008.

[82] W. F. Lunnon, P. A. B. Pleasants, and N. M. Stephens. Arithmetic properties of Bell numbers to a composite modulus. I. *Acta Arith.*, 35(1):1–16, 1979.

[83] I. G. Macdonald. The volume of a lattice polyhedron. *Proc. Cambridge Philos. Soc.*, 59:719–726, 1963.

[84] I. G. Macdonald. Polynomials associated with finite cell-complexes. *J. London Math. Soc. (2)*, 4:181–192, 1971.

[85] H. D. Macpherson. Growth rates in infinite graphs and permutation groups. *Proc. London Math. Soc. (3)*, 51(2):285–294, 1985.

[86] H. D. Macpherson. Orbits of infinite permutation groups. *Proc. London Math. Soc. (3)*, 51(2):246–284, 1985.

[87] A. Marcus and G. Tardos. Excluded permutation matrices and the Stanley-Wilf conjecture. *J. Combin. Theory Ser. A*, 107(1):153–160, 2004.

[88] D. Marinov and R. Radoičić. Counting 1324-avoiding permutations. *Electron. J. Combin.*, 9(2):Research paper 13, 9 pp., 2003.

[89] C. McDiarmid, A. Steger, and D. J. A. Welsh. Random graphs from planar and other addable classes. In *Topics in discrete mathematics*, volume 26 of *Algorithms Combin.*, pages 231–246. Springer, Berlin, 2006.

[90] M. Mishna. Automatic enumeration of regular objects. *J. Integer Seq.*, 10(5):Article 07.5.5, 18 pp., 2007.

[91] M. J. Mishna. *Une approche holonome à la combinatoire algébrique.* PhD thesis, Univ. Québec à Montréal, 2003.

[92] M. Morse and G. A. Hedlund. Symbolic Dynamics. *Amer. J. Math.*, 60(4):815–866, 1938.

[93] S. Norine, P. Seymour, R. Thomas, and P. Wollan. Proper minor-closed families are small. *J. Combin. Theory Ser. B*, 96(5):754–757, 2006.

[94] C. H. Papadimitriou. *Computational complexity.* Addison-Wesley Publishing Company, Reading, MA, 1994.

[95] M. Pouzet. *Sur la théorie des relations.* PhD thesis, Université Claude-Bernard, Lyon 1, 1978.

[96] M. Pouzet. Application de la notion de relation presque-enchaînable au dénombrement des restrictions finies d'une relation. *Z. Math. Logik Grundlag. Math.*, 27(4):289–332, 1981.

[97] M. Pouzet. The profile of relations. *Glob. J. Pure Appl. Math.*, 2(3):237–272, 2006.

[98] M. Pouzet and N. M. Thiéry. Some relational structures with polynomial growth and their associated algebras. arXiv:0601256v1 [math.CO].

[99] C. R. Read. Enumeration. In L. W. Beineke and R. J. Wilson, editors, *Graph connections*, pages 13–33, New York, 1997. The Clarendon Press Oxford University Press.

[100] N. Robertson and P. Seymour. Graph minors i–xx. *J. Combinatorial Theory Ser. B*, 1983–2004.

[101] A. Salomaa. *Formal languages.* Academic Press [Harcourt Brace Jovanovich Publishers], New York, 1973. ACM Monograph Series.

[102] A. Salomaa and M. Soittola. *Automata-theoretic aspects of formal power series.* Springer-Verlag, New York, 1978. Texts and Monographs in Computer Science.

[103] E. R. Scheinerman and J. Zito. On the size of hereditary classes of graphs. *J. Combin. Theory Ser. B*, 61(1):16–39, 1994.

[104] S. Shelah. Spectra of monadic second order sentences. *Sci. Math. Jpn.*, 59(2):351–355, 2004.

[105] S. Shelah and M. Doron. Bounded m-ary patch-width are equivalent for $m > 2$. arXiv:math/0607375v1 [math.LO].

[106] R. Simion. Noncrossing partitions. *Discrete Math.*, 217(1-3):367–409, 2000. Formal power series and algebraic combinatorics (Vienna, 1997).

[107] E. Specker. Application of logic and combinatorics to enumeration problems. In *Trends in theoretical computer science (Udine, 1984)*, volume 12 of *Principles Comput. Sci. Ser.*, pages 143–169. Computer Sci. Press, Rockville, MD, 1988.

[108] E. Specker. Modular counting and substitution of structures. *Combin. Probab. Comput.*, 14(1-2):203–210, 2005.

[109] J. Spencer. *The strange logic of random graphs*, volume 22 of *Algorithms and Combinatorics*. Springer-Verlag, Berlin, 2001.

[110] R. P. Stanley. Solution to problem E2546. *Amer. Math. Monthly*, 83(10):813–814, 1976.

[111] R. P. Stanley. *Enumerative combinatorics. Vol. 1*, volume 49 of *Cambridge Studies in Advanced Mathematics*. Cambridge University Press, Cambridge, 1997.

[112] R. P. Stanley. *Enumerative combinatorics. Vol. 2*, volume 62 of *Cambridge Studies in Advanced Mathematics*. Cambridge University Press, Cambridge, 1999.

[113] V. I. Trofimov. Growth functions of some classes of languages. *Cybernetics*, 17:727–731, 1982. translated from *Kibernetika* (1981), 9–12.

[114] V. Vatter. Permutation classes of every growth rate above 2.48188. *Mathematika*, 56:182–192, 2010.

[115] V. Vatter. Small permutation classes. arXiv:0712.4006v2 [math.CO].

[116] V. Vatter. Enumeration schemes for restricted permutations. *Combin. Probab. Comput.*, 17:137–159, 2008.

[117] H. S. Wilf. What is an answer? *Amer. Math. Monthly*, 89(5):289–292, 1982.

A survey of simple permutations

Robert Brignall

Department of Mathematics
University of Bristol
Bristol, BS8 1UJ England

Abstract

We survey the known results about simple permutations. In particular, we present a number of recent enumerative and structural results pertaining to simple permutations, and show how simple permutations play an important role in the study of permutation classes. We demonstrate how classes containing only finitely many simple permutations satisfy a number of special properties relating to enumeration, partial well-order and the property of being finitely based.

1 Introduction

An *interval* of a permutation π corresponds to a set of contiguous indices $I = [a, b]$ such that the set of values $\pi(I) = \{\pi(i) : i \in I\}$ is also contiguous. Every permutation of length n has intervals of lengths 0, 1 and n. If a permutation π has no other intervals, then π is said to be *simple*. For example, the permutation $\pi = 28146357$ is not simple as witnessed by the non-trivial interval 4635 $(= \pi(4)\pi(5)\pi(6)\pi(7))$, while $\sigma = 51742683$ is simple.†

While intervals of permutations have applications in biomathematics, particularly to genetic algorithms and the matching of gene sequences (see Corteel, Louchard, and Pemantle [21] for extensive references), simple permutations form the "building blocks" of permutation classes and have thus received intensive study in recent years. We will see in Section 3 the various ways in which simplicity plays a role in the study of permutation classes, but we begin this short survey by introducing the

---

† Simplicity may be defined for all relational structures; such structures have variously been called *prime* or *indecomposable*.

41

substitution decomposition in Subsection 1.1, and thence by reviewing the structural and enumerative results of simple permutations themselves in Section 2. The rest of this subsection will cover several basic definitions that we will require.

Two finite sequences of the same length, $\alpha = a_1 a_2 \cdots a_n$ and $\beta = b_1 b_2 \cdots b_n$, are said to be *order isomorphic* if, for all i, j, we have $a_i < a_j$ if and only if $b_i < b_j$. As such, each sequence of distinct real numbers is order isomorphic to a unique permutation. Similarly, any given subsequence (or *pattern*) of a permutation π is order isomorphic to a smaller permutation, σ say, and such a subsequence is called a *copy* of σ in π. We may also say that π *contains* σ (or, in some texts, π *involves* σ) and write $\sigma \leq \pi$. If, on the other hand, π does not contain a copy of some given σ, then π is said to *avoid* σ. For example, $\pi = 918572346$ contains 51342 because of the subsequence 91572 ($= \pi(1)\pi(2)\pi(4)\pi(5)\pi(6)$), but avoids 3142.

It will often be useful to view permutations and order isomorphism graphically. Two sets S and T of points in the plane are said to be order isomorphic if the axes for the set S can be stretched and shrunk in some manner to map the points of S bijectively onto the points of T, i.e., if there are strictly increasing functions $f, g : \mathbb{R} \to \mathbb{R}$ such that $\{(f(s_1), g(s_2)) : (s_1, s_2) \in S\} = T$. Note that this forms an equivalence relation since the inverse of a strictly increasing function is also strictly increasing. The *plot* of the permutation π is then the point set $\{(i, \pi(i))\}$, and every finite point set in the plane in which no two points share a coordinate (often called a *generic* or *noncorectilinear* set) is order isomorphic to the plot of a unique permutation. Note that, with a slight abuse of notation, we will say that a point set is order isomorphic to a permutation.

The pattern containment order forms a partial order on the set of all permutations. Downsets of permutations under this order are called *permutation classes*.† In other words, if $\mathcal{C}$ is a permutation class and $\pi \in \mathcal{C}$, then for any permutation σ with $\sigma \leq \pi$ we have $\sigma \in \mathcal{C}$. A given permutation class is often described in terms of its minimal avoidance set, or *basis*. More formally, the basis B of a permutation class $\mathcal{C}$ is the smallest set for which $\mathcal{C} = \{\pi \mid \beta \not\leq \pi \text{ for all } \beta \in B\}$. For a permutation class $\mathcal{C}$, we denote by $\mathcal{C}_n$ the set $\mathcal{C} \cap S_n$, i.e. the permutations in $\mathcal{C}$ of length n, and we refer to $f(x) = \sum |\mathcal{C}_n| x^n$ as the *generating function for $\mathcal{C}$*.

---

† In the past, permutation classes have also been called *closed classes* or *pattern classes*.

Analogues of pattern containment exist for other relational structures; sets of structures closed under taking induced substructures are known as *hereditary properties*. Hereditary properties of graphs have received considerable attention (see Bollobás [11] for a survey of some older results), while more recently attention has been given to hereditary properties of a variety of structures including tournaments, ordered graphs and posets (see, for example, Balogh *et al.* [6, 7, 8], and Bollobás's recent survey [12]).

1.1 Substitution Decomposition

The simple permutations form the elemental building blocks upon which all other permutations are constructed by means of the substitution decomposition.† Analogues of this decomposition exist for every relational structure, and it has frequently arisen in a wide variety of perspectives, ranging from game theory to combinatorial optimization — for references see Möhring [35] or Möhring and Radermacher [36]. Its first appearance seems to be in a 1953 talk by Fraïssé (though only the abstract of this talk [25] survives). It did not appear in an article until Gallai [26] (for an English translation, see [27]), who applied them particularly to the study of transitive orientations of graphs.

Given a permutation σ of length m and nonempty permutations α_1, ..., α_m, the *inflation* of σ by $\alpha_1, \ldots, \alpha_m$ — denoted $\sigma[\alpha_1, \ldots, \alpha_m]$ — is the permutation obtained by replacing each entry $\sigma(i)$ by an interval that is order isomorphic to α_i. For example, $2413[1, 132, 321, 12] = 479832156$. Conversely, a *deflation* of π is any expression of π as an inflation $\pi = \sigma[\pi_1, \pi_2, \ldots, \pi_m]$, and we will call σ a *quotient* of π. We then have the *substitution decomposition* of permutations:

Proposition 1.1 (Albert and Atkinson [1]). *Every permutation may be written as the inflation of a unique simple permutation. Moreover, if π can be written as $\sigma[\alpha_1, \ldots, \alpha_m]$ where σ is simple and $m \geq 4$, then the α_i's are unique.*

Non-unique cases arise when a permutation can be written as an inflation of either 12 or 21, and to recover uniqueness we may choose a particular decomposition in a variety of ways. The one we will use is as follows.

---

† This decomposition is also called the modular decomposition, disjunctive decomposition or X-join in other contexts.

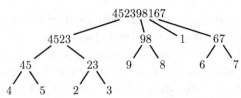

Fig. 1. The substitution decomposition tree of $\pi = 452398167$.

Proposition 1.2 (Albert and Atkinson [1]). *If π is an inflation of 12, then there is a unique sum indecomposable α_1 such that $\pi = 12[\alpha_1, \alpha_2]$ for some α_2, which is itself unique. The same holds with 12 replaced by 21 and "sum" replaced by "skew".*

The substitution decomposition tree for a permutation is obtained by recursively decomposing until we are left only with inflations of simple permutations by singletons. For example, the permutation $\pi = 452398167$ is decomposed as

$$
\begin{aligned}
452398167 &= 2413[3412, 21, 1, 12] \\
&= 2413[21[12, 12], 21[1, 1], 1, 12[1, 1]] \\
&= 2413[21[12[1, 1], 12[1, 1]], 21[1, 1], 1, 12[1, 1]]
\end{aligned}
$$

and its substitution decomposition tree is given in Figure 1.

Computation in Linear Time. The substitution decomposition is most frequently used in solving algorithmic problems, and consequently much attention has been given to its computation in optimal time.†
By its connection to the intervals of a permutation, a first approach to compute the substitution decomposition might be simply to compute all the intervals of our given permutation. Since there may be as many as $N = n(n-1)/2$ such intervals in a permutation of length n, listing these will not yield a linear $O(n)$ algorithm for the substitution decomposition. However, this computation has received significant attention through its connections with biomathematics, with an $O(n + N)$ time algorithm being given by Bergeron, Chauve, Montgolfier and Raffinot [10].‡
The first algorithm to compute the substitution decomposition of a

---

† In particular, graph decomposition has received significant attention, with the first $O(|V| + |E|)$ algorithms appearing in 1994 by McConnell and Spinrad [33] and Cournier and Habib [22].
‡ In fact, Bergeron *et al* show how to compute the "common intervals" — a generalisation of our notion of interval applied to sets of permutations.

permutation in linear time was given by Uno and Yagiura [41], while Bergeron *et al* have since given a simpler algorithm. A *strong interval* of a permutation π is an interval I for which every other interval J satisfies one of $J \subseteq I$, $I \subseteq J$ or $I \cap J = \emptyset$. For example, given $\pi = 234615$, the interval 234 ($= \pi(1)\pi(2)\pi(3)$) is a strong interval, but 23 is not, because it has non-trivial intersection with 34. A permutation can have at most $2n - 1$ strong intervals (note that the n singletons and the whole permutation are all strong intervals), and Bergeron *et al* give an optimal $O(n)$ algorithm to list them all. The substitution decomposition tree of the permutation follows immediately.

It is worth noticing that this algorithm does not give the simple quotients for each internal node of the decomposition tree — indeed, there are currently no linear time algorithms to do this. It is, however, straightforward to compute the label of any particular node in linear time, e.g. by finding a representative symbol for each strong interval lying below the node, and then computing the permutation order isomorphic to this sequence of representatives.

2 Enumeration and Structure

2.1 Enumeration and Asymptotics

The number of simple permutations of length $n = 1, 2, \ldots$ is $s_n = 1, 2, 0, 2, 6, 46, 338, 2926, 28146, \ldots$ (sequence A111111 of [40]). Albert, Atkinson and Klazar [2] showed, in a straightforward argument making use of the substitution decomposition, that the sequence $(s_n)_{n \geq 4}$ is given by

$$s_n = -\mathrm{Com}_n + (-1)^{n+1} \cdot 2,$$

where Com_n is the coefficient of x^n in the functional inverse of $f(x) = \sum_{n=1}^{\infty} n! x^n$ (sequence A059372 of [40]).†

Asymptotically, the sequence s_n may be counted using a probabilistic argument based on counting the intervals of a random permutation. Let the random variable X_k denote the number of intervals of length k in a random permutation π of length n. An interval of length k may be viewed as a mapping from a contiguous set of positions to a contiguous set of values, for which the set of positions must begin at one of the first $n - k + 1$ positions of π, and the lowest point in the set of values must be one of the lowest $n - k + 1$ values of π. Of the $\binom{n}{k}$ sets of values to

---

† The term Com_n is used since the function $f^{-1}(x)$ first appeared in an exercise in Comtet [20].

which the contiguous set of positions may be mapped, only one maps to the chosen contiguous set of values. Thus we have

$$\mathbb{E}[X_k] = \frac{(n-k+1)^2}{\binom{n}{k}} = \frac{(n-k+1)(n-k+1)!k!}{n!}.$$

Our first observation is that, as $n \to \infty$, $\mathbb{E}[X_2] = \frac{2(n-1)}{n} \to 2$. Thus, asymptotically, we should expect to find precisely two intervals of size two in a random permutation. We are seeking the asymptotics of the expected number of proper intervals, i.e. the sum $\sum_{k=2}^{n-1} \mathbb{E}[X_k]$, and want to demonstrate that $\sum_{k=3}^{n-1} \mathbb{E}[X_k] \to 0$ as $n \to 0$. We first consider the cases $k = 3$, $k = 4$, $k = n - 2$ (assuming $n \geq 4$) and $k = n - 1$ separately:

$$\mathbb{E}[X_3] = \frac{6(n-2)}{n(n-1)} \leq \frac{6}{n} \to 0$$

$$\mathbb{E}[X_4] = \frac{4!(n-3)}{n(n-1)(n-2)} \leq \frac{24}{n^2} \to 0$$

$$\mathbb{E}[X_{n-2}] = \frac{3 \cdot 3!}{n(n-1)} \leq \frac{24}{n^2} \to 0$$

$$\mathbb{E}[X_{n-1}] = \frac{4}{n} \to 0.$$

The remaining terms form a partial sum, which converges providing $\frac{\mathbb{E}[X_{k+1}]}{\mathbb{E}[X_k]} < 1$. Simplifying this equation gives $2k^2 - (3n+1)k + n^2 + n + 1 > 0$, a quadratic in k, which yields two roots. The smaller of these satisfies $0 < k^- \leq n$, the larger $k^+ > n$. Thus for $k \leq k^-$, $\mathbb{E}[X_k]$ is decreasing, while for $k^- < k < n$, $\mathbb{E}[X_k]$ is increasing, and hence $\mathbb{E}[X_k] \leq 24/n^2$ for $4 \leq k \leq n - 2$. Thus

$$\sum_{k=4}^{n-2} \mathbb{E}[X_k] \leq (n-5)\frac{24}{n^2} \leq \frac{24}{n} \to 0.$$

Subsequently, the only term of $\sum_{k=2}^{n-1} \mathbb{E}[X_k]$ which is non-zero in the limit $n \to \infty$ is $k = 2$.†

---

† A similar argument can be applied to graphs, but in this case we find that $\mathbb{E}[X_k] \to 0$ as $n \to 0$ for every $2 \leq k \leq n-1$, and thus, asymptotically, almost all graphs are indecomposable. The same applies to tournaments, posets, and (more generally) structures defined on a single asymmetric relation — see Möhring [34].

Ignoring larger intervals, occurrences of intervals of size 2 in a large random permutation π can roughly be regarded as independent events, and as we know the expectation of X_2 is 2, the occurrence of any specific interval is relatively rare. Heuristically, this suggests that X_2 is asymptotically Poisson distributed with parameter 2. Using this heuristic, we have $\Pr(X_2 = 0) \to e^{-2}$ as $n \to \infty$, and so there are approximately $\frac{n!}{e^2}$ simple permutations of length n.

A formal argument for this was implicitly given by Uno and Yagiura [41], and was made explicit by Corteel, Louchard, and Pemantle [21]. The method, however, essentially dates back to the 1940s with Kaplansky [31] and Wolfowitz [42], who considered "runs" within permutations — a *run* is a set of points with contiguous positions whose values are $i, i + 1, \ldots, i + r$ or $i + r, i + r - 1, \ldots, i$, in that order.† A non-probabilistic approach for these first order asymptotics based on Lagrange inversion can be obtained from a more general theorem of Bender and Richmond [9].

More precise asymptotics, meanwhile, have been found using a non-probabilistic method (but one relying on the work of Kaplansky) by Albert, Atkinson, and Klazar [2]. They obtain the following theorem, and note that higher order terms are calculable given sufficient computation:

Theorem 2.1 (Albert, Atkinson and Klazar [2]). *The number of simple permutations of length n is asymptotically given by*

$$\frac{n!}{e^2}\left(1 - \frac{4}{n} + \frac{2}{n(n-1)} + O(n^{-3})\right).$$

2.2 Exceptional Simple Permutations

Given a simple permutation π, one might ask what simple permutations are contained within π. In particular, is there a point that can be removed from π to leave a sequence order isomorphic to a simple permutation? This is not quite true, but allowing one- or two-point deletions suffices. The following theorem is a special case of a more general result for every relational structure whose relations are binary and irreflexive.

Theorem 2.2 (Schmerl and Trotter [39]). *Every simple permutation of length $n \geq 2$ contains a simple permutation of length $n - 1$ or $n - 2$.*

---

† Atkinson and Stitt [5] called permutations containing no runs *strongly irreducible*. Note that this is equivalent to a permutation containing no intervals of size two.

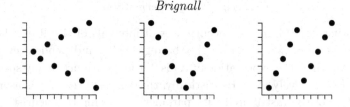

Fig. 2. The two permutations on the left are wedge alternations, the permutation on the right is a parallel alternation.

In most cases, however, a single point deletion is sufficient. If none of the one point deletions of a given simple permutation π is simple, then π is said to be *exceptional*. Schmerl and Trotter call such structures *critically indecomposable*, and present a complete characterisation in the analogous problem for partially ordered sets.

To consider the exceptional simple permutations, we first define a set of permutations called *alternations*. A *horizontal alternation* is a permutation in which every odd entry lies to the left of every even entry, or the reverse of such a permutation. Similarly, a *vertical alternation* is the group-theoretic inverse of a horizontal alternation. Of these alternations, we identify two families in which each "side" of the alternation forms a monotone sequence, namely the *parallel* and *wedge alternations* — see Figure 2 for definitions. While any parallel alternation is already simple or very nearly so, wedge alternations are not. Any wedge alternation may be extended to form a simple permutation by placing a single point in one of two places, thus forming wedge simple permutations of types 1 and 2 — see Figure 3.

The exceptional simple permutations turn out to be precisely the set of parallel alternations:

Theorem 2.3 (Albert and Atkinson [1]). *The only simple permutations that do not have a one point deletion are the simple parallel alternations, i.e. those of the form*

$$246\cdots(2m)135\cdots(2m-1)\quad(m\geq 2)$$

and every symmetry of this permutation.

Proof. Define a poset $(P_\pi, \prec_\pi)$ of the permutation π by $x \prec_\pi y$ if and only if $x < y$ and $\pi(x) < \pi(y)$. Note that the poset $(P_\pi, \prec_\pi)$ has dimension 2, and conversely that all posets of dimension 2 correspond to a unique permutation, up to permutation inverses. A permutation is simple if and only if its corresponding poset is also simple, i.e. P_π

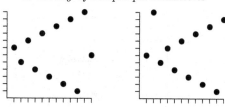

Fig. 3. The two types of wedge simple permutation, type 1 (left) and type 2 (right).

contains no proper nonsingleton subset I for which every pair x, y of I are ordered with respect to all elements of $P_\pi \setminus I$ in the same way. Furthermore, π is exceptional if and only if the corresponding poset is also exceptional (or, more usually, critically indecomposable). Schmerl and Trotter [39] classify all the critically indecomposable posets, from which the result follows. □

2.3 Pin Sequences and Decomposition

In the plot of a permutation, an interval is clearly identified as a set of points enclosed in an axes-parallel rectangle, with no points lying in the regions directly above, below, to the left or to the right. Conversely, since simple permutations contain no non-trivial intervals, any axes-parallel rectangle drawn over the plot of a simple permutation must be "separated" by at least one point in one of these four regions (unless the rectangle contains only one point or the whole permutation). Extending our rectangle to include this extra point puts us in a similar situation, and inductively our rectangle will eventually be extended to contain every point of the permutation. This is the motivating idea behind "pin sequences".

Given points $p_1, \ldots, p_m$ in the plane, we denote by $\mathrm{rect}(p_1, \ldots, p_m)$ the smallest axes-parallel rectangle containing them. A *pin* for the points $p_1, \ldots, p_m$ is any point p_{m+1} not contained in $\mathrm{rect}(p_1, \ldots, p_m)$ that lies either horizontally or vertically amongst them. Such a pin will have a *direction* — up, down, left or right — which records the position of p_{m+1} relative to $\mathrm{rect}(p_1, \ldots, p_m)$. A *pin sequence* is a sequence of points $p_1, p_2, \ldots$ such that p_i is a pin for the points $p_1, \ldots, p_{i-1}$ for every $i \geq 3$, while a *proper pin sequence* is a pin sequence satisfying two further conditions:

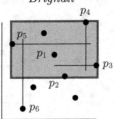

Fig. 4. A pin sequence of the simple permutation 713864295. The shaded box represents $\mathrm{rect}(p_1, p_2, p_3, p_4, p_5)$.

- *Maximality condition*: each pin must be maximal in its direction. That is, in the plot of a permutation, of all possible pins for $p_1, \ldots, p_{i-1}$ having the same direction as p_i $(i \geq 3)$, the pin p_i is furthest from $\mathrm{rect}(p_1, \ldots, p_{i-1})$.†
- *Separation condition*: p_{i+1} must separate p_i from $\mathrm{rect}(p_1, \ldots, p_{i-1})$ $(i \geq 2)$. That is, p_{i+1} must lie horizontally or vertically between p_i and $\mathrm{rect}(p_1, \ldots, p_{i-1})$.

See Figure 4 for an example. Proper pin sequences are intimately connected with simple permutations. In one direction, we have:

Theorem 2.4 (Brignall, Huczynska and Vatter [16]). *If $p_1, \ldots, p_m$ is a proper pin sequence of length $m \geq 5$ then one of the sets of points $\{p_1, \ldots, p_m\}$, $\{p_1, \ldots, p_m\} \setminus \{p_1\}$, or $\{p_1, \ldots, p_m\} \setminus \{p_2\}$ is order isomorphic to a simple permutation.*

This should come as no surprise — by our motivation, pin sequences encapsulate precisely what it means to be simple. Moreover, we also expect to be able to find long proper pin sequences within arbitrary simple permutations. If no such pin sequence exists, then we encounter the two families of alternations defined in the previous subsection.

Theorem 2.5 (Brignall, Huczynska and Vatter [16]). *Every simple permutation of length at least $2(256k^8)^{2k}$ contains a proper pin sequence of length $2k$, a parallel alternation of length $2k$, or a wedge simple permutation of length $2k$.*

Sketch of proof. Suppose that a simple permutation π of length n con-

---

† In certain situations, notably where pin sequences are used to construct permutations from scratch (rather than taking pin sequences from the points in the plot of a permutation), the maximality condition is replaced with the *externality condition*, requiring that p_{i+1} lies outside $\mathrm{rect}(p_1, \ldots, p_i)$.

tains neither a proper pin sequence of length $2k$ nor a parallel or wedge alternation of length $2k$. A pin sequence $p_1, p_2, \ldots, p_n$ of π is said to be *right-reaching* if p_n corresponds to the rightmost point of π. Moreover, from each pair of points in a simple permutation π there exists a right-reaching proper pin sequence. We now consider the collection of $\lfloor n/2 \rfloor$ proper right-reaching pin sequences of π beginning with the first and second points, the third and fourth points, and so on, reading from left to right.

Two pin sequences $p_1, p_2, \ldots$ and $q_1, q_2, \ldots$ are said to *converge at the point x* if there exists i and j such that $p_i = q_j = x$ but $\{p_1, \ldots, p_{i-1}\}$ and $\{q_1, \ldots, q_{i-1}\}$ are disjoint. Our $\lfloor n/2 \rfloor$ proper right-reaching pin sequences all end at the rightmost point of π, and so every pair of these must converge at some point.

If $16k$ proper pin sequences converge at a point x, then we may show there must exist an alternation of length at least $2k$. Such an alternation may not necessarily be parallel or wedge, but, by the Erdős-Szekeres Theorem, every alternation of length at least $2k^4$ must contain a parallel or wedge alternation of length at least $2k$. Accordingly, since π contains no parallel or wedge alternation of length $2k$, fewer than $16k^4$ of our pin sequences can converge at any point. Since no pin sequence contains $2k$ or more pins, the number of right-reaching pin sequences is bounded by how many can converge at each steps towards the rightmost pin, and there can be at most $2k - 1$ steps. Thus $\lfloor n/2 \rfloor < 2(16k^4)^{2k}$.

Finally, if this process has produced a wedge alternation, it is then necessary to demonstrate the existence of one additional point to form a wedge simple permutation of type 1 or 2. For a wedge alternation containing $4k^2$ points and oriented $<$ (as in Figure 3), consider a pin sequence starting on the two leftmost points of the alternation. At some stage, either the pin sequence attains length $2k$ (which we have assumed does not exist), or the pin has "jumped" a long way in the wedge alternation, giving the additional point required to form a wedge simple permutation of type 1 or 2 with length $2k$. Adjusting the bound $\lfloor n/2 \rfloor < 2(16k^4)^{2k}$ to take into account that a wedge alternation of length $4k^2$ must be avoided gives the stated bound $n < 2(256k^8)^{2k}$. □

An immediate consequence of this decomposition is that within a suitably long simple permutation σ we may find two "almost disjoint" simple subsequences by considering subsequences of the long proper pin sequence, parallel alternation or wedge simple permutation contained in σ. This result has consequences for certain permutation classes, which we will discuss in Subsection 3.5.

Corollary 2.6 (Brignall, Huczynska and Vatter [16]). *There is a function $f(k)$ such that every simple permutation of length at least $f(k)$ contains two simple subsequences, each of length at least k, sharing at most two entries.*

To observe that it may be necessary that the two simple subsequences share exactly two entries, consider the family of type 2 wedge simple permutations (i.e. those of the form $m(2m)(m-1)(m+1)(m-2)(m+2)\cdots 1(2m-1)$), in which the first two entries are required in every simple subsequence.

2.4 Simple Extensions

In addition to finding the underlying structures in simple permutations, we may ask how, for an arbitrary permutation π, we may embed π into a simple permutation, and how long this permutation has to be. The analogous problem for tournaments was solved by Erdős, Fried, Hajnal and Milner [23], where they demonstrated that every tournament may be embedded in a simple tournament containing at most two extra vertices. Furthermore, in another paper published in the same year, Erdős, Hajnal and Milner [24] listed the cases when one point was not sufficient. For permutations, however, adding two points is rarely going to be sufficient. In fact, in the case of an increasing permutation $12\cdots n$ of length n, an additional $\lceil (n+1)/2 \rceil$ points are both necessary and sufficient. The same bound can be obtained inductively for arbitrary permutations by means of the substitution decomposition, though it is not known when fewer points are sufficient.

Theorem 2.7 (Brignall, Ruškuc and Vatter [18]). *Every permutation π on n symbols has a simple extension with at most $\lceil (n+1)/2 \rceil$ additional points.*

Cases for graphs and posets are also obtainable by the same method. In the former, at most $\lceil \log_2(n+1) \rceil$ additional vertices are required, while for the latter, the bound is $\lceil (n+1)/2 \rceil$, as in the permutation case.

3 Permutation Classes with Finitely Many Simples

Considerable attention has been paid in recent years to the study of permutation classes which contain only finitely many simple permutations.

Much is known about such classes both in terms of enumeration and structure, and of course these two considerations are not unrelated — it is precisely the structure (given in terms of the substitution decomposition) of permutations lying in classes containing only finitely many simple permutations that gives these classes this well behaved enumeration.

3.1 Substitution Closures

As a preliminary to the results of this section, we consider permutation classes which may be described exactly by their (not necessarily finite) set of simple permutations. A class $\mathcal{C}$ of permutations is *substitution-closed* if $\sigma[\alpha_1, \ldots, \alpha_m] \in \mathcal{C}$ for all $\sigma, \alpha_1, \ldots, \alpha_m \in \mathcal{C}$. The *substitution-closure*† of a set X, $\langle X \rangle$, is defined as the smallest substitution-closed class containing X. (This concept is well-defined because the intersection of substitution-closed classes is substitution-closed, and the set of all permutations is substitution-closed.)

Letting $\text{Si}(\mathcal{C})$ denote the set of simple permutations in the class $\mathcal{C}$, we observe that $\text{Si}(\mathcal{C}) = \text{Si}(\langle \mathcal{C} \rangle)$: every permutation in $\mathcal{C}$ is an inflation of a member of $\text{Si}(\mathcal{C})$ so it follows (e.g., by induction) that $\mathcal{C} \subseteq \langle \text{Si}(\mathcal{C}) \rangle$. Thus $\langle \mathcal{C} \rangle \subseteq \langle \text{Si}(C) \rangle$, establishing that $\text{Si}(\mathcal{C}) = \text{Si}(\langle \mathcal{C} \rangle)$. As substitution-closed classes are uniquely determined by their sets of simple permutations, $\langle \mathcal{C} \rangle$ is the largest class with this property. For example, the substitution closure of $\text{Av}(132)$ is the largest class whose only simple permutations are 1, 12, and 21, which is precisely the class $\text{Av}(2413, 3142)$ of *separable permutations*.

It is quite easy to decide if a permutation class given by a finite basis is substitution-closed:

Proposition 3.1 (Albert and Atkinson [1]). *A permutation class is substitution-closed if and only if each of its basis elements is simple.*

One may also wish to compute the basis of $\langle \mathcal{C} \rangle$. This is routine for classes with finitely many simple permutations (see Proposition 3.10), but much less so in general. An example of a finitely based class whose substitution-closure is infinitely based is $\text{Av}(4321)$ — the basis of its substitution-closure contains an infinite family of permutations, one of which is shown in Figure 5.

The natural question is then:

---

† The substitution closure has extensively been called the *wreath closure* in the literature.

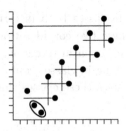

Fig. 5. A basis element of the substitution closure of Av(4321).

Question 3.2. Given a finite basis B, is it decidable whether $\langle \mathrm{Av}(B) \rangle$ is finitely based?†

The importance of substitution-closed classes will become increasingly evident throughout this section. In particular, it is often fruitful to prove results about a class containing only finitely many simple permutations by first considering its substitution closure, and then adding basis restrictions to recover the original class.

3.2 Algebraic Generating Functions

When a class is enumerated by an algebraic generating function, we intuitively expect to find some recursive description of the permutations in the class. The converse intuition is also good, i.e. that classes constructed via some set of recursions should have algebraic generating functions. In a class $\mathcal{C}$ which has only finitely many simple permutations, any long permutation must be constructed recursively via the substitution decomposition, starting from the finite set $\mathrm{Si}(\mathcal{C})$. Applying our intuition then immediately suggests the following result:

Theorem 3.3 (Albert and Atkinson [1]). *A permutation class with only finitely many simple permutations has a readily computable algebraic generating function.*

Theorem 3.3 was obtained by first proving it in the case of substitution-closed classes, and then demonstrating how adding an extra basis restriction does not affect the algebraicity of the generating function. A

---

† This question has recently been considered by Atkinson, Ruškuc and Smith [4], who settled the question in the case of singleton bases B. The analogous question for graphs was raised by Giakoumakis [28] and has received a sizable amount of attention, see for example Zverovich [43].

stronger result is true if, in addition to containing only finitely many simple permutations, our permutation class avoids a decreasing permutation of some length n, namely that such a class is enumerated by a rational generating function [1].

We may view permutation classes as subsets of the unique substitution-closed class containing the same set of simple permutations. When the substitution-closed class contains only finitely many simple permutations, Theorem 3.3 shows that subclasses — now thought of merely as particular subsets — have algebraic generating functions. These subsets are not alone in this property, however; there are many other subsets which are enumerated by algebraic generating functions.

Calling any set of permutations P a *property*, we may say that a permutation π *satisfies* P if $\pi \in P$. A set $\mathcal{P}$ of properties is said to be *query-complete* if, for every simple permutation σ and property $P \in \mathcal{P}$, one may determine whether $\sigma[\alpha_1, \ldots, \alpha_m]$ satisfies P simply by knowing which properties of $\mathcal{P}$ each α_i satisfies. For example, avoidance of a given pattern is query-complete, since the property $\mathrm{Av}(\beta)$ lies in the query-complete set $\{\mathrm{Av}(\delta) : \delta \leq \beta\}$. Moreover, since this query-complete set is finite, it is said to be *finite query-complete*.

Theorem 3.4 (Brignall, Huczynska and Vatter [17]). *Let $\mathcal{C}$ be a permutation class containing only finitely many simple permutations, $\mathcal{P}$ a finite query-complete set of properties, and $\mathcal{Q} \subseteq \mathcal{P}$. The generating function for the set of permutations in $\mathcal{C}$ satisfying every property in $\mathcal{Q}$ is algebraic over $\mathbb{Q}(x)$.*

Properties known to lie in finite query-complete sets include the set of alternating permutations, even permutations, Dumont permutations of the first kind, and those avoiding any number of blocked or barred permutations. For example, a permutation π of length n is *alternating* if, for all $i \in [2, n-1]$, $\pi(i)$ does not lie between $\pi(i-1)$ and $\pi(i+1)$. We can then explicitly state the finite query-complete set of properties:

Lemma 3.5. *The set of properties consisting of*

- $AL = \{alternating\ permutations\}$,
- $BR = \{permutations\ beginning\ with\ a\ rise,\ i.e.,\ permutations\ with\ \pi(1) < \pi(2)\}$,
- $ER = \{permutations\ ending\ with\ a\ rise\}$, *and*
- $\{1\}$.

is query-complete.

Proof. It is easy to show that $\{\{1\}, BR, ER\}$ is query-complete:

$$\sigma[\alpha_1, \ldots, \alpha_m] \in BR \quad \Longleftrightarrow \quad \alpha_1 \in BR \text{ or } (\alpha_1 = 1 \text{ and } \sigma \in BR),$$

$$\sigma[\alpha_1, \ldots, \alpha_m] \in ER \quad \Longleftrightarrow \quad \alpha_m \in ER \text{ or } (\alpha_m = 1 \text{ and } \sigma \in ER).$$

If $\pi = \sigma[\alpha_1, \ldots, \alpha_m]$ is an alternating permutation, we must have $\alpha_1, \ldots, \alpha_m \in AL$. After checking that all the entries of π up to and including the interval corresponding to $\sigma(i)$ are alternating, we must consider two cases: the first where $\sigma(i) > \sigma(i+1)$, the second $\sigma(i) < \sigma(i+1)$. We show only the former, the latter being analogous. Since $\sigma(i) > \sigma(i+1)$, π contains a descent between its $\sigma(i)$ interval and its $\sigma(i+1)$ interval. Thus α_i is allowed to be 1 (i.e., $\alpha_i \in \{1\}$) only if $i = 1$ or $\sigma(i-1) < \sigma(i)$, while if $\alpha_i \neq 1$ then we must have $\alpha_i \in ER$, and whether or not α_i is 1 we must have $\alpha_{i+1} \in BR \cup \{1\}$. $\square$

The intersection of two properties known to lie in finite query-complete sets again lies in a finite query-complete set, and thus any combination of the properties listed above will lie in a finite query complete-set. It is also possible to show that the number of involutions of a permutation class containing only finitely many simple permutations is enumerated by an algebraic generating function, as is the cyclic closure of such a permutation class, and these may again be combined with other properties. Thus, for example, the number of alternating even involutions in a permutation class with only finitely many simple permutations is enumerated algebraically.

3.3 Partial Well-Order

An *antichain* is a set of pairwise incomparable elements. Any partial order is *partially well-ordered* if it contains neither an infinite properly decreasing sequence nor an infinite antichain, but in the case of permutation classes we need only check the latter condition. Many of the results we are about to present may be applied to a variety of relational structures, though here we will restrict our attention to the permutation case.

Infinite antichains of permutations rely heavily on the structure of simple permutations to maintain their incomparability — two permutations of an antichain are incomparable either because their quotients are incomparable, or because there are incomparable blocks in the substitution decomposition — and thus it should not be surprising that infinite antichains arise only when we have infinitely many different quotients, *viz.* infinitely many simple permutations. This intuition is confirmed by studying classes containing only finitely many simple permutations — we will see that such classes are partially well-ordered.

To prove that a permutation class is not partially well-ordered, it is sufficient to show that the class contains an infinite antichain. In fact, we may restrict our search somewhat, namely to those antichains that are *minimal* under the following ordering: $A \preceq B$ if and only if for every $\beta \in B$ there exists $\alpha \in A$ such that $\alpha \leq \beta$. Essentially following Nash-Williams' "minimal bad sequence" argument [38], we obtain:

Proposition 3.6. *Every non-partially well-ordered permutation class contains an antichain minimal under $\preceq$.*

On the face of it, this result may not appear to be much of a gain. However, a helpful structural characterisation of antichains minimal under $\preceq$ exists and is due to Gustedt [29]; we state here only the part of the result that we need. The *closure* of a set A of permutations is the set $\mathrm{Cl}(A) = \{\pi : \pi \leq \alpha \text{ for some } \alpha \in A\}$, and accordingly the *proper closure* is the set $\mathrm{Cl}(A) \setminus A$.

Lemma 3.7 (Gustedt [29]). *The proper closure of an antichain minimal under $\preceq$ is partially well-ordered.*

Proving that a given permutation class is partially well-ordered is typically a harder task. The primary tool is a result due to Higman [30] which can be used to prove our anticipated Proposition 3.9 — see Albert and Atkinson's proof [1]. However, using our earlier observations on antichains, we will not in fact need Higman's Theorem and so have omitted it for brevity. Instead, we require one more straightforward observation:

Proposition 3.8. *The product $(P_1, \leq_1) \times \cdots \times (P_s, \leq_s)$ of a collection of partial orders is partially well ordered if and only if each of them is partially well-ordered.*

No further results are required. We follow the proof given by Gustedt; it is worth noting that Albert and Atkinson's proof also makes use of the substitution decomposition in a similar way.

Proposition 3.9 (Gustedt [29]; Albert and Atkinson [1]). *Every permutation class with only finitely many simple permutations is partially well-ordered.*

Proof. Suppose to the contrary that the class $\mathcal{C}$ contains an infinite antichain but only finitely many simple permutations. By Proposition 3.6, $\mathcal{C}$ contains an infinite antichain minimal under $\preceq$. Moreover, there is an infinite subset A of this antichain for which every element is an inflation of the same simple permutation, say σ. Let $\mathcal{D} = \mathrm{Cl}(A) \setminus A$ denote the proper closure of A, noting that $\mathcal{D}$ is partially well-ordered by Lemma 3.7. It is easy to see that the permutation containment order, when restricted to inflations of σ, is isomorphic to a product order: $\sigma[\alpha_1, \ldots, \alpha_m] \leq \sigma[\alpha'_1, \ldots, \alpha'_m]$ if and only if $\alpha_i \leq \alpha'_i$ for all $i \in [m]$. However, this implies that A is an infinite antichain in a product $\mathcal{D} \times \cdots \times \mathcal{D}$ of partially well-ordered posets, contradicting Proposition 3.8. $\qquad\square$

3.4 Finite Basis

That a class containing only finitely many simple permutations is finitely based arises by first considering its substitution closure. Our first task is to compute the basis of a substitution closed class containing only finitely many simple permutations, which is easily done using Theorem 2.2:

Proposition 3.10. *If the longest simple permutations in $\mathcal{C}$ have length k then the basis elements of $\langle \mathcal{C} \rangle$ have length at most $k + 2$.*

Proof. The basis of $\langle \mathcal{C} \rangle$ is easily seen to consist of the minimal (under the pattern containment order) simple permutations not contained in $\mathcal{C}$ (cf. Proposition 3.1). Let π be such a permutation of length n. Theorem 2.2 shows that π contains a simple permutation σ of length $n - 1$ or $n - 2$. If $n \geq k + 3$, then $\sigma \notin \mathcal{C}$, so $\sigma \notin \langle \mathcal{C} \rangle$ and thus π cannot lie in the basis of $\langle \mathcal{C} \rangle$. $\qquad\square$

For example, using this Proposition it can be computed that the wreath closure of 1, 12, 21, and 2413 is $\mathrm{Av}(3142, 25314, 246135, 362514)$.

The finite basis result for arbitrary permutation classes containing only finitely many simple permutations now follows by recalling that all such classes are partially well-ordered. Thus:

Theorem 3.11 (Murphy [37]; Albert and Atkinson [1]). *Every permutation class containing only finitely many simple permutations is finitely based.*

Proof. Let $\mathcal{C}$ be a class containing only finitely many permutations. By Proposition 3.10, $\langle \mathcal{C} \rangle$ is finitely based, and by Proposition 3.9 it is partially well ordered. The class $\mathcal{C}$ must therefore avoid all elements in the basis of $\langle \mathcal{C} \rangle$, together with the minimal elements of $\langle \mathcal{C} \rangle$ not belonging to $\mathcal{C}$, which form an antichain. By its partial well ordering any antichain in $\langle \mathcal{C} \rangle$ is finite, and so there can only be finitely many basis elements of $\mathcal{C}$. $\qquad\square$

3.5 Finding Finitely Many Simples

In order to find whether the above properties apply to a given permutation class defined by its (finite) basis, it is of course necessary to know which simple permutations lie in the class, or at least whether there are only finitely many of them. In the first instance, we may use Theorem 2.2 simply to look for the simple permutations of each length $n = 1, 2, \ldots$, since if we encounter two consecutive integers for which the class has no simple permutations of those lengths, then there can be no simple permutations of any greater length.

One of the simplest classes with only finitely many simple permutations is Av(132), containing only 1, 12 and 21. Bóna [13] and Mansour and Vainshtein [32] showed that that for every r, the class of all permutations containing at most r copies of 132 has an algebraic generating function. The reason for this algebraicity is as we might hope: these classes contain only finitely many simple permutations, and this follows from a more general result that relies on Corollary 2.6.

Denote by $\mathrm{Av}(\beta_1^{\leq r_1}, \beta_2^{\leq r_2}, \ldots, \beta_k^{\leq r_k})$ the set of permutations that have at most r_1 copies of β_1, r_2 copies of β_2, and so on. It should be clear that any such set forms a permutation class, although finding its basis is perhaps less obvious. Atkinson [3] showed that the basis elements of the class can have length at most $\max\{(r_i + 1)|\beta_i| : i \in [k]\}$. For example, $\mathrm{Av}(132^{\leq 1}) = \mathrm{Av}(1243, 1342, 1423, 1432, 2143, 35142, 354162, 461325, 465132)$. We then have the following result:

Theorem 3.12 (Brignall, Huczynska and Vatter [16]). *If the class* $\mathrm{Av}(\beta_1, \beta_2, \ldots, \beta_k)$ *contains only finitely many simple permutations then the class* $\mathrm{Av}(\beta_1^{\leq r_1}, \beta_2^{\leq r_2}, \ldots, \beta_k^{\leq r_k})$ *also contains only finitely many simple permutations for all choices of nonnegative integers* $r_1, r_2, \ldots, r_k$.

Proof. We need to show that for any choice of nonnegative integers $r_1, r_2, \ldots, r_k$, only finitely many simple permutations contain at most r_i copies of β_i for each $i \in [k]$. We may suppose that $|\beta_i| \geq 3$ for all $i \in [k]$ since if any β_i is of length 1 or 2 then the theorem follows easily. We now proceed by induction. The base case, arising when $r_i = 0$ for all i, follows trivially, so suppose that some $r_j > 0$ and set

$$g(r_1, r_2, \ldots, r_k) = f(g(r_1, r_2, \ldots, r_{j-1}, \lfloor r_j/2 \rfloor, r_{j+1}, \ldots, r_k)),$$

where f is the function from Corollary 2.6. By that result, every simple permutation π of length at least $g(r_1, r_2, \ldots, r_k)$ contains two simple subsequences of length at least $f(g(r_1, r_2, \ldots, r_{j-1}, \lfloor r_j/2 \rfloor, r_{j+1}, \ldots, r_k))$, and by induction each of these simple subsequences contains more than $\lfloor r_j/2 \rfloor$ copies of β_j. Moreover, because these simple subsequences share at most two entries, their copies of β_j are distinct, and thus π contains more than r_j copies of β_j, proving the theorem.   □

For a general answer to the decidability question, we turn to Theorem 2.5, which reduces the task to checking whether the permutation class in question contains arbitrarily long proper pin sequences, parallel alternations or wedge simple alternations of types 1 and 2. This may be done algorithmically:

Theorem 3.13 (Brignall, Ruškuc and Vatter [19]). *It is possible to decide if a permutation class given by a finite basis contains infinitely many simple permutations.*

Sketch of proof. To determine whether a permutation class $\mathrm{Av}(B)$ contains only finitely many parallel alternations oriented \\, we need only check that there is a permutation in B that is contained in such an alternation. More simply, it is sufficient to verify that B contains a permutation in the class $\mathrm{Av}(123, 2413, 3412)$. To check the other orientations, we need simply test the same condition for every symmetry of this class. Similarly, to check the wedge simple permutations of types 1 and 2, one needs to ensure that B contains an element of every symmetry of $\mathrm{Av}(1243, 1324, 1423, 1432, 2431, 3124, 4123, 4132, 4231, 4312)$ and $\mathrm{Av}(2134, 2143, 3124, 3142, 3241, 3412, 4123, 4132, 4231, 4312)$ respectively.

Thus it remains to determine whether $\mathrm{Av}(B)$ contains arbitrarily long proper pin sequences. This is done by encoding proper pin sequences as "strict pin words" over the four-letter alphabet of *directions* $\{L, R, U, D\}$ and subsequences of proper pin sequences as pin words over an eight-letter alphabet consisting of the four directions and four *numerals* $\{1, 2, 3, 4\}$. These numerals correspond to the four quadrants around an origin which is placed "close" to the first two points of the pin sequence in such a way that whenever a pin is not included in the subsequence the next pin that is included is encoded as a numeral corresponding to the quadrant in which it lies. The permutation containment order restricted to pin sequences and their subsequences corresponds to an order on these words.

For every $\beta \in B$, list all (non-strict) pin words corresponding to β. It may be shown that the set of strict pin words containing any non-strict pin word forms a regular language, and hence the union of all strict pin words containing any pin word corresponding to some β forms a regular language. The complement of this set of strict pin words corresponds to the proper pin sequences lying in $\mathrm{Av}(B)$, and also forms a regular language. It is decidable whether a regular language contains arbitrarily long words or not, from which the result follows. □

3.6 Algorithms

Linear Time Membership. Bose, Buss and Lubiw [14] showed that deciding whether a given permutation lies in some permutation class is in general NP-complete, but that one may use the substitution decomposition to decide whether a permutation is separable in polynomial time. Out of some of the recent machinery surveyed here comes an indication that, given a permutation class $\mathcal{C}$ that contains only finitely many simple permutations, it may be decided in linear time whether an arbitrary permutation π of length n lies in $\mathcal{C}$. The approach relies first and foremost on the fact that we may compute the substitution decomposition of any permutation in linear time, as mentioned in Subsection 1.1. We begin by first performing some precomputations specific to the class $\mathcal{C}$, all of which may be done essentially in constant time:

- Compute $\mathrm{Si}(\mathcal{C})$, the set of simple permutations in $\mathcal{C}$.
- Compute the basis B of $\mathcal{C}$, noting that permutations in B can be no longer than $\max\limits_{\sigma \in \mathrm{Si}(\mathcal{C})} |\sigma| + 2$ by the Schmerl-Trotter Theorem 2.2.
- For every β either lying in B or contained in a permutation lying in B, list all expressions of β as a *lenient inflation* (an inflation $\sigma[\gamma_1, \ldots, \gamma_m]$ in which the γ_i's are allowed to be empty) of each $\sigma \in \mathrm{Si}(\mathcal{C})$.

With this one-time work done, we now take our candidate permutation π of length n and compute its substitution decomposition, $\pi = \sigma[\alpha_1, \ldots, \alpha_m]$. Now, after first trivially checking that the quotient σ lies in $\mathcal{C}$, we look at all the expressions of each $\beta \in B$ as lenient inflations of σ. Note that if some $\beta \leq \pi$, there must exist an expression of β as a lenient inflation $\beta = \sigma[\gamma_1, \ldots, \gamma_m]$ so that $\gamma_i \leq \alpha_i$ for every $i = 1, \ldots, m$.

Thus, taking each lenient inflation $\beta = \sigma[\gamma_1, \ldots, \gamma_m]$ in turn, we look recursively at each block, testing to see if $\gamma_i \leq \alpha_i$ is true. Though this recursion makes the linear-time complexity non-obvious, note that the number of levels of recursion that are required cannot be more than the maximum depth of the substitution decomposition tree, which itself cannot have more than $2n$ nodes. The recursion will eventually reduce the problem to making only trivial comparisons, each of which is immediately answerable in constant time.

Longest Common Pattern. A possible generalisation of the pattern containment problem is that of computing the longest common pattern of two permutations. Bouvel and Rossin [15] demonstrate a general algorithm to find the longest common pattern between two permutations π_1 and π_2, relying on the substitution decomposition of either π_1 or π_2. For general π_1 and π_2 this does not run in polynomial time, but in the special case where π_1 comes from a permutation class whose simple permutations are of length at most d, their algorithm runs in $O(\min(n_1, n_2) n_1 n_2^{2d+2})$, where $n_1 = |\pi_1|$ and $n_2 = |\pi_2|$. In particular, they give an explicit $O(n^8)$ algorithm for finding the longest common pattern between two permutations of length n given one is separable.

4 Concluding Remarks

As we have seen, much is known about permutation classes containing only finitely many simple permutations. On the other hand, little is known in general about classes containing infinitely many simple permutations. The way in which many of the results presented in Section 3 are obtained suggests that a first step would be to restrict our attention to substitution-closed classes. There is one result to support this approach in the more general context: it is known that a substitution-closed class has an algebraic generating function if and only if the simple permutations of the class are also enumerated by an algebraic generating function [1].

Related to this is the question of partial well-order — we have seen that simple permutations are intimately related to the structure of fundamental antichains, but precisely how they are related remains unknown. In particular, if a permutation class is partially well-ordered, what can be said about its set of simple permutations?

Recalling that the substitution decomposition is defined analogously for all relational structures, there is no reason why some of the results reviewed here cannot be extended to other relational structures. While questions such as the enumeration of hereditary properties are not obviously extendable (since isomorphism between structures must be taken into account), questions relating to the decomposition of simple or indecomposable objects and to the study of partial well-order in hereditary properties are likely to behave in a similar way for all structures.

Acknowledgments. The author thanks Vince Vatter and the anonymous referee for their helpful and enriching comments. Thanks is also owed to Mathilde Bouvel for fruitful discussions.

References

[1] M. H. Albert and M. D. Atkinson. Simple permutations and pattern restricted permutations. *Discrete Math.*, 300(1-3):1–15, 2005.

[2] M. H. Albert, M. D. Atkinson, and M. Klazar. The enumeration of simple permutations. *J. Integer Seq.*, 6(4):Article 03.4.4, 18 pp., 2003.

[3] M. D. Atkinson. Restricted permutations. *Discrete Math.*, 195(1-3):27–38, 1999.

[4] M. D. Atkinson, N. Ruškuc, and R. Smith. Substitution-closed pattern classes. Preprint.

[5] M. D. Atkinson and T. Stitt. Restricted permutations and the wreath product. *Discrete Math.*, 259(1-3):19–36, 2002.

[6] J. Balogh, B. Bollobás, and R. Morris. Hereditary properties of partitions, ordered graphs and ordered hypergraphs. *European J. Combin.*, 27(8):1263–1281, 2006.

[7] J. Balogh, B. Bollobás, and R. Morris. Hereditary properties of combinatorial structures: posets and oriented graphs. *J. Graph Theory*, 56(4):311–332, 2007.

[8] J. Balogh, B. Bollobás, and R. Morris. Hereditary properties of tournaments. *Electron. J. Combin.*, 14(1):Research Paper 60, 25 pp., 2007.

[9] E. A. Bender and L. B. Richmond. An asymptotic expansion for the coefficients of some power series. II. Lagrange inversion. *Discrete Math.*, 50(2-3):135–141, 1984.

[10] A. Bergeron, C. Chauve, F. de Montgolfier, and M. Raffinot. Computing common intervals of k permutations, with applications to modular decomposition of graphs. In *Algorithms – ESA 2005*, volume 3669/2005 of *Lecture Notes in Computer Science*, pages 779–790. Springer Berlin, Heidelberg, 2005.

[11] B. Bollobás. Hereditary properties of graphs: asymptotic enumeration, global structure, and colouring. In *Proceedings of the International Congress of Mathematicians, Vol. III (Berlin, 1998)*, Extra Vol. III, pages 333–342, 1998.

[12] B. Bollobás. Hereditary and monotone properties of combinatorial structures. In A. Hilton and J. Talbot, editors, *Surveys in Combinatorics 2007*, number 346 in London Mathematical Society Lecture Note Series, pages 1–39. Cambridge University Press, 2007.

[13] M. Bóna. The number of permutations with exactly r 132-subsequences is P-recursive in the size! *Adv. in Appl. Math.*, 18(4):510–522, 1997.

[14] P. Bose, J. F. Buss, and A. Lubiw. Pattern matching for permutations. *Inform. Process. Lett.*, 65(5):277–283, 1998.

[15] M. Bouvel and D. Rossin. Longest common pattern between two permutations. arXiv:math/0611679v1 [math.CO].

[16] R. Brignall, S. Huczynska, and V. Vatter. Decomposing simple permutations, with enumerative consequences. *Combinatorica*, 28:385–400, 2008.

[17] R. Brignall, S. Huczynska, and V. Vatter. Simple permutations and algebraic generating functions. *J. Combin. Theory Ser. A*, 115(3):423–441, 2008.

[18] R. Brignall, N. Ruškuc, and V. Vatter. Simple extensions of combinatorial structures. arXiv:math/0911.4378v1 [math.CO].

[19] R. Brignall, N. Ruškuc, and V. Vatter. Simple permutations: decidability and unavoidable substructures. *Theoret. Comput. Sci.*, 391(1–2):150–163, 2008.

[20] L. Comtet. *Advanced combinatorics*. D. Reidel Publishing Co., Dordrecht, 1974.

[21] S. Corteel, G. Louchard, and R. Pemantle. Common intervals of permutations. *Discrete Math. Theor. Comput. Sci.*, 8(1):189–214, 2006.

[22] A. Cournier and M. Habib. A new linear algorithm for modular decomposition. In *Trees in algebra and programming—CAAP '94 (Edinburgh, 1994)*, volume 787 of *Lecture Notes in Comput. Sci.*, pages 68–84. Springer, Berlin, 1994.

[23] P. Erdős, E. Fried, A. Hajnal, and E. C. Milner. Some remarks on simple tournaments. *Algebra Universalis*, 2:238–245, 1972.

[24] P. Erdős, A. Hajnal, and E. C. Milner. Simple one-point extensions of tournaments. *Mathematika*, 19:57–62, 1972.

[25] R. Fraïssé. On a decomposition of relations which generalizes the sum of ordering relations. *Bull. Amer. Math. Soc.*, 59:389, 1953.

[26] T. Gallai. Transitiv orientierbare Graphen. *Acta Math. Acad. Sci. Hungar*, 18:25–66, 1967.

[27] T. Gallai. *A translation of T. Gallai's paper: "Transitiv orientierbare Graphen"*, pages 25–66. Wiley-Intersci. Ser. Discrete Math. Optim. Wiley, Chichester, 2001.

[28] V. Giakoumakis. On the closure of graphs under substitution. *Discrete Math.*, 177(1-3):83–97, 1997.

[29] J. Gustedt. Finiteness theorems for graphs and posets obtained by compositions. *Order*, 15:203–220, 1999.

[30] G. Higman. Ordering by divisibility in abstract algebras. *Proc. London Math. Soc. (3)*, 2:326–336, 1952.

[31] I. Kaplansky. The asymptotic distribution of runs of consecutive elements. *Ann. Math. Statistics*, 16:200–203, 1945.

[32] T. Mansour and A. Vainshtein. Counting occurrences of 132 in a permutation. *Adv. in Appl. Math.*, 28(2):185–195, 2002.

[33] R. M. McConnell and J. P. Spinrad. Linear-time modular decomposition and efficient transitive orientation of comparability graphs. In *Proceedings of the Fifth Annual ACM-SIAM Symposium on Discrete Algorithms (Arlington, VA, 1994)*, pages 536–545, New York, 1994. ACM.

[34] R. H. Möhring. On the distribution of locally undecomposable relations and independence systems. In *Colloquium on Mathematical Methods of Operations Research (Aachen, 1980)*, volume 42 of *Methods Oper. Res.*, pages 33–48. Athenäum/Hain/Hanstein, Königstein/Ts., 1981.

[35] R. H. Möhring. Algorithmic aspects of the substitution decomposition in optimization over relations, sets systems and Boolean functions. *Ann. Oper. Res.*, 4(1-4):195–225, 1985.

[36] R. H. Möhring and F. J. Radermacher. Substitution decomposition for discrete structures and connections with combinatorial optimization. In *Algebraic and combinatorial methods in operations research*, volume 95 of *North-Holland Math. Stud.*, pages 257–355. North-Holland, Amsterdam, 1984.

[37] M. M. Murphy. *Restricted Permutations, Antichains, Atomic Classes, and Stack Sorting*. PhD thesis, Univ. of St Andrews, 2002.

[38] C. S. J. A. Nash-Williams. On well-quasi-ordering finite trees. *Proc. Cambridge Philos. Soc.*, 59:833–835, 1963.

[39] J. H. Schmerl and W. T. Trotter. Critically indecomposable partially ordered sets, graphs, tournaments and other binary relational structures. *Discrete Math.*, 113(1-3):191–205, 1993.

[40] N. J. A. Sloane. The On-line Encyclopedia of Integer Sequences. Available

online at http://www.research.att.com/~njas/sequences/.

[41] T. Uno and M. Yagiura. Fast algorithms to enumerate all common intervals of two permutations. *Algorithmica*, 26(2):290–309, 2000.

[42] J. Wolfowitz. Note on runs of consecutive elements. *Ann. Math. Statistics*, 15:97–98, 1944.

[43] I. Zverovich. A finiteness theorem for primal extensions. *Discrete Math.*, 296(1):103–116, 2005.

Permuting machines and permutation patterns

M. D. Atkinson

Department of Computer Science
University of Otago
Dunedin New Zealand

Abstract

Permuting machines were the early inspiration of the theory of permutation pattern classes. Some examples are given which lead up to distilling the key properties that link them to pattern classes. It is shown how relatively simple ways of combining permuting machines can lead to quite complex behaviour and that the notion of regularity can sometimes be used to contain this complexity. Machines which are sensitive to their input data values are shown to be connected to a more general notion than pattern classes. Finally some open problems are presented.

1 Introduction

Although permutation patterns have only recently been studied in a systematic manner their history can be traced back many decades. It could be argued that the well-known lemma [12] of Erdős and Szekeres is really a result about pattern classes (a pattern class whose basis contains both an increasing and a decreasing permutation is necessarily finite). However it is perhaps more convincing to attribute the birth of the subject to the ground-breaking first volume of Donald Knuth's *Art of Computer Programming* series. In the main body of his text, and in some fascinating follow-up exercises, Knuth enumerated some pattern classes, and found some bases, while at the same time introducing some techniques on generating functions that, in due time, were codified as "the kernel method". His work attracted the attention of a few computer scientists [17, 20, 13] in the early 1970s who, in work on other data

67

structures, anticipated some of the ideas that now underpin the theory of pattern classes. However there was then an almost 15 year interval before the modern theory was kick-started by the work of Simion and Schmidt [19].

This early work was motivated by the idea of container data types as permuters or sorters of their input. In this article we set those ideas in a more general context by introducing a certain type of abstract machine called a permuting machine. We shall see that the theory of permuting machines inspires and is inspired by the theory of pattern classes.

In order to motivate some of the early definitions there is no better place to start than a reprise of the connection between stacks and permutations. A stack is just a container for some linear sequence that one is allowed to change by inserting new items at its tail and by removing tail items. Initially the stack is empty and then a sequence of insertions interleaved with removals is made. A list of input items is transformed thereby into a list of output items. We think of a stack as being an abstract machine with two types of instruction: insert and remove. Figure 1 shows a simple example:

Evidently the behaviour of the stack depends on the way in which the insertions and removals are interleaved, and hardly at all on the actual input items. In fact, we may as well assume that the input items are $1, 2, \ldots$ (in that order); then the sequence of output items will be some permutation of these items. These permutations are called *stack permutations*.

It is clear that the set of stack permutations is closed under pattern containment, since removing an item from the input and ignoring the insertion and removal operations on it gives a computation that applies to the shorter input. Since the set of stack permutations is a pattern class it can be defined by a basis. In fact, we have [15]

Theorem 1.1. σ *is a stack permutation if and only if it does not involve the permutation* 312.

Another result from [15] is

Theorem 1.2. *There are*

$$\frac{\binom{2n}{n}}{n+1}$$

stack permutations of length n.

Rather than repeat any of the many proofs of these theorems we prefer

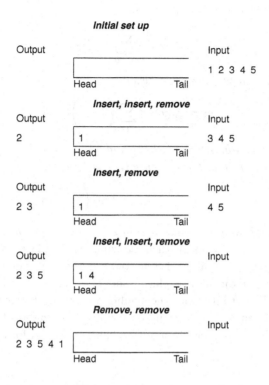

Fig. 1. Generating 23541 with a stack

to consider a small variation on the stack structure. A k-bounded stack is a stack that cannot contain more than k items at any time. Thus, if such a stack is "full" (contains k items) then the next operation must be a removal not an insertion. The permutations that such a stack can produce are called k-stack permutations and, again, they form a pattern class F_k. The following theorem gives the basis, enumeration and growth rate of F_k and is related to the treatment given in [10].

Theorem 1.3. *A permutation belongs to F_k if and only if it does not involve both 312 and $k, k-1, \ldots, 2, 1$ Furthermore F_k contains t_n permutations of length n ($n = 0, 1, 2, \ldots$) where*

(i) *the ordinary generating function*

$$f_k = f_k(x) = \sum_{n=0}^{\infty} t_n x^n = \frac{q_{k-1}}{q_k}$$

is a rational function of x, *where*

$$q_k = q_k(x) = \sum_i \binom{k-i}{i}(-x)^i$$

is a polynomial of degree $\lceil k/2 \rceil$ *and*

(ii) *the growth rate of the sequence* (t_n) *is*

$$\lim_{n \to \infty} \sqrt[n]{t_n} = 2 + 2\cos\left(\frac{2\pi}{k+1}\right)$$

Proof. Consider how a particular permutation σ of length n is generated from an input $12\cdots n$. Clearly, the first symbol to be placed in the stack is symbol 1. Until the symbol 1 is removed the stack behaves as a $(k-1)$-stack and therefore during this phase some $(k-1)$-stack permutation of $2, 3, \ldots, t$ will be output. Once the symbol 1 is output the remainder of the process will output a k-stack permutation of the symbols $t+1, \ldots, n$. This tells us that a k-stack permutation has the general form:

$$\sigma = \sigma_1 \, 1 \, \sigma_2 \tag{1}$$

where σ_1 is the sequence of output produced while symbol 1 resided in the stack, and σ_2 is the sequence of output produced after symbol 1 had been removed. The sequences σ_1 and σ_2 are order isomorphic to permutations in F_{k-1} and F_k and $\sigma_1 < \sigma_2$. Conversely, every permutation structured like this is a k-stack permutation.

The first part of the theorem now follows by induction. For the second part we pass to the equation between generating functions that follows from the structural decomposition (1) and we obtain

$$f_k(x) = f_{k-1}(x) x f_k(x) + 1$$

(the last term because we have to account for the single permutation of length 0) and we write this as

$$f_k(x) = \frac{1}{1 - x f_{k-1}(x)} \tag{2}$$

Obviously, $f_k(x)$ is a rational function of x and, if we write it as

$$f_k(x) = \frac{p_k(x)}{q_k(x)}$$

where $p_k = p_k(x)$ and $q_k = q_k(x)$ are polynomials with unit leading term and with no non-trivial common factor, we can rewrite the recurrence (2) as

$$\frac{p_k}{q_k} = \frac{q_{k-1}}{q_{k-1} - xp_{k-1}}$$

which yields

$$p_k = q_{k-1}$$
$$q_k = q_{k-1} - xp_{k-1}$$

From these equations we find that

$$q_k = q_{k-1} - xq_{k-2} \tag{3}$$

Clearly, we have

$$q_0 = 1$$
$$q_1 = 1 - x$$

Solving the recurrence (3) for q_k we find that

$$q_k = \frac{(1 + \sqrt{1 - 4x})^{k+1} - (1 - \sqrt{1 - 4x})^{k+1}}{2^{k+1}\sqrt{1 - 4x}}$$

and then, by binomial expansion,

$$q_k = \sum_i \binom{k - i}{i}(-x)^i$$

The growth rate of the class is now easily found by analysing the singularities of $f_k(x)$ and these occur at the zeros $q_k(x)$. Therefore we solve $q_k(x) = 0$ for its largest root and take the reciprocal. The growth rate turns out to be

$$2 + 2\cos\left(\frac{2\pi}{k + 1}\right)$$

□

Stacks are just one example of structures whose output is a permutation of their input and it was not long before this scenario was extended. The first extensions were to other abstract data types such as deques and queues and various ways in which they could be combined. But, in general, we can consider *any* devices that accept some stream of input values, process them in some way, and then output them. If the output of the device is simply some rearrangement of the input we call it a

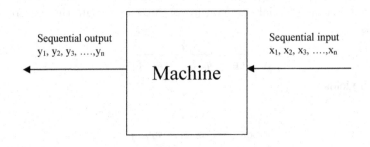

Fig. 2. A permuting machine

Fig. 3. A road bridge and parking bay

permuting machine (see Figure 2) and to understand its behaviour we have to determine how the input and output are related.

There are many situations that fall under this umbrella. We give two further examples.

In the first of these examples consider the bridge shown in Figure 3.

Cars enter the bridge in single file and one car can be overtaken only if that car pulls into the bay shown (which has room for one car only), subsequently rejoining the traffic after possibly several cars have passed it. The original stream of cars $1, 2, \ldots, n$ is transformed into an output stream and the resulting permutations are most easily describable in cycle form as

$$(1, 2, \ldots)(\ldots) \ldots (\ldots)$$

in which the cycle entries are the numbers $1, 2, \ldots$ in that order. This set of permutations is a pattern class with basis $\{321, 312\}$ and it has 2^{n-1} permutations of each length n.

For the second example consider Figure 4 which represents a small art gallery.

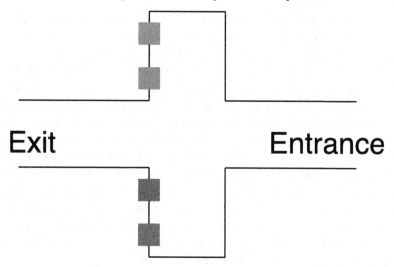

Fig. 4. A small art gallery

In this art gallery patrons have to move carefully there being not much room. In fact there is room only for at most 4 persons:– in front of the 4 paintings illustrated, and patrons cannot squeeze by one another in the 4 quadrants of the gallery. Patrons enter, move among the 4 allowed paintings and eventually exit. It is not hard to see that the resulting permutations of the input form a pattern class. Surprisingly, this simple scenario is quite hard to analyse and we shall return to it in Section 3.

The two previous examples give rise to pattern classes of permutations because their behaviour does not depend on the values of the items that pass through them and input items can be omitted without changing the essential behaviour of the machines. But this is not always the case as shown by the following examples.

In the "Lucky third" machine patrons $1, 2, \ldots, n$ line up in order at a ticket booth, buy a ticket, and leave the line; however every lucky third customer is promoted to the head of the line, buys their ticket and leaves. The order in which n customers depart is unique $(3, 1, 2, 6, 4, 5, 9, 7, 8 \ldots)$. This collection of permutations is clearly not closed under involvement since "lucky third" is not respected by subsequences.

A more important example is the priority queue machine. This is similar to a stack differing only in that the removal rule is that the least element, rather than the most recently inserted element, is removed. There is one very significant difference compared to stacks. Since priority queues are sensitive to the values in the input stream, different inputs can be permuted in different ways. For example, the input stream 123 can give rise only to the output stream 123, but the input 321 can produce 5 possible outputs. That means that we cannot understand a priority queue structure simply by analysing one input stream of each length only, as was the case with stacks. Nevertheless, priority queues have some properties that allow progress to be made. Two input streams whose items are arranged order isomorphically have similar behaviours; for example, there is an obvious correspondence between behaviours on the input streams $4\,2\,5\,1\,3$ and $40\,20\,50\,10\,30$. Furthermore, if we take a subsequence of the input and just track how it is transformed, then we can realise this same behaviour if the subsequence is presented as input in its own right. These properties certainly recall pattern class properties and we shall see in section 4 exactly what generalisation of permutation patterns is required to study them.

The permutational behaviour of a permuting machine M is represented by its set $A(M)$ of *allowable pairs*: those pairs (σ, τ) such that τ is a possible output if M is presented with σ as input. The examples above suggest the properties we require of permuting machines and their allowable pairs in order that we can use them to obtain pattern classes.

Subsequence property If

$$(x_1 x_2 \cdots x_n, y_1 y_2 \cdots y_n) \in A(M)$$

and $x_{i_1} x_{i_2} \cdots x_{i_m}$ is a subsequence of $x_1 x_2 \cdots x_n$ whose terms appear as the subsequence $y_{j_1} y_{j_2} \cdots y_{j_m}$ in $y_1 y_2 \cdots y_n$ then

$$(x_{i_1} x_{i_2} \cdots x_{i_m}, y_{j_1} y_{j_2} \cdots y_{j_m}) \in A(M)$$

Relative property If

$$(x_1 x_2 \cdots x_n, y_1 y_2 \cdots y_n) \in A(M)$$

and ρ is any order-preserving bijection with $x_i' = \rho(x_i)$ then (putting $y_i' = \rho(y_i)$) we have

$$(x_1' x_2' \cdots x_n', y_1' y_2' \cdots y_n') \in A(M)$$

Oblivious property If

$$(x_1 x_2 \cdots x_n, y_1 y_2 \cdots y_n) \in A(M)$$

and ρ is *any* bijection with $x_i' = \rho(x_i)$ then (putting $y_i' = \rho(y_i)$)

we have

$$(x_1' x_2' \cdots x_n', y_1' y_2' \cdots y_n') \in A(M)$$

Clearly, the oblivious property implies the relative property. The stack, the roadbridge, and the art gallery are machines that have all of these properties; the priority queue satisfies the subsequence and relative properties only; and the "third time lucky" machine has the relative and oblivious properties but not the subsequence property.

Suppose we have a machine M that satisfies the relative and oblivious properties. Then we may partition the permutations of length n in the set $A(M)$ as

$$P(\sigma) = \bigcup_\tau \{(\sigma, \tau) \in A(M)\}$$

The oblivious property guarantees that the $n!$ subsets of this union are all bijective with each other. Furthermore, the bijections are given by simple renamings of the symbols. Thus an understanding of this set can be obtained from any one of the subsets in the partition. If we take the subset corresponding to $\sigma = 12 \cdots n$ then we can analyse the set $A(M)$ by analysing the set of pairs

$$P(12 \cdots n) = \{(12 \cdots n, \tau) \in A(M)\}$$

We call the set of permutations τ that arise as the second component of such pairs the pattern class *generated by* M. The subsequence property then implies that this set is indeed a pattern class.

There is a dual point of view that is sometimes useful. We can partition the permutations of length n in $A(M)$ according to their second component. This gives rise to another pattern class: the class of all permutations that M can transform into $12 \cdots n$; this is the pattern class *sorted* by M.

It is not difficult to see that the pattern class sorted by M consists of the inverses of the permutations of the class generated by M.

Pattern classes that arise out of permuting machines are very numerous. Indeed there is a sense in which *all* pattern classes arise in this way, for given a pattern class X we can simply define a machine to have the behaviour that it receives a sequence $1, 2, \ldots$ and (non-deterministically) outputs a permutation in X. That may seem to be somewhat of a cheat since we do not describe a mechanism whereby the output permutations are generated symbol by symbol. Nevertheless, it is a powerful conceptual tool and this will become more evident later on.

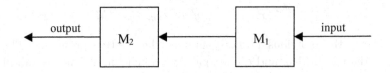

Fig. 5. Serial composition of two machines

2 Composition of machines

Lemma 2.1. *Let L_1, L_2 be pattern classes. Then the set of all permutation products (left to right composition)*

$$L_2 \circ L_1 = \{\tau_2 \circ \tau_1 \mid \tau_1 \in L_1, \tau_2 \in L_2, |\tau_1| = |\tau_2|\}$$

is also a pattern class.

Proof. Let M_1, M_2 be the permuting machines associated with L_1, L_2 and consider the machine shown in Figure 5. In this machine the output of M_1 is fed to M_2 as input. So, if we have $1, 2, \ldots, n$ as input then there will be a permutation $t_1 t_2 \cdots t_n = \tau_1 \in L_1$ that is the input to M_2 and M_2 will transform this via a permutation $\tau_2 \in L_2$ into

$$t_{\tau_2(1)} t_{\tau_2(2)} t_{\tau_2(3)} \cdots t_{\tau_2(n)}$$

which is the permutation $\tau_2 \circ \tau_1$.

Conversely, any permutation $\tau_2 \circ \tau_1$ can be obtained in this way. Finally, it is obvious that the machine is oblivious and has the subsequence property so $L_2 \circ L_1$ is a pattern class. □

$L_2 \circ L_1$ is usually a very much more complicated class than either L_1 or L_2. Consider, for example, the class of all products $\sigma \circ \tau$ where $\sigma, \tau \in \mathrm{Av}(312)$. Since $\mathrm{Av}(312)$ is the class of stack permutations this class is simply the set of permutations generated by two stacks in series as shown in Figure 6. The enumeration and basis of this class are unknown. It is known that the basis is infinite [16] and also, if the first stack is 2-bounded, the basis is finite [11]. Incidentally, it is worth noting that there are some more restricted ways of chaining stacks together and a survey of some of these may be found in [9].

A more tractable example is the class $V = \mathrm{Av}(132, 231)$. It consists of all permutations which can be partitioned into two segments, the first decreasing, and the second increasing (like a 'V'). Its enumeration

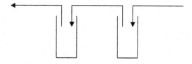

Fig. 6. Two stacks in series

function is easily seen to be 2^{n-1} because the permutations of V can be viewed as being generated from $1, 2, \ldots, n$ by a machine that processes the input symbols in order placing each one at the beginning or end of an output stream; for all symbols except the last there are two choices. But how about the class V^2 of all products $\nu_1 \nu_2$ with $\nu_1, \nu_2 \in V$?

We can think of the class V as being generated from its input by the following machine M, a minor variation on the machine above. First M chooses a subsequence of the input, and places it in reverse as the initial segment of the output; the remaining elements are then appended to the output. Certainly, this machine produces exactly the permutations of V if presented with $1, 2, \ldots$ as input. To understand V^2 we must apply this machine to a permutation σ of V itself. The first step chooses a subsequence σ_1 of σ, reverses it and places it as the initial segment of the output; the second step takes the remaining items, a subsequence σ_2 and appends them to the output. Since all of σ_1, and its reverse, and σ_2 are also V-shaped the output will be shaped as two V's side by side: a W-shape (which may be degenerate if not all the arms are non-empty).

This proves that the permutations of V^2 have the form $\nu_1 \nu_2 \nu_3 \nu_4$ where ν_1, ν_3 are decreasing, and ν_2, ν_4 are increasing.

To prove that every permutation of this type lies in V^2 we consider the machine that undoes the permutations created by M. This machine, M' say, divides its input into an initial segment and a final segment, reverses the initial segment, and merges it with the final segment. Suppose that M' is presented with a permutation $\nu_1 \nu_2 \nu_3 \nu_4$ where ν_1, ν_3 are decreasing, and ν_2, ν_4 are increasing. Then it can split it into $\nu_1 \nu_2$ followed by $\nu_3 \nu_4$ and merge $(\nu_1 \nu_2)^R = \nu_2^R \nu_1^R$ with $\nu_3 \nu_4$ in such a way that ν_2^R and ν_3 are merged into a decreasing sequence and ν_1^R and ν_4 are merged into an increasing sequence; the result, a decreasing segment followed by an increasing segment, can then be processed by M' again to yield $12 \cdots n$.

A more general result of this type is proved in [1].

Once we have the idea of putting two machines together by serial composition it is natural to consider the parallel composition of two ma-

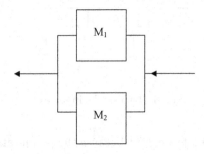

Fig. 7. Machines in parallel

chines. The parallel composition of M_1 and M_2 accepts input symbols into one of the two machines, which are then output into a common output stream, see Figure 7. Thus the set of output permutations consists of interleavings of two subsequences, one generated by M_1, the other by M_2.

It would be nice if the parallel composition generated all such interleavings; unfortunately that is false. The reason that all interleavings might not be generated is that M_1 may be required to discharge some output before reading symbols it will eventually process; these latter symbols prevent symbols that follow being output via M_2 before the initial symbols output by M_1. However, if M_1 and M_2 are endowed with output queues where symbols may wait until they are sent to the output stream we do indeed generate all interleavings.

Despite the great increase in complexity caused by forming serial and parallel combinations there is, nevertheless, a large family of permuting machines that are relatively tractable. This family, called the family of regular permuting machines, is defined in terms of classical finite automata and we discuss it later in the next section. In very rough terms, we define an encoding of a set X of permutations as strings over some finite alphabet; when this set of encoded strings is regular (in the sense of finite automata) then we can often prove properties of X and define natural permuting machines that generate X.

3 Regularity

We turn now to a concept borrowed from the theory of automata that has many uses in the study of pattern classes. Finite state automata are usually defined as machines that respond to an input sequence that causes a sequence of state changes. However, it is rather more useful here to define them as (non-deterministic) machines where each state change is accompanied by outputting a symbol from a finite alphabet (or the null symbol ϵ).

For our purposes, a finite state machine consists of

(i) A finite set Q of states, one of which is a start state, and some of which are final states,

(ii) A finite vocabulary Σ of output symbols,

(iii) A ternary transition relation $T \subseteq Q \times Q \times \Sigma \cup \{\epsilon\}$.

A transition of the machine from state q to state q' producing output symbol x can occur if $(q, q', x) \in T$. A computation of the machine is a sequence of transitions from its starting state to one of its final states; the output of the computation is the sequence of symbols generated at each transition.

The language defined by the machine is the set of words over Σ that it can output.

The general idea is to associate a pattern class with a set of words over the alphabet Σ, each permutation being encoded by a word. Furthermore, the operation of a permuting machine that generates a class X is then associated with the operation of some finite state machine that defines a language Y, and the words in Y correspond, via the encoding, to the permutations of X.

The *rank encoding* of permutations by words over a finite alphabet is the one most commonly used but the *insertion encoding* [4] is another powerful alternative. The rank encoding can be used in any permuting machine with a "finite capacity" that is, a machine that cannot accept more than a limited number k of symbols before it has to output one. More generally, the next symbol to output must be among the next k symbols so far not output. This condition means that every permutation $\pi = p_1 p_2 \cdots p_n$ can be encoded over the k-letter alphabet $\{1, 2, \ldots, k\}$ as $c_1 c_2 \cdots c_n$ where c_i is the rank of p_i in the set $\{p_i, p_{i+1}, \cdots, p_n\}$.

Example 3.1. The rank encoding of 48273156 is 47252111.

To illustrate the applications of finite state machines we give some examples. The first example is the art gallery machine. At any point in the operation of the machine the possible next action depends on what patrons are currently in the four locations of the gallery, and the next action leads to another disposition of patrons. When a patron is output

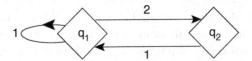

Fig. 8. A simple finite state machine; q_1 is the initial and final state

the output symbol can be encoded by its rank among the remaining symbols and this rank can only be 1, 2, 3 or 4. So if we regard the state of the gallery as being the relative order of the patrons within it there will then be just a finite number of possible states, and the transitions between these states reflect how the patrons move within the gallery. Thus the system can be modeled by a finite state machine in which the output symbols rank encode the permutations that are being generated. This example generalises to *token passing networks* [6, 3, 21]. Here the machine is just a finite graph in which each node is a register (a storage location that can contain one token at a time). Symbols can move between registers if there is a connecting edge. The state of such a machine is just the relative order of the symbols within the registers and output symbols are rank-encoded by one of a fixed number of ranks.

Consider next the finite state machine F defined by Figure 8. The diagram indicates that the machine has two states q_1, q_2, and the possible transitions between them. The machine undergoes a series of transitions, beginning and ending in state q_1. At each transition it emits the symbol marked on the arrow. It is easy to see that the set of strings that this machine can output consists of those strings of 1s and 2s ending in 1 with no two consecutive 2s. Such strings are counted by the Fibonacci numbers.

From this machine we construct a permuting machine G. The permuting machine also has two states q_1, q_2 and two internal registers R_1, R_2 each able to hold one symbol. The arrow from R_2 to R_1 (see Figure 9) indicates that, whenever R_1 is empty and R_2 contains a symbol then that symbol is transferred to R_1. The input symbols $1, 2, \ldots$ enter the machine at register R_2 provided that R_2 is empty (so, if register R_1 happens to be empty, they then pass immediately to R_1). Output can only occur if R_1 is occupied (and R_2 may or may not be occupied). The operation of G follows that of F in that output operations trigger state changes and whenever F would have emitted the symbol 1 (respec-

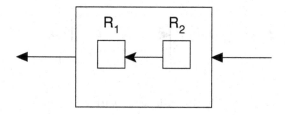

Fig. 9. A regular permuting machine

tively 2) the content of register R_1 (respectively R_2) is emitted (this is consistent in that, if R_2 is empty, the machine is necessarily in state q_1).

Certainly every computation of F is mirrored by a computation of G and, in each, distinct computations produce distinct results. In particular, the number of strings of length n generated by F equals the number of permutations of length n generated by G. These numbers are therefore the Fibonacci numbers. In fact it is quite straightforward to prove that this set of permutations is the pattern class $\text{Av}(321, 312, 231)$.

The output strings of F and the permutations generated by G are connected via the rank encoding. For the permutations generated by G have the property that each term is either the smallest or second smallest of all terms that are equal or later than it; these two possibilities correspond to the symbols 1 or 2 in the alphabet of F.

The machine G has two registers only which are connected in a rather simple fashion. This suggests two directions in which we might generalise. First we can allow there to be any (finite) number k of registers. Second we can allow the contents of the registers to change in a less restrictive way. We shall insist that the registers contain (in some order) the k smallest symbols that have not yet been output (the symbols greater, and later, than these have not yet entered the machine). The particular way in which these next k symbols occupy the registers is called a *disposition*. As before the machine, M say, is in one of a finite number of states but now the state change rules depend not just on the current state but on the current disposition also; and they specify not only which state to enter and which register symbol to output but also a new disposition of the next k symbols among the registers. Such a permuting machine is called a *regular* permuting machine; all token-passing networks are regular in this sense.

From such a machine we can construct a finite state machine N that

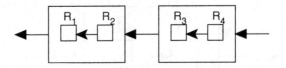

Fig. 10. G in series with G

outputs the rank-encoded forms of the permutations generated by M. The states of N are ordered pairs (q, d) where q is a state of M and d is a disposition. The state change rules are the obvious ones: if M, in state q and disposition d, can make a transition to state q' and disposition d' then N can change state from (q, d) to (q', d'). The output rule when N changes state is to output the rank of the symbol output by M; this rank is completely determined by the register whose contents M discharges, and the disposition of M (which tells us the rank of the symbol residing in that register).

If t_n denotes the number of permutations of length n producible from G (and therefore the number of output strings of length n producible from N) then, by the general theory of finite state machines, the generating function

$$\sum_{n=0}^{\infty} u_n x^n$$

is a rational function.

To clarify these constructions we consider the machine G above and its serial composition with itself (see Figure 10). The state of this serial composition is given by the pair of states for the two components.

Input $1, 2, \ldots, n$ enters on the right. The right hand copy of G discharges its symbols (either from register R_3 or R_4) into the left hand copy. Although the registers R_3 and R_4 contain symbols r, s with $r < s$ always it is now possible that the symbols u, v in R_1 and R_2 might have $u > v$. So in this machine there are two possible dispositions with all 4 registers occupied, in which the symbols of R_1, R_2, R_3, R_4 are ordered as 1234 or 2134; and, if not all the registers are occupied, there are some dispositions where the empty registers are denoted by "E".

For example, if R_2 happened to be empty, and the contents of R_1, R_3, R_4 were ordered as 123 (disposition $1E23$), then possible actions that could now occur would be:

(i) move the symbol in R_1 to the output stream (where it is encoded as 1),

(ii) move the symbol in R_3 to R_2,

(iii) move the symbol in R_4 to R_2

The new dispositions of R_1, R_2, R_3, R_4 in these three cases are (respectively) $E123$, $12E3$, $132E$.

It is evident from the discussions above that this composite machine is regular and generates the products $\sigma\tau$ where σ, τ are permutations generated by G. It is also evident that these permutations can be rank-encoded in a 4-symbol alphabet.

This construction is fairly general and proves

Theorem 3.2. *Serial composition preserve regularity.*

For parallel composition we also have

Theorem 3.3. *Parallel composition preserves regularity.*

Unfortunately regularity is not preserved by machines endowed with output queues. This is evident because, if regularity were preserved, then the set of interleavings of the class of increasing permutations with itself would be regular — but that is impossible since this class does not have a rational generating function.

4 Non-oblivious machines

In this section we lay the foundations for the theory of machines that possess only the subsequence and relative properties. Here we have an extra layer of complexity because the machine may behave quite differently on different inputs. That means that we have to consider the set of *all* allowable (input, output) pairs for the machine rather than choosing the input to be the special sequence $1, 2, \dots$. We need to extend the idea of pattern involvement to pairs of permutations.

Definition 4.1. The pair of permutations (α, β) of length m is involved in the pair (σ, τ) of length n if there exist subsequences σ_0 and τ_0 of σ and τ such that

(i) σ_0 is a rearrangement of τ_0, and

(ii) σ_0 is order isomorphic to α and τ_0 is order isomorphic to β.

The set of allowable pairs for a machine that has the subsequence and relative properties is closed under this involvement relation.

Motivated by the ideas behind pattern classes we can then define a set of pairs of permutations to be a pattern class if it is closed under this notion of pair involvement. Just as for ordinary pattern classes we have a notion of basis: the basis of a pattern class of pairs is the set of pairs minimal with respect to lying outside the class.

The priority queue machine is the best-known example of a machine with the subsequence and relative properties but not the oblivious property. The basis for its class of allowable pairs is known to be [5]

$$\{(12, 21), (321, 132)\}$$

Enumeration is also an issue for pattern classes of pairs but, rather than enumerate permutations, we have to enumerate the allowable pairs themselves. For the priority queue machine the number of allowable pairs is known to be [7]

$$(n + 1)^{n-1}$$

and [14] relates the priority queue allowable pairs to parking functions. We give one further example of enumeration.

Proposition 4.2. *Let u_n be the number of pairs that avoid the pairs of the set*

$$\Gamma = \{(12, 21), (321, 132), (231, 123), (321, 123)\}$$

Then

$$U(x) = \sum u_n x^n / n! = \frac{1}{1 + \log(1 - x)}$$

Proof. Let (σ, τ) be one of the pairs that avoid the permutations of Γ. Put $\sigma = s_1 s_2 \cdots s_n$ and write $\tau = \lambda s_1 \mu$. Note first that, because of the $(12, 21)$ avoidance, every term of λ is smaller than s_1. Consider any term s_j that appears in λ and any other s_i with $1 < i < j$. Then s_i appears in λ and occurs before s_j. To see this consider the possible positioning of $\{s_1, s_i, s_j\}$ in τ.

   (i) $\tau = \cdots s_j \cdots s_1 \cdots s_i \cdots$. Then $s_j < s_i$ because of the $(12, 21)$ avoidance. But then we have either $s_j < s_1 < s_i$ or $s_j < s_i < s_1$. In the former case $(s_1 s_i s_j, s_j s_1 s_i)$ is isomorphic to $(231, 123)$ and in the latter case $(s_1 s_i s_j, s_j s_1 s_i)$ is isomorphic to $(321, 132)$. So this case cannot arise.

(ii) $\tau = \cdots s_j \cdots s_i \cdots s_1 \cdots$. Again $s_j < s_i$ because of the $(12, 21)$ avoidance and $s_i < s_1$ for the same reason. But now the pair $(s_1 s_i s_j, s_j s_i s_1)$ is isomorphic to $(321, 123)$ which is also impossible.

We now know that

$$
\begin{aligned}
\sigma &= s_1 s_2 \cdots s_i \alpha \\
\tau &= s_2 s_3 \cdots s_i s_1 \beta
\end{aligned}
$$

and that

(i) All of $s_2, s_3, \ldots, s_i$ are smaller than s_1, and
(ii) (α, β) is a pair that avoids the pairs of Γ and take their terms from some set Δ of size $n - i$.

It is easily verified that these conditions in turn imply that (σ, τ) avoids the pairs of Γ. We can now count these pairs using induction on n. The set Δ may be chosen in $\binom{n}{n-i}$ ways and, once chosen, there are u_{n-i} choices for (α, β) and $(i - 1)!$ choices for $s_1 s_2 \cdots s_i$. Since (σ, τ) determines the value of i uniquely we have

$$
u_n = \sum_{i=1}^{n} (i - 1)! \binom{n}{n - i} u_{n-i}
$$

which may be rewritten as

$$
\frac{u_n}{n!} = \sum_{i=1}^{n} \frac{u_{n-i}}{(n - i)!} \frac{1}{i}
$$

Expressing this in terms of the generating function $U(x)$ we have

$$
\begin{aligned}
U(x) &= 1 + U(x) \left(x + \frac{x^2}{2} + \frac{x^3}{3} + \cdots \right) \\
&= 1 - U(x) \log(1 - x)
\end{aligned}
$$

giving the result. $\qquad\square$

This result is actually a result about machines since the pairs that avoid Γ are precisely the (input, output) pairs for a priority queue of size 2. The allowable pairs for priority queues of other bounded sizes have not been enumerated.

For ordinary pattern classes it is helpful to regard a set of n points in general position in the plane as determining a permutation. Reading the points in the horizontal direction establishes names $1, 2, \ldots$ for them and then reading these names in the vertical direction defines the permutation itself; conversely, the graph of a permutation (the set of points $\{(i, p_i) \mid 1 \leq i \leq n\}$) determines a set of points. This viewpoint makes clear that the 8 symmetries used to map pattern classes into one another arise from symmetries of a square. It also allows the notion of one pattern being contained in another to be represented by one point set being contained in another. For pattern classes of pairs there is a similar viewpoint in 3 dimensions.

For every pair $(a_1 a_2 \ldots a_n, b_1 b_2 \ldots b_n)$ of permutations, we define a $(0-1)$ $n \times n \times n$ tensor τ_{ijk} where

$$\tau_{ijk} = 1 \text{ if and only if } j = a_i \text{ and } k = b_i$$

and we associate it with a set of n points in an $n \times n \times n$ cube in an obvious way. The symmetry group of the cube acts on these tensors and induces 48 symmetries among pattern classes of pairs. Furthermore a pair of permutations is involved in another pair if and only if the first associated tensor defines a 3-dimensional point set that is (isomorphic) to the point set of the second associated tensor.

In contrast to the situation for ordinary pattern classes very few pattern classes of pairs have been enumerated. Even the enumeration of the class W defined by the single basis pair $(12, 21)$ is unknown. For $(a_1 a_2 \ldots a_n, b_1 b_2 \ldots b_n)$ to be in W it must be that every pair of symbols a_i, a_j with $i < j$ and $a_i < a_j$ will occur in that order in $b_1 b_2 \ldots b_n$. This is exactly the condition that $(a_1 a_2 \ldots a_n) \leq (b_1 b_2 \ldots b_n)$ in the weak order [8] and, frustratingly, the number of pairs of permutations of length n that are related in the weak order is unknown.

Using again the notion of combining two machines in series we have the following result which is analogous to Lemma 2.1.

Lemma 4.3. *Let* L_1, L_2 *be pattern classes of pairs. Then the relational composition*

$$L_2 \circ L_1 = \{(\alpha, \gamma) \mid (\alpha, \beta) \in L_2, (\beta, \gamma) \in L_1 \text{ for some permutation } \beta\}$$

is also a pattern class.

Some examples of classes derived through this relational composition are given in [5].

5 Concluding Remarks

We have seen that permuting machines were one of the main originators in the modern theory of permutation patterns. They provide many examples of pattern classes and, in turn, the machine viewpoint can be a powerful tool in proving results about pattern classes. However, many problems remain and we conclude by mentioning two general areas that it would be useful to develop and one specific unsolved problem.

Section 3 demonstrated the utility of bringing the classical theory of finite automata to bear on some permuting machine questions. Can we use, in a significant way, other families of classical machine such as pushdown automata? The hope here would be to devise tools that could deal with pattern classes whose generating functions were algebraic (in the same way that finite state machines may sometimes be used to derive rational generating functions).

The theory of pattern classes of pairs of permutations is in its infancy and much work needs to be done if it is to become a useful tool for studying non-oblivious machines. Early indications are that exponential generating functions will have to be used in place of ordinary generating functions.

Finally, returning to the original work of Knuth [15], one problem has resisted attack for over 40 years. How many permutations of length n can be generated by a deque (a linear list that allows insertions and deletions at both ends)? Virtually nothing was known about this problem until very recently. Of course, deque permutations do form a pattern class and the basis of this class is known (and is infinite) [17]; moreover, there is a $O(n)$ recognition procedure for deque permutations [18]. But for the enumeration problem even the rate of growth, Δ say, is unknown. Recently it has been proved [2] that

$$7.88966 \leq \Delta \leq 8.51951$$

The lower bound is obtained by approximating a deque with a bounded deque and using regularity techniques. The upper bound is obtained by counting operation sequences and finding rules that imply equivalences between these sequences. Both bounds require extensive machine computation. With more extensive computing resources they could be improved but the methods are not capable of delivering exact enumerations.

References

[1] M. H. Albert, R. E. L. Aldred, M. D. Atkinson, H. P. van Ditmarsch, C. C. Handley, D. A. Holton, and D. J. McCaughan. Compositions of pattern restricted sets of permutations. *Australas. J. Combin.*, 37:43–56, 2007.

[2] M. H. Albert, M. D. Atkinson, and S. Linton. Permutations generated by stacks and deques. *Ann. Comb.*, 14(1):3–16, 2010.

[3] M. H. Albert, S. Linton, and N. Ruškuc. On the permutational power of token passing networks. In *this volume*, 317–338.

[4] M. H. Albert, S. Linton, and N. Ruškuc. The insertion encoding of permutations. *Electron. J. Combin.*, 12(1):Research paper 47, 31 pp., 2005.

[5] R. E. L. Aldred, M. D. Atkinson, H. P. van Ditmarsch, C. C. Handley, D. A. Holton, and D. J. McCaughan. Permuting machines and priority queues. *Theoret. Comput. Sci.*, 349(3):309–317, 2005.

[6] M. D. Atkinson, M. J. Livesey, and D. Tulley. Permutations generated by token passing in graphs. *Theoret. Comput. Sci.*, 178(1-2):103–118, 1997.

[7] M. D. Atkinson and M. Thiyagarajah. The permutational power of a priority queue. *BIT*, 33(1):2–6, 1993.

[8] A. Björner. Orderings of Coxeter groups. In *Combinatorics and algebra (Boulder, Colo., 1983)*, volume 34 of *Contemp. Math.*, pages 175–195. Amer. Math. Soc., Providence, RI, 1984.

[9] M. Bóna. A survey of stack-sorting disciplines. *Electron. J. Combin.*, 9(2):Article 1, 16 pp., 2003.

[10] T. Chow and J. West. Forbidden subsequences and Chebyshev polynomials. *Discrete Math.*, 204(1-3):119–128, 1999.

[11] M. Elder. Permutations generated by a stack of depth 2 and an infinite stack in series. *Electron. J. Combin.*, 13:Research paper 68, 12 pp., 2006.

[12] P. Erdős and G. Szekeres. A combinatorial problem in geometry. *Compos. Math.*, 2:463–470, 1935.

[13] S. Even and A. Itai. Queues, stacks, and graphs. In *Theory of machines and computations (Proc. Internat. Sympos., Technion, Haifa, 1971)*, pages 71–86. Academic Press, New York, 1971.

[14] J. D. Gilbey and L. H. Kalikow. Parking functions, valet functions and priority queues. *Discrete Math.*, 197/198:351–373, 1999. 16th British Combinatorial Conference (London, 1997).

[15] D. E. Knuth. *The art of computer programming. Vol. 1: Fundamental algorithms.* Addison-Wesley Publishing Co., Reading, Mass., 1969.

[16] M. M. Murphy. *Restricted Permutations, Antichains, Atomic Classes, and Stack Sorting.* PhD thesis, Univ. of St Andrews, 2002.

[17] V. R. Pratt. Computing permutations with double-ended queues, parallel stacks and parallel queues. In *STOC '73: Proceedings of the fifth annual ACM symposium on Theory of computing*, pages 268–277, New York, NY, USA, 1973. ACM Press.

[18] P. Rosenstiehl and R. E. Tarjan. Gauss codes, planar Hamiltonian graphs, and stack-sortable permutations. *J. Algorithms*, 5(3):375–390, 1984.

[19] R. Simion and F. W. Schmidt. Restricted permutations. *European J. Combin.*, 6(4):383–406, 1985.

[20] R. Tarjan. Sorting using networks of queues and stacks. *J. Assoc. Comput. Mach.*, 19:341–346, 1972.

[21] S. Waton. *On Permutation Classes Defined by Token Passing Networks, Gridding Matrices and Pictures: Three Flavours of Involvement.* PhD thesis, Univ. of St Andrews, 2007.

On three different notions of monotone subsequences

Miklós Bóna

Department of Mathematics
University of Florida
Gainesville, FL 32611 USA

Abstract

We review how the monotone pattern compares to other patterns in terms of enumerative results on pattern avoiding permutations. We consider three natural definitions of pattern avoidance, give an overview of classic and recent formulas, and provide some new results related to limiting distributions.

1 Introduction

Monotone subsequences in a permutation $p = p_1 p_2 \cdots p_n$ have been the subject of vigorous research for over sixty years. In this paper, we will review three different lines of work. In all of them, we will consider increasing subsequences of a permutation of length n that have a *fixed* length k. This is in contrast to another line of work, started by Ulam more than sixty years ago, in which the distribution of the *longest* increasing subsequence of a random permutation has been studied. That direction of research has recently reached a high point in the article [4] of Baik, Deift, and Johansson.

The three directions we consider are distinguished by their definition of monotone subsequences. We can simply require that k entries of a permutation increase from left to right, or we can in addition require that these k entries be in consecutive positions, or we can even require that they be consecutive integers *and* be in consecutive positions.

2 Monotone Subsequences with No Restrictions

The classic definition of pattern avoidance for permutations is as follows. Let $p = p_1 p_2 \cdots p_n$ be a permutation, let $k < n$, and let $q = q_1 q_2 \cdots q_k$ be another permutation. We say that p *contains* q as a pattern if there exists a subsequence $1 \leq i_1 < i_2 < \cdots < i_k \leq n$ so that for all indices j and r, the inequality $q_j < q_r$ holds if and only if the inequality $p_{i_j} < p_{i_r}$ holds. If p does not contain q, then we say that p *avoids* q. In other words, p contains q if p has a subsequence of entries, not necessarily in consecutive positions, which relate to each other the same way as the entries of q do.

Example 2.1. The permutation 3174625 contains the pattern 123. Indeed, consider the first, fourth, and seventh entries.

As in this paper, the monotone pattern $12 \cdots k$ plays a special role, we introduce the special notation

$$\alpha_k = 12 \cdots k. \tag{1}$$

In particular, p contains α_k if and only if p contains an increasing subsequence of length k. The elements of this increasing subsequence do not have to be in consecutive positions.

The enumeration of permutations avoiding a given pattern is a fascinating subject. Let $S_n(q)$ denote the number of permutations of length n (or, in what follows, n-permutations) that avoid the pattern q.

2.1 Patterns of Length Three

Among patterns of length three, there is no difference between the monotone pattern and other patterns as far as $S_n(q)$ is concerned. This is the content of our first theorem.

Theorem 2.2. *Let q be any pattern of length three, and let n be any positive integer. Then $S_n(q) = C_n = \binom{2n}{n}/(n+1)$. In other words, $S_n(q)$ is the nth Catalan number.*

Proof. If p avoids q, then the reverse of p avoids the reverse of q, and the complement of p avoids the complement of q. Therefore, $S_n(123) = S_n(321)$ and $S_n(132) = S_n(231) = S_n(213) = S_n(312)$.

The fact that $S_n(132) = S_n(123)$ is proved using the well-known Simion-Schmidt bijection [23]. In a permutation, let us call an entry a *left-to-right* minimum if it is smaller than every entry on its left. For instance, the left-to-right minima of 4537612 are the entries 4, 3, and 1.

Take an n-permutation p of length n that avoids 132, keep its left-to-right minima fixed, and arrange all other entries in decreasing order in the positions that do not belong to left-to-right minima, to get the permutation $f(p)$. For instance, if $p = 34125$, then $f(p) = 35142$. Then $f(p)$ is a union of two decreasing sequences, so it is 123-avoiding. Furthermore, f is a bijection between the two relevant set of permutations. Indeed, if r is a permutation counted by $S_n(123)$, then $f^{-1}(r)$ is obtained by keeping the left-to-right minima of r fixed, and rearranging the remaining entries so that moving from left to right, each slot is filled by the smallest remaining entry that is larger than the closest left-to-right minimum on the left of that position.

In order to prove that $S_n(132) = C_n$, just note that in a 132-avoiding n-permutation, any entry to the left of n must be smaller than any entry to the right of n. Therefore, if n is in the ith position, then there are $S_{i-1}(132)S_{n-i}(132)$ permutations of length n that avoid 132. Summing over all i, we get the recurrence

$$S_n(132) = \sum_{i=0}^{n-1} S_{i-1}(132)S_{n-i}(132),$$

which is the well-known recurrence for Catalan numbers. □

2.2 Patterns of Length Four

When we move to longer patterns, the situation becomes much more complicated and less well understood. In his doctoral thesis [27], Julian West published the following numerical evidence.

- for $S_n(1342)$, and $n = 1, 2, \cdots, 8$, we have 1, 2, 6, 23, 103, 512, 2740, 15485
- for $S_n(1234)$, and $n = 1, 2, \cdots, 8$, we have 1, 2, 6, 23, 103, 513, 2761, 15767
- for $S_n(1324)$, and $n = 1, 2, \cdots, 8$, we have 1, 2, 6, 23, 103, 513, 2762, 15793.

These data are startling for at least two reasons. First, the numbers $S_n(q)$ are no longer independent of q; there are some patterns of length four that are easier to avoid than others. Second, the monotone pattern 1234, special as it is, does not provide the minimum or the maximum value for $S_n(q)$. We point out that for each q of the other 21 patterns of length four, it is known that the sequence $S_n(q)$ is identical to one of the three sequences $S_n(1342)$, $S_n(1234)$, and $S_n(1324)$. See [7], Chapter 4, for more details.

Exact formulas are known for two of the above three sequences. For the monotone pattern, Ira Gessel gave a formula using symmetric functions.

Theorem 2.3 (Gessel [13]). *For all positive integers n, the identity*

$$S_n(1234) \;=\; 2 \cdot \sum_{k=0}^{n} \binom{2k}{k} \binom{n}{k}^2 \frac{3k^2 + 2k + 1 - n - 2nk}{(k+1)^2(k+2)(n-k+1)} \quad (2)$$

$$= \frac{1}{(n+1)^2(n+2)} \sum_{k=0}^{n} \binom{2k}{k} \binom{n+1}{k+1} \binom{n+2}{k+1}. \quad (3)$$

The formula for $S_n(1342)$ is due to the present author [6], and is quite surprising.

Theorem 2.4. *For all positive integers n, we have*

$$S_n(1342) \;=\; (-1)^{n-1} \cdot \frac{(7n^2 - 3n - 2)}{2}$$

$$+ \; 3 \sum_{i=2}^{n} (-1)^{n-i} \cdot 2^{i+1} \cdot \frac{(2i-4)!}{i!(i-2)!} \cdot \binom{n-i+2}{2}.$$

This result is unexpected for two reasons. First, it shows that $S_n(1342)$ is not simply less than $S_n(1234)$ for every $n \geq 6$; it is *much less*, in a sense that we will explain in Subsection 2.4. For now, we simply state that while $S_n(1234)$ is "roughly" 9^n, the value of $S_n(1342)$ is "roughly" 8^n. Second, the formula is, in some sense, simpler than that for $S_n(1234)$. Indeed, it follows from Theorem 2.4 that the ordinary generating function of the sequence $S_n(1342)$ is

$$H(x) = \sum_{i \geq 0} F^i(x) = \frac{1}{1 - F(x)} = \frac{32x}{-8x^2 + 20x + 1 - (1-8x)^{3/2}}.$$

This is an *algebraic* power series. On the other hand, it is known (Problem Plus 5.10 in [7] that the ordinary generating function of the sequence $S_n(1234)$ is *not* algebraic. So permutations avoiding the monotone pattern are not even the *nicest* among permutations avoiding a given pattern, in terms of the generating functions that count them.

There is no known formula for the third sequence, that of the numbers $S_n(1324)$. However, the following inequality is known [5].

Theorem 2.5. *For all integers $n \geq 7$, the inequality*

$$S_n(1234) < S_n(1324)$$

holds.

Proof. Let us call an entry of a permutation a *right-to-left maximum* if it is larger than all entries on its right. Then let us say that two n-permutations are in the same class if they have the same left-to-right minima, and they are in the same positions, and they have the same right-to-left maxima, and they are in the same positions as well. For example, 51234 and 51324 are in the same class, but $z = 24315$ and $v = 24135$ are not, as the third entry of z is not a left-to-right minimum, whereas that of v is.

It is straightforward to see that each non-empty class contains exactly one 1234-avoiding permutation, the one in which the subsequence of entries that are neither left-to-right minima nor right-to-left maxima is decreasing.

It is less obvious that each class contains *at least one* 1324-avoiding permutation. Note that if a permutation contains a 1324-pattern, then we can choose such a pattern so that its first element is a left-to-right minimum and its last element is a right-to-left maximum. Take a 1324-avoiding permutation, and take one of its 1324-patterns of the kind described in the previous sentence. Interchange its second and third element. Observe that this will keep the permutation within its original class. Repeat this procedure as long as possible. The procedure will stop after a finite number of steps since each step decreases the number of inversions of the permutation. When the procedure stops, the permutation at hand avoids 1324.

This shows that $S_n(1234) \leq S_n(1324)$ for all n. If $n \geq 7$, then the equality cannot hold since there is at least one class that contains more than one 1324-avoiding permutation. For $n = 7$, this is the class $3 * 1 * 7 * 5$, which contains 3612745 and 3416725. For larger n, this class can be prepended by $n(n-1) \cdots 8$ to get a suitable class. $\quad\square$

It turns out again that $S_n(1324)$ is *much* larger than $S_n(1234)$. We will give the details in Subsection 2.4.

2.3 Patterns of Any Length

For general k, there are some good estimates known for the value of $S_n(\alpha_k)$. The first one can be proved by an elementary method.

Theorem 2.6. *For all positive integers n and k > 2, we have*

$$S_n(123\cdots k) \le (k-1)^{2n}.$$

Proof. Let us say that an entry x of a permutation is of rank i if it is the end of an increasing subsequence of length i, but there is no increasing subsequence of length $i+1$ that ends in x. Then for all i, elements of rank i must form a decreasing subsequence. Therefore, a q-avoiding permutation can be decomposed into the union of $k-1$ decreasing subsequences. Clearly, there are at most $(k-1)^n$ ways to partition our n entries into $k-1$ blocks. Then we have to place these blocks of entries somewhere in our permutation. There are at most $(k-1)^n$ ways to assign each position of the permutation to one of these blocks, completing the proof. □

Indeed, Theorem 2.6 has a stronger version, obtained by Amitaj Regev [20]. It needs heavy analytic machinery, and therefore will not be proved here. We mention the result, however, as it shows that no matter what k is, the constant $(k-1)^2$ in Theorem 2.6 cannot be replaced by a smaller number, so the elementary estimate of Theorem 2.6 is optimal in some strong sense. We remind the reader that functions $f(n)$ and $g(n)$ are said to be *asymptotically equal* if $\lim_{n\to\infty}\frac{f(n)}{g(n)} = 1$.

Theorem 2.7. *[20] For all n, $S_n(1234\cdots k)$ asymptotically equals*

$$\lambda_k \frac{(k-1)^{2n}}{n^{(k^2-2k)/2}}.$$

Here

$$\lambda_k = \gamma_k^2 \int\limits_{x_1 \ge x_2 \ge \cdots \ge x_k} \int \cdots \int [D(x_1, x_2, \cdots, x_k) \cdot e^{-(k/2)x^2}]^2 dx_1 dx_2 \cdots dx_k,$$

where $D(x_1, x_2, \cdots, x_k) = \Pi_{i<j}(x_i - x_j)$, and $\gamma_k = (1/\sqrt{2\pi})^{k-1} \cdot k^{k^2/2}$.

2.4 Stanley-Wilf Limits

The following celebrated result of Adam Marcus and Gábor Tardos [18] shows that in general, it is very difficult to avoid any given pattern q.

Theorem 2.8 (Marcus and Tardos [18]). *For all patterns q, there exists a constant c_q so that*

$$S_n(q) \le c_q^n. \tag{4}$$

It is not difficult [2] to show using Fekete's lemma that the sequence $(S_n(q))^{1/n}$ is monotone increasing. The previous theorem shows that it is bounded from above, leading to the following.

Corollary 2.9. *For all patterns q, the limit*

$$L(q) = \lim_{n \to \infty} (S_n(q))^{1/n}$$

exists.

The real number $L(q)$ is called the *Stanley-Wilf limit*, or *growth rate* of the pattern q. In this terminology, Theorem 2.7 implies that $L(\alpha_k) = (k-1)^2$. In particular, $L(1234) = 9$, while Theorem 2.4 implies that $L(1342) = 8$. So it is not simply easier to avoid 1234 than 1342, it is *exponentially* easier to do so.

Numerical evidence suggests that in the multiset of $k!$ real numbers $S_n(q)$, the numbers $S_n(\alpha_k)$ are much closer to the maximum than to the minimum. This led to the plausible conjecture that for any pattern q of length k, the inequality $L(q) \leq (k-1)^2$ holds. This would mean that while there are patterns of length k that are easier to avoid than α_k, there are none that are much easier to avoid, in the sense of Stanley-Wilf limits. However, this conjecture has been disproved by the following result of Michael Albert et al.

Theorem 2.10 (Albert et al. [1]). *The inequality $L(1324) \geq 11.35$ holds.*

In other words, it is not simply harder to avoid 1234 than 1324, it is *exponentially* harder to do so.

2.5 *Asymptotic Normality*

In this section we change direction and prove that the distribution of the number of copies of α_k in a randomly selected n-permutation converges in distribution to a normal distribution. (For the rest of this paper, when we say random permutation of length n, we always assume that each n-permutation is selected with probability $1/n!$.) Note that in the special case of $k = 2$, this is equivalent to the classic result that the distribution of inversions in random permutations is asymptotically normal. See [12] and its references for various proofs of that result, or [8] for a generalization.

We need to introduce some notation for transforms of the random variable Z. Let $\bar{Z} = Z - E(Z)$, let $\tilde{Z} = \bar{Z}/\sqrt{\operatorname{Var}(Z)}$, and let $Z_n \to$

$N(0,1)$ mean that Z_n converges in distribution to the standard normal variable.

Our main tool in this section will be a theorem of Svante Janson [16]. In order to be able to state that theorem, we need the following definition.

Definition 2.11. Let $\{Y_{n,k}|k = 1, 2, \cdots, N_n\}$ be an array of random variables. We say that a graph G is a *dependency graph* for $\{Y_{n,k}|k = 1, 2 \cdots, N_n\}$ if the following two conditions are satisfied:

(i) There exists a bijection between the random variables $Y_{n,k}$ and the vertices of G, and

(ii) If V_1 and V_2 are two disjoint sets of vertices of G so that no edge of G has one endpoint in V_1 and another one in V_2, then the corresponding sets of random variables are independent.

Note that the dependency graph of a family of variables is not unique. Indeed if G is a dependency graph for a family and G is not a complete graph, then we can get other dependency graphs for the family by simply adding new edges to G.

Now we are in position to state Janson's theorem, the famous *Janson dependency criterion*.

Theorem 2.12 (Janson [16]). *Let $Y_{n,k}$ be an array of random variables such that for all n, and for all $k = 1, 2, \cdots, N_n$, the inequality $|Y_{n,k}| \leq A_n$ holds for some real number A_n, and that the maximum degree of a dependency graph of $\{Y_{n,k}|k = 1, 2, \cdots, N_n\}$ is Δ_n.*

Set $Y_n = \sum_{k=1}^{N_n} Y_{n,k}$ and $\sigma_n^2 = Var(Y_n)$. If there is a natural number m so that

$$N_n \Delta_n^{m-1} \left(\frac{A_n}{\sigma_n} \right)^m \to 0, \tag{5}$$

as n goes to infinity, then

$$\tilde{Y}_n \to N(0,1).$$

Let us order the $\binom{n}{k}$ subsequences of length k of the permutation $p_1 p_2 \cdots p_n$ linearly in some way. For $1 \leq i \leq \binom{n}{k}$, let $X_{n,i}$ be the indicator random variable of the event that in a randomly selected permutation of length n, the ith subsequence of length k in the permutation $p = p_1 p_2 \cdots p_n$ is a $12 \cdots k$-pattern. We will now verify that the family of the $X_{n,i}$ satisfies all conditions of the Janson Dependency Criterion.

First, $|X_{n,i}| \leq 1$ for all i and all n, since the $X_{n,i}$ are indicator random variables. So we can set $A_n = 1$. Second, $N_n = \binom{n}{k}$, the total number of subsequences of length k in p. Third, if $a \neq b$, then $X_{n,a}$ and $X_{n,b}$ are independent unless the corresponding subsequences intersect. For that, the bth subsequence must intersect the ath subsequence in j entries, for some $1 \leq j \leq k - 1$. For a fixed ath subsequence, the number of ways that can happen is $\sum_{j=1}^{k-1} \binom{k}{j}\binom{n-k}{k-j} = \binom{n}{k} - \binom{n-k}{k} - 1$, where we used the well-known Vandermonde identity to compute the sum. Therefore,

$$\Delta_n \leq \binom{n}{k} - \binom{n-k}{k} - 1. \tag{6}$$

In particular, note that (6) provides an upper bound for Δ_n in terms of a polynomial function of n that is of degree $k - 1$ since terms of degree k will cancel.

There remains the task of finding a lower bound for σ_n that we can then use in applying Theorem 2.12. Let $X_n = \sum_{i=1}^{\binom{n}{k}} X_{n,i}$. We will show the following.

Proposition 2.13. *There exists a positive constant c so that for all n, the inequality*

$$Var(X_n) \geq cn^{2k-1}$$

holds.

Proof. By linearity of expectation, we have

$$Var(X_n) \;=\; E(X_n^2) - (E(X_n))^2 \tag{7}$$

$$= \; E\left(\left(\sum_{i=1}^{\binom{n}{k}} X_{n,i}\right)^2\right) - \left(E\left(\sum_{i=1}^{\binom{n}{k}} X_{n,i}\right)\right)^2 \tag{8}$$

$$= \; E\left(\left(\sum_{i=1}^{\binom{n}{k}} X_{n,i}\right)^2\right) - \left(\sum_{i=1}^{\binom{n}{k}} E(X_{n,i})\right)^2 \tag{9}$$

$$= \; \sum_{i_1,i_2} E(X_{n,i_1} X_{n,i_2}) - \sum_{i_1,i_2} E(X_{n,i_1})E(X_{n,i_2}). \tag{10}$$

Let I_1 (resp. I_2) denote the k-element subsequence of p indexed by i_1, (resp. i_2). Clearly, it suffices to show that

$$\sum_{|I_1 \cap I_2| \leq 1} E(X_{n,i_1} X_{n,i_2}) - \sum_{i_1,i_2} E(X_{n,i_1})E(X_{n,i_2}) \geq cn^{2k-1}, \tag{11}$$

since the left-hand side of (11) is obtained from the (10) by removing the sum of some positive terms, that is, the sum of all $E(X_{n,i_1} X_{n,i_2})$ where $|I_1 \cap I_2| > 1$.

As $E(X_{n,i}) = 1/k!$ for each i, the sum with negative sign in (10) is

$$\sum_{i_1, i_2} E(X_{n,i_1}) E(X_{n,i_2}) = \binom{n}{k}^2 \cdot \frac{1}{k!^2},$$

which is a polynomial function in n, of degree $2k$ and of leading coefficient $\frac{1}{k!^4}$. As far as the summands in (10) with a positive sign go, *most* of them are also equal to $\frac{1}{k!^2}$. More precisely, $E(X_{n,i_1} X_{n,i_2}) = \frac{1}{k!^2}$ when I_1 and I_2 are disjoint, and that happens for $\binom{n}{k}\binom{n-k}{k}$ ordered pairs (i_1, i_2) of indices. The sum of these summands is

$$d_n = \binom{n}{k}\binom{n-k}{k}\frac{1}{k!^2}, \tag{12}$$

which is again a polynomial function in n, of degree $2k$ and with leading coefficient $\frac{1}{k!^4}$. So summands of degree $2k$ will cancel out in (10). (We will see in the next paragraph that the summands we have not yet considered add up to a polynomial of degree $2k - 1$.) In fact, considering the two types of summands we studied in (10) and (12), we see that they add up to

$$\binom{n}{k}\binom{n-k}{k}\frac{1}{k!^2} - \binom{n}{k}^2\frac{1}{k!^2} \tag{13}$$

$$= n^{2k-1}\frac{2\binom{k}{2} - \binom{2k-1}{2}}{k!^4} + O(n^{2k-2}) \tag{14}$$

$$= n^{2k-1}\frac{-k^2}{k!^4} + O(n^{2k-2}). \tag{15}$$

Next we look at ordered pairs of indices (i_1, i_2) so that the corresponding subsequences I_1 and I_2 intersect in exactly one entry, the entry x. Let us say that counting from the left, x is the ath entry in I_1, and the bth entry in I_2. See Figure 1 for an illustration.

Observe that $X_{n,i_1} X_{n,i_2} = 1$ if and only if all of the following independent events occur.

(a) In the $(2k - 1)$-element set of entries that belong to $I_1 \cup I_2$, the entry x is the $(a+b-1)$th smallest. This happens with probability $1/(2k - 1)$.

(b) The $a + b - 2$ entries on the left of x in $I_1 \cup I_2$ are all smaller than the $2k - a - b$ entries on the right of x in $I_1 \cup I_2$. This happens with probability $\frac{1}{\binom{2k-2}{a+b-2}}$.

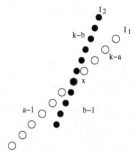

Fig. 1. In this example, $k = 11$, $a = 7$, and $b = 5$.

(c) The subsequences of I_1 on the left of x and on the right of x, and the subsequences of I_2 on the left of x and on the right of x are all monotone increasing. This happens with probability $\frac{1}{(a-1)!(b-1)!(k-a)!(k-b)!}$.

Therefore, if $|I_1 \cap I_2| = 1$, then

$$P(X_{i_1} X_{i_2} = 1) \tag{16}$$

$$= \frac{1}{(2k-1)\binom{2k-2}{a+b-2}(a-1)!(b-1)!(k-a)!(k-b)!} \tag{17}$$

$$= \frac{1}{(2k-1)!} \cdot \binom{a+b-2}{a-1}\binom{2k-a-b}{k-a}. \tag{18}$$

How many such ordered pairs (I_1, I_2) are there? There are $\binom{n}{2k-1}$ choices for the underlying set $I_1 \cup I_2$. Once that choice is made, the $a + b - $ 1st smallest entry of $I_1 \cup I_2$ will be x. Then the number of choices for the set of entries other than x that will be part of I_1 is $\binom{a+b-2}{a-1}\binom{2k-a-b}{k-a}$. Therefore, summing over all a and b and recalling (16),

$$p_n = \sum_{|I_1 \cap I_2|=1} P(X_{i_1} X_{i_2} = 1) = \sum_{|I_1 \cap I_2|=1} E(X_{i_1} X_{i_2}) \tag{19}$$

$$= \frac{1}{(2k-1)!}\binom{n}{2k-1}\sum_{a,b}\binom{a+b-2}{a-1}^2\binom{2k-a-b}{k-a}^2. \tag{20}$$

The expression we just obtained is a polynomial of degree $2k - 1$, in the variable n. We claim that its leading coefficient is larger than $k^2/k!^4$. If we can show that, the proposition will be proved since (15) shows that the summands not included in (19) contribute about $-\frac{k^2}{k!^4}n^{2k-1}$ to the left-hand side of (11).

Recall that by the Cauchy-Schwarz inequality, if $t_1, t_2, \cdots, t_m$ are non-negative real numbers, then

$$\frac{\left(\sum_{i=1}^{m} t_i\right)^2}{m} \leq \sum_{i=1}^{m} t_i^2, \tag{21}$$

where equality holds if and only if all the t_i are equal.

Let us apply this inequality with the numbers $\binom{a+b-2}{a-1}^2 \binom{2k-a-b}{k-a}^2$ playing the role of the t_i, where a and b range from 1 to k. We get that

$$\sum_{1 \leq a,b \leq k} \binom{a+b-2}{a-1}^2 \binom{2k-a-b}{k-a}^2$$
$$> \frac{\left(\sum_{1 \leq a,b \leq k} \binom{a+b-2}{a-1}\binom{2k-a-b}{k-a}\right)^2}{k^2}. \tag{22}$$

We will use Vandermonde's identity to compute the right-hand side. To that end, we first compute the sum of summands with a *fixed* $h = a + b$. We obtain

$$\sum_{1 \leq a,b \leq k} \binom{a+b-2}{a-1}\binom{2k-a-b}{k-a} \tag{23}$$

$$= \sum_{h=2}^{2k} \sum_{a=1}^{k} \binom{h-2}{a-1}\binom{2k-h}{k-a} \tag{24}$$

$$= \sum_{h=2}^{2k} \binom{2k-2}{k-1} \tag{25}$$

$$= (2k-1) \cdot \binom{2k-2}{k-1}. \tag{26}$$

Substituting the last expression into the right-hand side of (22) yields

$$\sum_{1 \leq a,b \leq k} \binom{a+b-2}{a-1}^2 \binom{2k-a-b}{k-a}^2 > \frac{1}{k^2} \cdot (2k-1)^2 \cdot \binom{2k-2}{k-1}^2. \tag{27}$$

Therefore, (19) and (27) imply that

$$p_n > \frac{1}{(2k-1)!}\binom{n}{2k-1}\frac{(2k-1)^2}{k^2}\binom{2k-2}{k-1}^2.$$

As we pointed out after (19), p_n is a polynomial of degree $2k - 1$ in the variable n. The last displayed inequality shows that its leading coefficient is larger than

$$\frac{1}{(2k-1)!^2} \cdot \frac{1}{k^2} \cdot \frac{(2k-2)!^2}{(k-1)!^4} = \frac{k^2}{k!^4}$$

as claimed.

Comparing this with (15) completes the proof of our Proposition.   □

We can now return to the application of Theorem 2.12 to our variables $X_{n,i}$. By Proposition 2.13, there is an absolute constant C so that $\sigma_n > Cn^{k-0.5}$ for all n. So (5) will be satisfied if we show that there exists a positive integer m so that

$$\binom{n}{k}(dn^{k-1})^{m-1} \cdot (n^{-k+0.5})^m < dn^{-0.5m} \to 0.$$

Clearly, any positive integer m is a good choice. So we have proved the following theorem.

Theorem 2.14. *Let k be a fixed positive integer, and let X_n be the random variable counting occurrences of α_k in permutations of length n. Then $\tilde{X}_n \to N(0,1)$. In other words, X_n is asymptotically normal.*

3 Monotone Subsequences with Entries in Consecutive Positions

In 2001, Sergi Elizalde and Marc Noy [11] considered similar problems using another definition of pattern containment. Let us say that the permutation $p = p_1 p_2 \cdots p_n$ *tightly* contains the permutation $q = q_1 q_2 \cdots q_k$ if there exists an index $0 \le i \le n-k$ so that $q_j < q_r$ if and only if $p_{i+j} < p_{i+r}$. (We point out that this definition is a very special case of the one introduced by Babson and Steingrímsson in [3] and called *generalized pattern avoidance*, but we will not need that much more general concept in this paper.)

Example 3.1. While permutation 246351 contains 132 (take the second, third, and fifth entries), it does not *tightly contain* 132 since there are no three entries in consecutive positions in 246351 that would form a 132-pattern.

If p does not tightly contain q, then we say that p *tightly avoids q*. Let $T_n(q)$ denote the number of n-permutations that tightly avoid q. An intriguing conjecture of Elizalde and Noy [11] is the following.

Conjecture 3.2. For any pattern q of length k and for any positive integer n, the inequality

$$T_n(q) \leq T_n(\alpha_k)$$

holds.

This is in stark contrast with the situation for traditional patterns, where, as we have seen in the previous section, the monotone pattern is not the easiest or the hardest to avoid, even in the sense of growth rates.

3.1 Tight Patterns of Length Three

Conjecture 3.2 is proved in [11] in the special case of $k = 3$. As it is clear by taking reverses and complements that $T_n(123) = T_n(321)$ and that $T_n(132) = T_n(231) = T_n(213) = T_n(312)$, it suffices to show that $T_n(132) < T_n(123)$ if $n \geq n$. The authors achieve that by a simple injection.

It turns out that the numbers $T_n(123)$ are not simply larger than the numbers $T_n(132)$; they are larger even in the sense of logarithmic asymptotics. The following results contain the details.

Theorem 3.3 (Elizalde and Noy [11]). *Let $A_{123}(x) = \sum_{n\geq 0} T_n(123)\frac{x^n}{n!}$ be the exponential generating function of the sequence $\{T_n(123)\}_{n\geq 0}$. Then*

$$A_{123}(x) = \frac{\sqrt{3}}{2} \cdot \frac{e^{x/2}}{\cos\left(\frac{\sqrt{3}}{2}x + \frac{\pi}{6}\right)}.$$

Furthermore,

$$T_n(123) \sim \gamma_1 \cdot (\rho_1)^n \cdot n!,$$

where $\rho_1 = \frac{3\sqrt{3}}{2\pi}$ and $\gamma_1 = e^{3\sqrt{3}\pi}$.

Theorem 3.4 (Elizalde and Noy [11]). *Let $A_{132}(x) = \sum_{n\geq 0} T_n(132)\frac{x^n}{n!}$ be the exponential generating function of the sequence $\{T_n(132)\}_{n\geq 0}$. Then*

$$A_{132}(x) = \frac{1}{1 - \int_0^x e^{-t^2/2}\,dt}.$$

Furthermore,

$$T_n(132) \sim \gamma_2 \cdot (\rho_2)^n \cdot n!,$$

where ρ_2^{-1} is the unique positive root of the equation $\int_0^x e^{-t^2/2}\,dt = 1$, and $\gamma_2 = e^{(\rho_2)^{-2}/2}$.

3.2 Tight Patterns of Length Four

For tight patterns, the case of length four is even more complex than it is for traditional patterns. Indeed, it is not true that each of the 24 sequences $T_n(q)$, where q is a tight pattern of length four, is identical to one of $T_n(1342)$, $T_n(1234)$, and $T_n(1324)$. In fact, in [11], Elizalde and Noy showed that there are exactly seven distinct sequences of this kind. They have also proved the following results.

Theorem 3.5. *We have*

(i) $T_n(1342) \sim \gamma_1(\rho_1)^n \cdot n!$,
(ii) $T_n(1234) \sim \gamma_2(\rho_2)^n \cdot n!$, *and*
(iii) $T_n(1243) \sim \gamma_3(\rho_3)^n \cdot n!$,

where ρ_1^{-1} is the smallest positive root z of the equation $\int_0^z e^{-t^3/6} dt = 1$, ρ_2^{-1} is the smallest positive root of $\cos z - \sin z + e^{-z} = 0$, and ρ_3 is the solution of a certain equation involving Airy functions.

The approximate values of these constants are

- $\rho_1 = 0.954611$, $\gamma_1 = 1.8305194$,
- $\rho_2 = 0.963005$, $\gamma_2 = 2.2558142$,
- $\rho_3 = 0.952891$, $\gamma_3 = 1.6043282$.

These results are interesting for several reasons. First, we see that again, $T_n(\alpha_4)$ is larger than the other $T_n(q)$, even in the asymptotic sense. Second, $T_n(1234) \neq T_n(1243)$, in contrast to the traditional case, where $S_n(1234) = S_n(1243)$. Third, the tight pattern 1342 is *not* the hardest to avoid, unlike in the traditional case, where $S_n(1342) \leq S_n(q)$ for any pattern q of length four.

3.3 Longer Tight Patterns

For tight patterns that are longer than four, the only known results concern monotone patterns. They have been found by Richard Warlimont, and, independently, also by Sergi Elizalde and Marc Noy.

Theorem 3.6 (Elizalde, Noy, and Warlimont [11, 26, 25]). *For all integers $k \geq 3$, the identity*

$$\sum_{n \geq 0} T_n(\alpha_k) \frac{x^n}{n!} = \left(\sum_{i \geq 0} \frac{x^{ik}}{(ik)!} - \sum_{i \geq 0} \frac{x^{ik+1}}{(ik+1)!} \right)^{-1}$$

holds.

Theorem 3.7 (Warlimont [25]). *Let $k \geq 3$, let $f_k(x) = \sum_{i \geq 0} \frac{x^{ik}}{(ik)!} - \sum_{i \geq 0} \frac{x^{ik+1}}{(ik+1)!}$, and let ω_k denote the smallest positive root of $f_k(x)$. Then*

$$\omega_k = 1 + \frac{1}{m!}\left(1 + O(1)\right),$$

and

$$\frac{T_n(\alpha_k)}{n!} \sim c_m \omega_k^{-n}.$$

3.4 Growth Rates

The form of the results in Theorems 3.3 and 3.4 is not an accident. They are special cases of the following general theorem.

Theorem 3.8 (Elizalde [10]). *For all patterns q, there exists a constant w_q so that*

$$\lim_{n \to \infty} \left(\frac{T_n(q)}{n!}\right)^{1/n} = w_q.$$

Compare this with the result of Corollary 2.9. That Corollary and the fact that the sequence $(S_n(q)^{1/n}$ is increasing, show that the numbers $S_n(q)$ are roughly as large as $L(q)^n$, for some constant $L(q)$. Clearly, it is much easier to avoid a tight pattern than a traditional pattern. However, Theorem 3.8 shows how much easier it is. Indeed, this time it is not the *number* of pattern avoiding permutations is simply exponential; it is their *ratio* to all permutations that is exponential.

The fact that $T_n(q)/n! < C_q^n$ for *some* C_q is straightforward. Indeed, $T_n(q)/n! < \left(\frac{k!-1}{k!}\right)^{\lfloor n/k \rfloor}$ by simply looking at $\lfloor n/k \rfloor$ distinct subsequences of k consecutive entries. Interestingly, Theorem 3.8 shows that this straightforward estimate is optimal in some (weak) sense. Note that there is no known way to get a result similarly close to the truth for traditional patterns.

3.5 Asymptotic Normality

Our goal now is to prove that the distribution of tight copies of α_k is asymptotically normal in randomly selected permutations of length n. Note that in the special case of $k = 2$, our problem is reduced to the classic result stating that descents of permutations are asymptotically normal. (Just as in the previous section, see [12] and its references for various proofs of this fact, or [8] for a generalization.) Our method is very similar to the one we used in Subsection 2.5. For fixed n and $1 \le i \le n - k + 1$, let $Y_{n,i}$ denote the indicator random variable of the event that in $p = p_1 p_2 \cdots p_n$, the subsequence $p_i p_{i+1} \cdots p_{i+k-1}$ is increasing. Set $Y_n = \sum_{i=1}^{n-k+1} Y_{n,i}$. We want to use Theorem 2.12. Clearly, $|Y_{n,i}| \le 1$ for every i, and $N_n = n - k + 1$. Furthermore, the graph with vertex set $\{1, 2, \cdots, n - k + 1\}$ in which there is an edge between i and j if and only if $|i - j| \le k - 1$ is a dependency graph for the family $\{Y_{n,i} | 1 \le i \le n - k + 1\}$. In this graph, $\Delta_n = 2k - 2$. We will prove the following estimate for $\mathrm{Var}(Y)$.

Proposition 3.9. *There exists a positive constant C so that $\mathrm{Var}(Y) \ge cn$ for all n.*

Proof. By linearity of expectation, we have

$$\mathrm{Var}(Y_n) = E(Y_n^2) - (E(Y_n))^2 \tag{28}$$

$$= E\left(\left(\sum_{i=1}^{n-k+1} Y_{n,i}\right)^2\right) - \left(E\left(\sum_{i=1}^{n-k+1} Y_{n,i}\right)\right)^2 \tag{29}$$

$$= E\left(\left(\sum_{i=1}^{n-k+1} Y_{n,i}\right)^2\right) - \left(\sum_{i=1}^{n-k+1} E(Y_{n,i})\right)^2 \tag{30}$$

$$= \sum_{i_1, i_2} E(Y_{n,i_1} Y_{n,i_2}) - \sum_{i_1, i_2} E(Y_{n,i_1}) E(Y_{n,i_2}). \tag{31}$$

In (31), all the $(n - k + 1)^2$ summands with a negative sign are equal to $1/k!^2$. Among the summands with a positive sign, the $(n - 2k + 1)(n - 2k + 2)$ summands in which $|i_1 - i_2| \ge k$ are equal to $1/k!^2$, the $n - k + 1$ summands in which $i_1 = i_2$ are equal to $1/k!$, and the $2(n - 2k + 2)$ summands in which $|i_1 - i_2| = k - 1$ are equal to $1/(k+1)!$. All remaining summands are non-negative. This shows that

$$\mathrm{Var}(Y_n) \ge \frac{n(1 - 2k) + 3k^2 - 2k + 1}{k!^2} + \frac{n - k + 1}{k!} + \frac{2(n - k + 2)}{(k+1)!}$$

$$\ge \left(\frac{1}{k!} + \frac{2}{(k+1)!} - \frac{2k - 1}{k!^2}\right) n + d_k,$$

where d_k is a constant that depends only on k. As the coefficient $\frac{1}{k!} + \frac{2}{(k+1)!} - \frac{2k-1}{k!^2}$ of n in the last expression is positive for all $k \ge 2$, our claim is proved. $\square$

The main theorem of this subsection is now immediate.

Theorem 3.10. *Let Y_n denote the random variable counting tight copies of α_k in a randomly selected permutation of length n. Then $\tilde{Y}_n \rightarrow N(0,1)$.*

Proof. Use Theorem 2.12 with $m = 3$, and let C be the constant of Proposition 3.9. Then (5) simplifies to

$$(n - k + 1) \cdot (2k - 2)^2 \cdot \frac{C^3}{n^{1.5}},$$

which converges to 0 as n goes to infinity. □

4 Consecutive Entries in Consecutive Positions

Let us take the idea of Elizalde and Noy one step further, by restricting the notion of pattern containment further as follows. Let $p = p_1 p_2 \cdots p_n$ be a permutation, let $k < n$, and let $q = q_1 q_2 \cdots q_k$ be another permutation. We say that p *very tightly* contains q if there is an index $0 \leq i \leq n - k$ and an integer $0 \leq a \leq n - k$ so that $q_j < q_r$ if and only if $p_{i+j} < p_{i+r}$, and,

$$\{p_{i+1}, p_{i+2}, \cdots, p_{i+k}\} = \{a + 1, a + 2, \cdots, a + k\}.$$

That is, p very tightly contains q if p tightly contains q and the entries of p that form a copy of q are not just in consecutive positions, but they are also consecutive as integers (in the sense that their set is an interval). We point out that this definition was used by Amy Myers [19] who called it *rigid* pattern avoidance. However, in order to keep continuity with our previous definitions, we will refer to it as very tight pattern avoidance.

For example, 15324 tightly contains 132 (consider the first three entries), but does not very tightly contain 132. On the other hand, 15324 very tightly contains 213, as can be seen by considering the last three entries. If p does not very tightly contain q, then we will say that p *very tightly avoids q*.

4.1 Enumerative Results

Let $V_n(q)$ be the number of permutations of length n that very tightly avoid the pattern q. The following early results on $V_n(\alpha_k)$ are due to David Jackson and others. They generalize earlier work by Riordan [21] concerning the special case of $k = 3$.

Theorem 4.1 (Jackson, Read, and Reilly [14, 15]). *For all positive integers n, and any $k \leq n$, the value of $V_n(\alpha_k)$ is equal to the coefficient of x^n in the formal power series*

$$\sum_{m \geq 0} m! x^m \left(\frac{1 - x^{k-1}}{1 - x^k} \right)^m.$$

Note that in particular, this implies that for $k \leq n < 2k$, the number of permutations of length $k + r$ *containing* a very tight copy of α_k is $r!(r^2 + r + 1)$.

4.2 An Extremal Property of the Monotone Pattern

Recall that we have seen in Section 2 that in the multiset of the $k!$ numbers $S_n(q)$ where q is of length k, the number $S_n(\alpha_k)$ is neither minimal nor maximal. Also recall that in Section 3 we mentioned that in the multiset of the $k!$ numbers $T_n(q)$, where q is of length k, the number $T_n(\alpha_k)$ is *conjectured* to be maximal. While we cannot prove that we prove that in the multiset of the $k!$ numbers $V_n(q)$, where q is of length k, the number $V_n(\alpha_k)$ is maximal, in this Subsection we prove that for almost all very tight patterns q of length k, the inequality $V_n(q) \leq V_n(\alpha_k)$ does hold.

4.2.1 An Argument Using Expectations

Let q be any pattern of length k. For a fixed positive integer n, let $X_{n,q}$ be the random variable counting the very tight copies of q in a randomly selected n-permutation. It is straightforward to see that by linearity of expectation,

$$E(X_{n,q}) = \frac{(n - k + 1)^2}{\binom{n}{k} k!}. \tag{32}$$

In particular, $E(X_{n,q})$ does not depend on q, just on the length k of q.

Let $p_{n,i,q}$ be the probability that a randomly selected n-permutation contains *exactly* i very tight copies of q, and let $P(n, i, q)$ be the probability that a randomly selected n-permutation contains *at least* i very tight copies of q. Note that $V_n(q) = (1 - P(n, 1, q))n!$, for any given pattern q.

Now note that by the definition of expectation

$$E(X_n, q)$$

$$= \sum_{i=1}^{m} i p_{n,i,q}$$

$$= \sum_{j=0}^{m-1} \sum_{i=0}^{j} p_{n,m-i,q}$$

$$= p_{n,m,q} + (p_{n,m,q} + p_{n,m-1,q}) + \cdots + (p_{n,m,q} + \cdots + p_{n,1,q})$$

$$= \sum_{i=1}^{m} P(n, i, q).$$

We know from (32) that $E(X_{n,q}) = E(X_{n,\alpha_k})$, and then the previous displayed equation implies that

$$\sum_{i=1}^{m} P(n, i, q) = \sum_{i=1}^{m} P(n, i, \alpha). \qquad (33)$$

So if we can show that for $i \geq 2$, the inequality

$$P(n, i, q) \leq P(n, i, \alpha_k) \qquad (34)$$

holds, then (33) will imply that $P(n, 1, q) \geq P(n, 1, \alpha_k)$, which is equivalent to $V_n(q) \leq V_n(\alpha_k)$, which we set out to prove.

4.2.2 Extendible and Non-extendible Patterns

Now we are going to describe the set of patterns q for which we will prove that $V_n(q) \leq V_n(\alpha_k)$.

Let us assume that the permutation $p = p_1 p_2 \cdots p_n$ very tightly contains two *non-disjoint* copies of the pattern $q = q_1 q_2 \cdots q_k$. Let these two copies be $q^{(1)}$ and $q^{(2)}$, so that $q^{(1)} = p_{i+1} p_{i+2} \cdots p_{i+k}$ and $q^{(2)} = p_{i+j+1} p_{i+j+2} \cdots p_{i+j+k}$ for some $j \in [1, k-1]$. Then $|q^{(1)} \cap q^{(2)}| = k - j + 1 =: s$. Furthermore, since the set of entries of $q^{(1)}$ is an interval, and the set of entries of $q^{(2)}$ is an interval, it follows that the set of entries of $q^{(1)} \cap q^{(2)}$ is also an interval. So the rightmost s entries of q, and the leftmost s entries of q must form identical patterns, and the respective sets of these entries must both be intervals.

If q' is the reverse of the pattern q, then clearly $V_n(q) = V_n(q')$. Therefore, we can assume without loss of generality that that the first entry of q is less than the last entry of q. For shortness, we will call such patterns *rising* patterns.

We claim that if p very tightly contains two non-disjoint copies $q^{(1)}$

and $q^{(2)}$ of the rising pattern q, and s is defined as above, then the *rightmost s* entries of q must also be the *largest s* entries of q. This can be seen by considering $q^{(1)}$. Indeed, the set of these entries of $q^{(1)}$ is the intersection of two intervals of the same length, and therefore, must be an ending segment of the interval that starts on the left of the other. An analogous argument, applied for $q^{(2)}$, shows that the leftmost s entries of q must also be the *smallest s* entries of q. So we have proved the following.

Proposition 4.2. *Let p be a permutation that very tightly contains copies $q^{(1)}$ and $q^{(2)}$ of the pattern $q = q_1 q_2 \cdots q_k$. Let us assume without loss of generality that q is rising. Then $q^{(1)}$ and $q^{(2)}$ are disjoint unless all of the following hold.*

There exists a positive integer $s \leq k - 1$ so that

(i) *the rightmost s entries of q are also the largest s entries of q, and the leftmost s entries of q are also the smallest s entries of q, and*

(ii) *the pattern of the leftmost s entries of q is identical to the pattern of the rightmost s entries of q.*

If q satisfies both of these criteria, then two very tightly contained copies of q in p may indeed intersect. For example, the pattern $q = 2143$ satisfies both of the above criteria with $s = 2$, and indeed, 214365 very tightly contains two intersecting copies of q, namely 2143 and 4365.

The following definition is similar to one in [19].

Definition 4.3. *Let $q = q_1 q_2 \cdots q_k$ be a rising pattern that satisfies both conditions of Proposition 4.2. Then we say that q is extendible.*

If q is rising and not extendible, then we say that q is non-extendible.

Note that the notions of extendible and non-extendible patterns are only defined for rising patterns here.

Example 4.4. The extendible patterns of length four are as follows:

- 1234, 1324 (here $s = 1$),
- 2143 (here $s = 2$).

Now we are in a position to prove the main result of this Subsection.

Theorem 4.5. *Let q be any pattern of length k so that either q or its reverse q' is non-extendible. Then for all positive integers n,*

$$V_n(q) \leq V_n(\alpha_k).$$

Proof. We have seen in Subsubsection 4.2.1 that it suffices to prove (34). On the one hand,

$$\frac{(n-k-i+2)!}{n!} \leq P(n,i,\alpha_k), \tag{35}$$

since the number of n-permutations very tightly containing i copies of α is at least as large as the number of n-permutations very tightly containing the pattern $12\cdots(i+k-1)$. The latter is at least as large as the number of n-permutations that very tightly contain a $12\cdots(i+k-1)$-pattern in their first $i+k-1$ positions.

On the other hand,

$$P(n,i,q) \leq \binom{n-i(k-1)}{i}^2 (n-ik)!\frac{1}{n!}. \tag{36}$$

This can be proved by noting that if S is the i-element set of starting positions of i (necessarily disjoint) very tight copies of q in an n-permutation, and A_S is the event that in a random permutation $p = p_1\cdots p_n$, the subsequence $p_j p_{j+1}\cdots p_{j+k-1}$ is a very tight q-subsequence for all $j \in S$, then $P(A_S) = \binom{n-i(k-1)}{i}(n-ik)!\frac{1}{n!}$. The details can be found in [9].

Comparing (35) and (36), the claim of the theorem follows. Again, the reader is invited to consult [9] for details. $\qquad\square$

It is not difficult to show [9] that the ratio of extendible permutations of length k among all permutations of length k converges to 0 as k goes to infinity. So Theorem 4.5 covers almost all patterns of length k.

4.3 The Limiting Distribution of the Number of Very Tight Copies

In the previous two sections, we have seen that the limiting distribution of the number of copies of α_k, as well as the limiting distribution of the number of tight copies of α_k, is normal. Very tight copies behave differently. We will discuss the special case of $k = 2$, that is, the case of the very tight pattern 12.

Theorem 4.6. *Let Z_n be the random variable that counts very tight copies of 12 in a randomly selected permutation of length n. Then Z_n converges a Poisson distribution with parameter $\lambda = 1$.*

A version of this result was proved, in a slightly different setup, by

Wolfowitz in [28] and by Kaplansky in [17]. They used the *method of moments*, which is the following.

Lemma 4.7 (Rucinski [22]). *Let U be a random variable so that*

(i) *for every positive integer k, the moment $E(U^k)$ exists, and*

(ii) *the variable U is completely determined by its moments, that is, there is no other variable with the same sequence of moments.*

Let $U_1, U_2, \cdots$ be a sequence of random variables, and let us assume that for all positive integers k,

$$\lim_{n \to \infty} U_n^k = U^k.$$

Then $U_n \to U$ in distribution.

Proof. (of Theorem 4.6.) It is well-known [24] that the Poisson distribution (with any parameter) is determined by its moments, so the method of moments can be applied to prove convergence to a Poisson distribution. Let $Z_{n,i}$ be the indicator random variable of the event that in a randomly selected n-permutation $p = p_1 p_2 \cdots p_n$, the inequality $p_i + 1 = p_{i+1}$. Then $E(Z_{n,i}) = 1/n$, and the probability that p has a very tight copy of α_k for $k > 2$ is $O(1/n)$. Therefore, we have

$$\lim_{n \to \infty} E(Z_n^j) = \lim_{n \to \infty} E\left(\left(\sum_{i=1}^{n-1} Z_{n,i} \right)^j \right)$$

$$= \lim_{n \to \infty} E\left(\left(\sum_{i=1}^{n-1} V_{n,i} \right)^j \right), \tag{37}$$

where the $V_{n,i}$ are *independent* random variables and each of them takes value 0 with probability $(n-1)/n$, and value 1 with probability $1/n$. (See [28] for more details.) The rightmost limit in the above displayed equation is not difficult to compute. Let t be a fixed non-negative integer. Then the probability that exactly t variables $V_{n,i}$ take value 1 is $\binom{n-1}{t} n^{-t} (\frac{n-1}{n})^{n-t} \sim \frac{e^{-1}}{t!}$. Once we know the t-element set of the $V_{n,i}$ that take value 1, each of the t^j strings of length j formed from those t variables contributes 1 to $E(V^j)$. Summing over all t, this proves that

$$\lim_{n \to \infty} E\left(\left(\sum_{i=1}^{n-1} V_{n,i} \right)^j \right) = e^{-1} \sum_{t \geq 0} \frac{t^j}{j!}.$$

On the other hand, it is well-known that $e^{-1} \sum_{t \geq 1} \frac{t^j}{j!}$, the jth Bell number, is also the jth moment of the Poisson distribution with parameter 1. Comparing this to (37), we see that the sequence $E(Z_n^j)$ converges to the jth moment of the Poisson distribution with parameter 1. Therefore, by the method of moments, our claim is proved. $\qquad\square$

5 Added In Proof

While this is a survey on monotone patterns, it is worth pointing out that Theorem 2.14, and its proof, survive even if we replace α_k by an *arbitrary* pattern. Most of the proof carries through without modification. All that has to be changed are the independent events (b) and (c) considered following equation (15).

Recall that we are in the special case when I_1 and I_2 both form q-patterns, and $I_1 \cap I_2 = x$ is the ath smallest entry in I_1 and the bth smallest entry in I_2. Given q, the pair (a, b) describes the location of x in I_1 and in I_2 as well. Let I_1' (resp. I_2') denote the set of $a - 1$ positions in I_1 (resp. $b - 1$ positions in I_2) which must contain entries smaller than x given that I_1 (resp. I_2) forms a q-pattern. Similarly, let I_1'' (resp. I_2'') denote the set of $k - a$ positions in I_1 (resp. $k - b$ positions I_2) which must contain entries larger than x given that I_1 (resp. I_2) forms a q-pattern.

Now leave condition (a) unchanged, and change conditions (b) and (c) as follows.

(b') The $a + b - 2$ entries in positions belonging to $I_1' \cup I_2'$ must all be smaller than the $2k - a - b$ entries in positions belonging to $I_1'' \cup I_2''$. This happens with probability $\frac{1}{\binom{2k-2}{a+b-2}}$.

(c') • the subsequence I_1' is a pattern that is isomorphic to the pattern formed by the $a - 1$ smallest entries of q,
   • the subsequence I_2' is a pattern that is isomorphic to the pattern formed by the $b - 1$ smallest entries of q,
   • the subsequence I_1'' is a pattern that is isomorphic to the pattern formed by the $k - a$ largest entries of q, and
   • the subsequence I_2'' is a pattern that is isomorphic to the pattern formed by the $k - b$ largest entries of q.
   This happens with probability $\frac{1}{(a-1)!(b-1)!(k-a)!(k-b)!}$.

The rest of the proof is unchanged. It is worth pointing out that while the expectation of the number of copies of a pattern of a given length k does not depend on the pattern (it is $\binom{n}{k}/k!$), the variance of these numbers does. However, it follows easily from our work that the variance is a polynomial function of n that has degree $2n - 1$, and that the *leading coefficient* of this polynomial does not depend on q. It is the terms of lower degree that depend on q.

References

[1] M. H. Albert, M. Elder, A. Rechnitzer, P. Westcott, and M. Zabrocki. On the Wilf-Stanley limit of 4231-avoiding permutations and a conjecture of Arratia. *Adv. in Appl. Math.*, 36(2):95–105, 2006.

[2] R. Arratia. On the Stanley-Wilf conjecture for the number of permutations avoiding a given pattern. *Electron. J. Combin.*, 6:Note, N1, 4 pp., 1999.

[3] E. Babson and E. Steingrímsson. Generalized permutation patterns and a classification of the Mahonian statistics. *Sém. Lothar. Combin.*, 44:Article B44b, 18 pp., 2000.

[4] J. Baik, P. Deift, and K. Johansson. On the distribution of the length of the longest increasing subsequence of random permutations. *J. Amer. Math. Soc.*, 12(4):1119–1178, 1999.

[5] M. Bóna. *Exact and asymptotic enumeration of permutations with subsequence conditions.* PhD thesis, M.I.T., 1997.

[6] M. Bóna. Exact enumeration of 1342-avoiding permutations: a close link with labeled trees and planar maps. *J. Combin. Theory Ser. A*, 80(2):257–272, 1997.

[7] M. Bóna. *Combinatorics of permutations.* Discrete Mathematics and its Applications (Boca Raton). Chapman & Hall/CRC, Boca Raton, FL, 2004.

[8] M. Bóna. Generalized descents and normality. *Electron. J. Combin.*, 15(1):Note 21, 8, 2008.

[9] M. Bóna. Where the monotone pattern (mostly) rules. *Discrete Math.*, 308(23):5782–5788, 2008.

[10] S. Elizalde. Asymptotic enumeration of permutations avoiding generalized patterns. *Adv. in Appl. Math.*, 36(2):138–155, 2006.

[11] S. Elizalde and M. Noy. Consecutive patterns in permutations. *Adv. in Appl. Math.*, 30(1-2):110–125, 2003.

[12] J. Fulman. Stein's method and non-reversible Markov chains. In *Stein's method: expository lectures and applications*, volume 46 of *IMS Lecture Notes Monogr. Ser.*, pages 69–77. Inst. Math. Statist., Beachwood, OH, 2004.

[13] I. M. Gessel. Symmetric functions and P-recursiveness. *J. Combin. Theory Ser. A*, 53(2):257–285, 1990.

[14] D. M. Jackson and R. C. Read. A note on permutations without runs of given length. *Aequationes Math.*, 17(2-3):336–343, 1978.

[15] D. M. Jackson and J. W. Reilly. Permutations with a prescribed number of p-runs. *Ars Combinatoria*, 1(1):297–305, 1976.

[16] S. Janson. Normal convergence by higher semi-invariants with applications to sums of dependent random variables and random graphs. *Ann. Probab.*, 16(1):305–312, 1988.

[17] I. Kaplansky. The asymptotic distribution of runs of consecutive elements. *Ann. Math. Statistics*, 16:200–203, 1945.

[18] A. Marcus and G. Tardos. Excluded permutation matrices and the Stanley-Wilf conjecture. *J. Combin. Theory Ser. A*, 107(1):153–160, 2004.

[19] A. N. Myers. Counting permutations by their rigid patterns. *J. Combin. Theory Ser. A*, 99(2):345–357, 2002.

[20] A. Regev. Asymptotic values for degrees associated with strips of Young diagrams. *Adv. in Math.*, 41(2):115–136, 1981.

[21] J. Riordan. Permutations without 3-sequences. *Bull. Amer. Math. Soc.*, 51:745–748, 1945.

[22] A. Ruciński. Proving normality in combinatorics. In *Random graphs, Vol. 2 (Poznań, 1989)*, Wiley-Intersci. Publ., pages 215–231. Wiley, New York, 1992.

[23] R. Simion and F. W. Schmidt. Restricted permutations. *European J. Combin.*, 6(4):383–406, 1985.

[24] R. von Mises. über die wahrscheinlichkeit seltener ereignisse. *Z. Angew. Math. Mech.*, 1(2):121–124, 1921.

[25] R. Warlimont. Permutations avoiding consecutive patterns. *Ann. Univ. Sci. Budapest. Sect. Comput.*, 22:373–393, 2003.

[26] R. Warlimont. Permutations avoiding consecutive patterns. II. *Arch. Math. (Basel)*, 84(6):496–502, 2005.

[27] J. West. *Permutations with forbidden subsequences and stack-sortable permutations*. PhD thesis, M.I.T., 1990.

[28] J. Wolfowitz. Note on runs of consecutive elements. *Ann. Math. Statistics*, 15:97–98, 1944.

A survey on partially ordered patterns

Sergey Kitaev

The Mathematics Institute
Reykjavík University
IS-103 Reykjavík, Iceland

Abstract

The paper offers an overview over selected results in the literature on partially ordered patterns (POPs) in permutations, words and compositions. The POPs give rise in connection with co-unimodal patterns, peaks and valleys in permutations, Horse permutations, Catalan, Narayana, and Pell numbers, bi-colored set partitions, and other combinatorial objects.

1 Introduction

An occurrence of a *pattern* τ in a permutation π is defined as a subsequence in π (of the same length as τ) whose letters are in the same relative order as those in τ. For example, the permutation 31425 has three occurrences of the pattern 1-2-3, namely the subsequences 345, 145, and 125. *Generalized permutation patterns (GPs)* being introduced in [1] allow the requirement that some adjacent letters in a pattern must also be adjacent in the permutation. We indicate this requirement by removing a dash in the corresponding place. Say, if pattern 2-31 occurs in a permutation π, then the letters in π that correspond to 3 and 1 are adjacent. For example, the permutation 516423 has only one occurrence of the pattern 2-31, namely the subword 564, whereas the pattern 2-3-1 occurs, in addition, in the subwords 562 and 563. Placing "[" on the left (resp., "]" on the right) next to a pattern p means the requirement that p must begin (resp., end) from the leftmost (resp., rightmost) letter. For example, the permutation 32415 contains two occurrences of the pattern [2-13, namely the subwords 324 and 315 and no occurrences

115

of the pattern 3-2-1]. We refer to [3] and [18] for more information on patterns and GPs.

A further generalization of the GPs (see [17]) is *partially ordered patterns (POPs)*, where the letters of a pattern form a partially ordered set (poset), and an occurrence of such a pattern in a permutation is a linear extension of the corresponding poset in the order suggested by the pattern (we also pay attention to eventual dashes and brackets). For instance, if we have a poset on three elements labeled by $1'$, 1, and 2, in which the only relation is $1 < 2$ (see Figure 1), then in an occurrence of $p = 1'$-12 in a permutation π the letter corresponding to the $1'$ in p can be either larger or smaller than the letters corresponding to 12. Thus, the permutation 31254 has three occurrences of p, namely 3-12, 3-25, and 1-25.

Fig. 1. A poset on three elements with the only relation $1 < 2$.

Let $\mathcal{S}_n(p_1, \ldots, p_k)$ denote the set of n-permutations simultaneously avoiding each of the patterns $p_1, \ldots, p_k$.

The POPs were introduced in [15]† as an auxiliary tool to study the maximum number of non-overlapping occurrences of *segmented* GPs (*SGPs*), also known as *consecutive* GPs, that is, the GPs, occurrences of which in permutations form contiguous subwords (there are no dashes). However, the most useful property of POPs known so far is their ability to "encode" certain sets of GPs which provides a convenient notation for those sets and often gives an idea how to treat them. For example, the original proof of the fact that $|\mathcal{S}_n(123, 132, 213)| = \binom{n}{\lfloor n/2 \rfloor}$ took 3 pages ([14]); on the other hand, if one notices that $|\mathcal{S}_n(123, 132, 213)| = |\mathcal{S}_n(11'2)|$, where the letters 1, $1'$, and 2 came from the same poset as above, then the result is easy to see. Indeed, we may use the property that the letters in odd and even positions of a "good" permutation do not affect each other because of the form of $11'2$. Thus we choose the letters in odd positions in $\binom{n}{\lfloor n/2 \rfloor}$ ways, and we must arrange them in decreasing order. We then must arrange the letters in even positions in decreasing order too.

The POPs can be used to encode certain combinatorial objects by re-

† The POPs in this paper, as well as in [17], are the same as the POGPs in [15], which is an abbreviation for Partially Ordered Generalized Patterns.

stricted permutations. Examples of that are Theorem 2.1, Propositions 3.5, 3.6, 4.2, and 4.5, as well as several other results in the literature (see, e.g., [5]). Such encoding is interesting from the point of view of finding bijections between the sets of objects involved, but it also may have applications for enumerating certain statistics. The idea is to encode a set of objects under consideration as a set of permutations satisfying certain restrictions (given by certain POPs); under appropriate encodings, this allows us to transfer the interesting statistics from the original set to the set of permutations, where they are easy to handle. For an illustration of how encoding by POPs can be used, see [20, Theorem 2.4] which deals with POPs in *compositions* (discussed in Section 5) rather than in permutations, though the approach remains the same.

As a matter of fact, some POPs appeared in the literature before they were actually introduced. Thus the notion of a POP allows us to collect under one roof (to provide a uniform notation for) several combinatorial structures such as *peaks, valleys, modified maxima* and *modified minima* in permutations, *Horse permutations* and *p-descents* in permutations discussed in Section 2.

There are several other ways to define occurrences of patterns in permutations (and other combinatorial objects like words and compositions) for which POPs can be defined and studied (see, e.g., [27] where certain POPs are studied in connection with *cyclic occurrence of patterns*). However, this survey deals with occurrences of patterns in the sense specified above.

The paper is organized as follows. Section 2 deals with *co-unimodal patterns* and some of their variations. In particular, this involves considering *peaks* and *valleys* in permutations, as well as so called *V- and Λ−patterns*. Sections 3 and 4 discuss POPs with, and without, dashes involved, respectively. In particular, Section 3 deals with *Horse permutations* and *multi-patterns*, while Section 4 presents results on *flat posets, non-overlapping SPOPs* in permutations and words, and *q-analogues* for non-overlapping SPOPs (*SPOP* abbreviates Segmented POP). Further, Section 5 states some of results on POPs in *compositions*, which can be viewed as a generalization for certain results on POPs in words. Finally, in Section 6, we state a couple of concluding remarks.

In what follows we need the following notations. Let σ and τ be two POPs of length greater than 0. We write $\sigma < \tau$ to indicate that any letter of σ is less than any letter of τ. We write $\sigma <> \tau$ when no letter in σ is comparable with any letter in τ. The *GF* (*EGF; BGF*) denotes the (*exponential; bivariate*) *generating function*. If $\pi = a_1 a_2 \cdots a_n \in \mathcal{S}_n$,

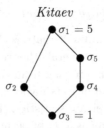

Fig. 2. A poset for co-unimodal pattern in the case $j = 3$ and $k = 5$.

then the *reverse* of π is $\pi^r := a_n \cdots a_2 a_1$, and the *complement* of π is a permutation π^c such that $\pi_i^c = n + 1 - a_i$, where $i \in [n] = \{1, \ldots, n\}$. We call π^r, π^c, and $(\pi^r)^c = (\pi^c)^r$ *trivial bijections*.

2 Co-unimodal patterns and their variations

For a permutation $\pi = \pi_1 \pi_2 \cdots \pi_n \in \mathcal{S}_n$, the *inversion index,* $\mathrm{inv}(\pi)$, is the number of ordered pairs (i, j) such that $1 \le i < j \le n$ and $\pi_i > \pi_j$. The major index, $\mathrm{maj}(\pi)$, is the sum of all i such that $\pi_i > \pi_{i+1}$. Suppose σ is a SPOP and

$$\mathrm{place}_\sigma(\pi) = \{i \mid \pi \text{ has an occurrence of } \sigma \text{ starting at } \pi_i\}.$$

Let $\mathrm{maj}_\sigma(\pi)$ be the sum of the elements of $\mathrm{place}_\sigma(\pi)$.

If σ is *co-unimodal*, meaning that $k = \sigma_1 > \sigma_2 > \cdots > \sigma_j < \cdots < \sigma_k$ for some $2 \le j \le k$ (see Figure 2 for a corresponding poset in the case $j = 3$ and $k = 5$), then the following formula holds [2]:

$$\sum_{\pi \in \mathcal{S}_n} t^{\mathrm{maj}_\sigma(\pi^{-1})} q^{\mathrm{maj}(\pi)} = \sum_{\pi \in \mathcal{S}_n} t^{\mathrm{maj}_\sigma(\pi^{-1})} q^{\mathrm{inv}(\pi)}.$$

If $k = 2$ we deal with usual descents. Thus a co-unimodal pattern can be viewed as a generalization of the notion of a descent. This may be a reason why a co-unimodal pattern p is called *p-descent* in [2]. Also, setting $t = 1$ we get a well-known result by MacMahon on equidistribution of maj and inv.

The notion of co-unimodal patterns was refined and generalized in [26], where the authors use symmetric functions along with λ-*brick tabloids* and *weighted λ-brick tabloids* to obtain their (new) results as well as some known results. Moreover, in all the cases in [26], it is possible to extend the results to q-analogues, where the powers of q count the inversion statistic. See [25] for basic techniques and ideas used in [26].

2.1 Peaks and valleys in permutations

A permutation π has exactly k *peaks* (resp., *valleys*), also known as *maxima* (resp., *minima*), if $|\{j \mid \pi_j > \max\{\pi_{j-1}, \pi_{j+1}\}\}| = k$ (resp., $|\{j \mid \pi_j < \min\{\pi_{j-1}, \pi_{j+1}\}\}| = k$). Thus, an occurrence of a peak in a permutation is an occurrence of the SPOP $1'21''$, where relations in the poset are $1' < 2$ and $1'' < 2$. Similarly, occurrences of valleys correspond to occurrences of the SPOP $2'12''$, where $2' > 1$ and $2'' > 1$. See Figure 3 for the posets corresponding to the peaks and valleys. So, any research done on the peak (or valley) statistics can be regarded as research on (S)POPs (e.g., see [30]).

Fig. 3. Posets corresponding to peaks and valleys.

Also, results related to *modified maxima* and *modified minima* can be viewed as results on SPOPs. For a permutation $\sigma_1 \ldots \sigma_n$ we say that σ_i is a *modified maximum* if $\sigma_{i-1} < \sigma_i > \sigma_{i+1}$ and a *modified minimum* if $\sigma_{i-1} > \sigma_i < \sigma_{i+1}$, for $i = 1, \ldots, n$, where $\sigma_0 = \sigma_{n+1} = 0$. Indeed, we can view a pattern p as a function from the set of all symmetric groups $\cup_{n \geq 0} \mathcal{S}_n$ to the set of natural numbers such that $p(\pi)$ is the number of occurrences of p in π, where π is a permutation. Thus, studying the distribution of modified maxima (resp., minima) is the same as studying the function $ab] + 1'21'' + [dc$ (resp., $ba] + 2'12'' + [cd$) where $a < b$, $c < d$ and the other relations between the patterns' letters are taken from Figure 3. Also, recall that placing "[" (resp., "]") next to a pattern p means the requirement that p must begin (resp., end) with the leftmost (resp., rightmost) letter.

A specific result in this direction is problem 3.3.46(c) on page 195 in [9]: We say that σ_i is a *double rise* (resp., *double fall*) if $\sigma_{i-1} < \sigma_i < \sigma_{i+1}$ (resp., $\sigma_{i-1} > \sigma_i > \sigma_{i+1}$); The number of permutations in $\mathcal{S}_n$ with i_1 modified minima, i_2 modified maxima, i_3 double rises, and i_4 double falls is

$$\left[u_1^{i_1} u_2^{i_2 - 1} u_3^{i_3} u_4^{i_4} \frac{x^n}{n!} \right] \frac{e^{\alpha_2 x} - e^{\alpha_1 x}}{\alpha_2 e^{\alpha_1 x} - \alpha_1 e^{\alpha_2 x}}$$

where $\alpha_1 \alpha_2 = u_1 u_2$, $\alpha_1 + \alpha_2 = u_3 + u_4$.

In Corollary 4.15 one has an explicit generating function for the distribution of peaks (valleys) in permutations. This result is an analogue to a result in [7] where the circular case of permutations is considered, that is, when the first letter of a permutation is thought to be to the right of the last letter in the permutation. In [7] it is shown that if $M(n,k)$ denotes the number of circular permutations in S_n having k maxima, then

$$\sum_{n\geq 1}\sum_{k\geq 0} M(n,k)y^k \frac{x^n}{n!} = \frac{zx(1-z\tanh xz)}{z-\tanh xz}$$

where $z = \sqrt{1-y}$.

2.2 V- and Λ-patterns

A variation of co-unimodal patterns is when we do not require in a co-unimodal pattern the first element to be the largest one. More precisely, we say that a factor $\pi_{i-k}\cdots\pi_i\cdots\pi_{i+\ell}$ of a permutation $\pi_1\cdots\pi_n$ is an occurrence of the pattern $V(k,\ell)$ (resp. $\Lambda(k,\ell)$) if $\pi_{i-k} > \pi_{i-k+1} > \cdots > \pi_i < \pi_{i+1} < \cdots < \pi_{i+\ell}$ (resp. $\pi_{i-k} < \pi_{i-k+1} < \cdots < \pi_i > \pi_{i+1} > \cdots > \pi_{i+\ell}$). Such patterns are a refinement of the concept of peaks and valleys.

A general approach to study avoidance of V- and Λ−patterns is suggested in [21] (see [21, Subsec. 2.2]). Below, we list explicit enumerative results in [21] starting with the one having a combinatorial interpretation for avoidance of a certain V-pattern.

Let K_n' denote the *corona* of the complete graph K_n and the complete graph K_1; in other words, K_n' is the graph constructed from K_n by adding for each vertex v a new vertex v' and the edge vv'. The following theorem provides a combinatorial property involving the pattern $V(1,2)$.

Theorem 2.1. ([21, Theorem 7]) *The set of $(n+1)$-permutations simultaneously avoiding the patterns 213 and $V(1,2)$ is in one-to-one correspondence with the set of all matchings of K_n'. Thus, the EGF for the number of permutations avoiding the patterns 213 and $V(1,2)$ is given by*

$$A(x) = 1 + \int_0^x e^{2t+t^2/2}\, dt.$$

Theorem 2.2. ([21, Theorem 1]) *The EGF $A(x)$ for the number of permutations avoiding $V(2,1)$ is given by*

$$1 + \exp\left(\frac{3x}{2}\right)\sec\left(\frac{\sqrt{3}x}{2}+\frac{\pi}{6}\right)\int_0^x \exp\left(-\frac{3u}{2}\right)\cos\left(\frac{\sqrt{3}u}{2}+\frac{\pi}{6}\right)du.$$

Theorem 2.3. ([21, Theorem 2]) *The EGF $A(x)$ for the number of permutations simultaneously avoiding the patterns $V(1,2)$, $V(2,1)$, and $\Lambda(1,2)$ is given by*

$$\frac{1}{2}(e^x + (\tan x + \sec x)(e^x + 1) - (1 + 2x + x^2)).$$

Theorem 2.4. ([21, Theorem 2]) *The EGF $A(x)$ for the number of permutations simultaneously avoiding the patterns $V(1,2)$ and $\Lambda(1,2)$ is given by*

$$1 + x + (\tan x + \sec x - 1)(e^x - 1).$$

Theorem 2.5. ([21, Corollary 5]) *The EGF $A(x)$ for the number of permutations simultaneously avoiding the patterns $V(1,2)$ and $\Lambda(2,1)$ is given by*

$$\frac{\sqrt{3}}{2} \exp\left(\frac{x}{2}\right) \sec\left(\frac{\sqrt{3}}{2}x + \frac{\pi}{6}\right) + e^x - \left(1 + x + \frac{x^2}{2}\right).$$

Theorem 2.6. ([21, Theorem 6]) *The number of n-permutations simultaneously avoiding $V(1,2)$ and $V(2,1)$ is given by*

$$A_n = \sum_{\substack{i,j \geq 1 \\ i+j \leq n+1}} A_{i,j}^n$$

with

$$A_{i,j}^n = \sum_{\substack{i,j \geq 1 \\ i+j \leq n+1 \\ n-i-j \text{ odd}}} A_{i,j}^n$$

where E_n is the number of alternating permutations.

3 POPs involving dashes

In this section we consider some of the results on POPs involving at least one dash.

3.1 *Patterns containing $\square$-symbol*

In [13] the authors study simultaneous avoidance of the patterns 1-3-2 and 1$\square$23. A permutation π avoids 1$\square$23 if there is no $\pi_i < \pi_j < \pi_{j+1}$ with $i < j - 1$. Thus the $\square$ symbol has the same meaning as "-" except for $\square$ does not allow the letters separated by it to be adjacent in an occurrence of the corresponding pattern. In the POP-terminology, $1\square 23$ is the pattern 1-1'-23, or 1-1'23, or 11'-23, where 1' is incomparable with the letters 1, 2, and 3 which, in turn, are ordered naturally: $1 < 2 < 3$. The permutations avoiding 1-3-2 and $1\square 23$ are called *Horse permutations*. The reason for the name came from the fact that these permutations are in one to one correspondence with *Horse paths*, which are the lattice paths from (0,0) to (n, n) containing the steps $(0, 1)$, $(1, 1)$, $(2, 1)$, and $(1, 2)$ and not passing the line $y = x$. According to [13], the generating function for the horse permutations is

$$\frac{1 - x - \sqrt{1 - 2x - 3x^2 - 4x^3}}{2x^2 (1 + x)}.$$

Moreover, in [13] the generating functions for Horse permutations avoiding, or containing (exactly) once, certain patterns are given.

In [8], patterns of the form x-$y\square z$ are studied, where $xyz \in \mathcal{S}_3$. Such a pattern can be written in the POP-notation as, for example, x-y-a-z where a is not comparable to x, y, and z. A bijection between permutations avoiding the pattern 1-2$\square$3, or 2-1$\square$3, and the set of *odd-dissection convex polygons* is given. Moreover, generating functions for permutations avoiding 1-3$\square$2 and certain additional patterns are obtained in [8].

3.2 *Patterns of the form σ-m-τ*

Let σ and τ be two SGPs (the results below work for SPOPs as well). We consider the POP $\alpha = \sigma$-m-τ with $m > \sigma$, $m > \tau$, and $\sigma <> \tau$, that is, each letter of σ is incomparable with any letter of τ and m is the largest letter in α. The POP α is an instance of so called *shuffle patterns* (see [15, Sec 4]).

Theorem 3.1. ([15, Theorem 16]) *Let $A(x)$, $B(x)$ and $C(x)$ be the EGF for the number of permutations that avoid σ, τ and α respectively. Then $C(x)$ is the solution to the following differential equation with $C(0) = 1$:*

$$C'(x) = (A(x) + B(x))C(x) - A(x)B(x).$$

If τ is the empty word then $B(x) = 0$ and we get the following result for segmented GPs:

Corollary 3.2. ([15, Theorem 13],[22]) *Let $\alpha = \sigma$-m, where σ is a SGP on $[k - 1]$. Let $A(x)$ (resp., $C(x)$) be the EGF for the number of permutations that avoid σ (resp., α). Then $C(x) = e^{F(x, A)}$, where $F(x, A) = \int_0^x A(y)\, dy$.*

Example 3.3. ([15, Ex 15]) Suppose $\alpha = $ 12-3. Here $\sigma = 12$, whence $A(x) = e^x$, since there is only one permutation that avoids σ. So

$$C(x) = e^{F(x,\exp)} = e^{e^x - 1}.$$

We get [6, Proposition 4] since $C(x)$ is the EGF for the Bell numbers.

Corollary 3.4. ([15, Corollary 19]) *Let* $\alpha = \sigma$-*m*-τ *is as described above. We consider the pattern* $\varphi(\alpha) = \varphi_1(\sigma)$-*m*-$\varphi_2(\tau)$, *where* φ_1 *and* φ_2 *are any trivial bijections. Then* $|\mathcal{S}_n(\alpha)| = |\mathcal{S}_n(\varphi(\alpha))|$.

3.3 Patterns of the form m-σ-m

This subsection contains results on patterns in which two largest *incomparable* elements of the corresponding poset embrace the other elements building a consecutive POP (SPOP).

Proposition 3.5. ([10]) *Suppose the elements* $1, 2, 3', 3''$ *build the poset with the relations* $1 < 2 < 3'$ *and* $2 < 3''$ ($3'$ *is incomparable with* $3''$). *Then permutations avoiding the POP* $3'$-12-$3''$ *are in one-to-one correspondence with* bi-colored set partitions.

Proposition 3.6. ([10]) *Suppose the elements* $1', 1'', 2, 3', 3''$ *build the poset with the relations* $1', 1'' < 2 < 3', 3''$ ($1'$ *is incomparable with* $1''$, *and* $3'$ *is incomparable with* $3''$). *Then permutations avoiding the POP* $3'$-$1'21''$-$3''$ *are in one-to-one correspondence with* Dowling partitions. *Moreover, the EGF for such permutations is given by*

$$1 + \int_0^x \exp\left(\frac{e^t + 2t - 1}{2}\right) dt.$$

3.4 Multi-patterns

Suppose $\{\sigma_1, \sigma_2, \ldots, \sigma_k\}$ is a set of segmented GPs and $p = \sigma_1$-σ_2-$\cdots$-σ_k where each letter of σ_i is incomparable with any letter of σ_j whenever $i \neq j$ ($\sigma_i <> \sigma_j$). We call such POPs *multi-patterns*. Clearly, the Hasse diagram for such a pattern is k disjoint chains similar to that in Figure 4.

Theorem 3.7. ([15, Theorem 23 and Corollary 24]) *The number of permutations avoiding the pattern* $p = \sigma_1$-σ_2-$\cdots$-σ_k *is equal to that avoiding a multi-pattern obtained from* p *by an arbitrary permutation of* σ_i's *as well as by applying to* σ_i's *any of trivial bijections.*

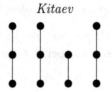

Fig. 4. A poset corresponding to a multi-pattern.

The following theorem is the basis for calculating the number of permutations that avoid a multi-pattern.

Theorem 3.8. ([15, Theorem 28]) *Let* $p = \sigma_1\text{-}\sigma_2\text{-}\cdots\text{-}\sigma_k$ *be a multi-pattern and let* $A_i(x)$ *be the EGF for the number of permutations that avoid* σ_i. *Then the EGF* $A(x)$ *for the number of permutations that avoid* p *is*

$$A(x) = \sum_{i=1}^{k} A_i(x) \prod_{j=1}^{i-1} ((x-1)A_j(x) + 1).$$

Corollary 3.9. ([15, Corollary 26]) *Let* $p = \sigma_1\text{-}\sigma_2\text{-}\cdots\text{-}\sigma_k$ *be a multi-pattern, where* $|\sigma_i| = 2$ *for all* i. *That is, each* σ_i *is either 12 or 21. Then the EGF for the number of permutations that avoid* p *is given by*

$$A(x) = \frac{1 - (1 + (x-1)e^x)^k}{1 - x}.$$

Remark 3.10. Although the results in Theorems 3.7 and 3.8 are stated in [15] for σ_i's which are SGPs, they are true for σ_is which are SPOPs ([17, Remark 7]).

4 Segmented POPs (SPOPs)

Patterns in Section 2 are also examples of SPOPs. In fact, the most of known results on POPs are related to SPOPs.

4.1 Segmented patterns of length four

In this subsection we provide the known results related to SPOPs of length four. Theorem 2.2, Proposition 4.9, and Corollary 4.14 give extra results on such patterns. In this subsection, $A(x) = \sum_{n \geq 0} A_n x^n / n!$ is the EGF for the number of permutations in question. The patterns in the subsection are built on the poset in Figure 6 and the letter $1''$ is not comparable to any other letter.

Theorem 4.1. ([15, Theorem 30]) *For the SPOP* 122′1′, *we have that*

$$A(x) = \frac{1}{2} + \frac{1}{4} \tan x (1 + e^{2x} + 2e^x \sin x) + \frac{1}{2} e^x \cos x.$$

Proposition 4.2. ([16, Propositions 8 and 9]) *There are*

$$\binom{n-1}{\lfloor (n-1)/2 \rfloor} \binom{n}{\lfloor n/2 \rfloor}$$

permutations in $\mathcal{S}_n$ *that avoid the SPOP* 12′21′. *The* (n+1)-*permutations avoiding* 12′21′ *are in one-to-one correspondence with different walks of* n *steps between lattice points, each in a direction N, S, E or W, starting from the origin and remaining in the positive quadrant.*

Proposition 4.3. ([16, Propositions 4, 5 and 6]) *For the SPOP* 11′1″2, *one has*

$$A_n = \frac{n!}{\lfloor n/3 \rfloor! \lfloor (n+1)/3 \rfloor! \lfloor (n+2)/3 \rfloor!},$$

and for the SPOP 11′21″ *and* $n \geq 1$, *we have* $A_n = n \cdot \binom{n-1}{\lfloor (n-1)/2 \rfloor}$. *Moreover, for the SPOPs* 1′1″12 *and* 1′121″, *we have* $A_0 = A_1 = 1$, *and, for* $n \geq 2$, $A_n = n(n-1)$.

Proposition 4.4. ([16, Proposition 7]) *For the SPOP* 1231′, *we have*

$$A(x) = xe^{x/2} \left(\cos \frac{\sqrt{3}x}{2} - \frac{\sqrt{3}}{3} \sin \frac{\sqrt{3}x}{2} \right)^{-1} + 1,$$

and for the SPOPs 1321′ *and* 2131′, *we have*

$$A(x) = x(1 - \int_0^x e^{-t^2/2} \, dt)^{-1} + 1.$$

We end this subsection with a result on multi-avoidance of SPOPs that has a combinatorial interpretation.

Proposition 4.5. ([5, Propositions 2.1 and 2.2]) *There are* $2\binom{n}{\lfloor n/2 \rfloor}$ *permutations in* $\mathcal{S}_n$ *that avoid the SPOPs* 11′22′ *and* 22′11′ *simultaneously. For* $n \geq 3$, *there is a bijection between such* n-*permutations and the set of all* (n + 1)-*step walks on the x-axis with the steps* $a = (1, 0)$ *and* $\bar{a} = (-1, 0)$ *starting from the origin but not returning to it.*

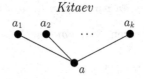

Fig. 5. A flat poset.

4.2 SPOPs built on flat posets

In this subsection, we consider flat posets built on $k+1$ elements a, a_1, ..., a_k with the only relations $a < a_i$ for all i. A Hasse diagram for the flat poset is in Figure 5. Theorem 4.11 and Corollary 4.15 are the main results in the subsection.

The following proposition generalizes [6, Proposition 6]. Indeed, letting $k = 2$ in the proposition we deal with involutions and permutations avoiding 1-23 and 1-32. Note that even though Proposition 4.6 and Corollary 4.7 contain dashes in the patterns, those results are actually on SPOPs due to Proposition 4.8. (We stated the results with dashes to be consistent with [6, Proposition 6].)

Proposition 4.6. ([17, Proposition 14]) *The permutations in $\mathcal{S}_n$ having cycles of length at most k are in one-to-one correspondence with permutations in $\mathcal{S}_n$ that avoid a-$a_1 \cdots a_k$.*

Corollary 4.7. ([17, Corollary 15]) *The EGF for the number of permutations avoiding a-$a_1 \cdots a_k$ is given by* $\exp\left(\sum_{i=1}^{k} x^i/i\right)$.

Proposition 4.8. ([17, Proposition 16]) *One has*

$$\mathcal{S}_n(a\text{-}a_1 \cdots a_k) = \mathcal{S}_n(aa_1 \cdots a_k),$$

and thus the EGF for the number of permutations avoiding $aa_1 \cdots a_k$ is

$$\exp\left(\sum_{i=1}^{k} x^i/i\right).$$

Proposition 4.9. ([17, Corollary 17]) *The EGF for the number of permutations avoiding $aa_1a_2a_3$ is given by* $\exp(x + x^2/2 + x^3/3)$.

Theorem 4.10. (Distribution of $aa_1a_2 \cdots a_k$, [17, Theorem 18]) *Let*

$$P := P(x,y) = \sum_{n \geq 0} \sum_{\pi \in \mathcal{S}_n} y^{e(\pi)} x^n/n!$$

be the BGF on permutations, where $e(\pi)$ is the number of occurrences of the SPOP $p = aa_1a_2 \cdots a_k$ in π. Then P is the solution to

$$\frac{\partial P}{\partial x} = yP^2 + \frac{(1-y)(1-x^k)}{1-x}P \tag{1}$$

with the initial condition $P(0, y) = 1$.

Note, that if $y = 0$ in Theorem 4.10, then the function in Corollary 4.7, due to Proposition 4.8, is supposed to be the solution to (1), which is true. If $k = 1$ in Theorem 4.10, then as the solution to (1) we get nothing else but the distribution of descents in permutations: $(1-y)(e^{(y-1)x} - y)^{-1}$. Thus Theorem 4.10 can be thought as a generalization of the result on the descent distribution.

The following theorem generalizes Theorem 4.10. Indeed, Theorem 4.10 is obtained from Theorem 4.11 by plugging in $\ell = 0$ and observing that obviously $aa_1 \cdots a_k$ and $a_1 \cdots a_k a$ are equidistributed.

Theorem 4.11. (Distribution of $a_1a_2 \cdots a_k aa_{k+1}a_{k+2} \cdots a_{k+\ell}$, [17, Theorem 19])) *Let*

$$P := P(x, y) = \sum_{n \geq 0} \sum_{\pi \in \mathcal{S}_n} y^{e(\pi)} x^n / n!$$

be the BGF of permutations where $e(\pi)$ is the number of occurrences of the SPOP $p = a_1a_2 \cdots a_k aa_{k+1}a_{k+2} \cdots a_{k+\ell}$ in π. Then P is the solution to

$$\frac{\partial P}{\partial x} = y\left(P - \frac{1-x^k}{1-x}\right)\left(P - \frac{1-x^\ell}{1-x}\right) + \frac{2 - x^k - x^\ell}{1-x}P$$
$$- \frac{1 - x^k - x^\ell + x^{k+\ell}}{(1-x)^2} \tag{2}$$

with the initial condition $P(0, y) = 1$.

If $y = 0$ in Theorem 4.11 then we get the following corollary:

Corollary 4.12. ([17, Corollary 20]) *The EGF $A(x)$ for the number of permutations avoiding the SPOP $p = a_1a_2 \cdots a_k aa_{k+1}a_{k+2} \cdots a_{k+\ell}$ satisfies the differential equation*

$$A'(x) = \frac{2 - x^k - x^\ell}{1-x}A(x) - \frac{1 - x^k - x^\ell + x^{k+\ell}}{(1-x)^2}$$

with the initial condition $A(0) = 1$.

The following corollaries to Corollary 4.12 are obtained by plugging in $k = \ell = 1$ and $k = 1$ and $\ell = 2$ respectively.

Corollary 4.13. ([14]) The EGF for the number of permutations avoiding $a_1 a a_2$ is $(\exp(2x) + 1)/2$ and thus $|\mathcal{S}_n (a_1 a a_2)| = 2^{n-1}$.

Corollary 4.14. ([17, Corollary 22]) *The EGF for the number of permutations avoiding $a_1 a a_2 a_3$ is*

$$1 + \sqrt{\pi/2} \left(\operatorname{erf}(x/\sqrt{2} + \sqrt{2}) - \operatorname{erf}(\sqrt{2}) \right) e^{x(x+4)/2+2}$$

where $\operatorname{erf}(x) = \dfrac{2}{\sqrt{\pi}} \displaystyle\int_0^x e^{-t^2}\, dt$ is the error function.

If $k = 1$ and $\ell = 1$, then our pattern $a_1 a a_2$ is nothing else but the valley statistic. In [28] a recursive formula for the generating function of permutations with exactly k valleys is obtained, which however does not seem to allow (at least easily) finding the corresponding BGF. As a corollary to Theorem 4.11 we get the following BGF by solving (2) for $k = 1$ and $\ell = 1$:

Corollary 4.15. ([17, Corollary 23]) *The BGF for the distribution of peaks (valleys) in permutations is given by*

$$1 - \frac{1}{y} + \frac{1}{y}\sqrt{y - 1} \cdot \tan \left(x\sqrt{y - 1} + \arctan \left(\frac{1}{\sqrt{y - 1}} \right) \right).$$

4.3 Distribution of SPOPs on flat posets with additional restrictions

The results from this subsection are in a similar direction as that in the papers [4], [23], [24], and several other papers, where the authors study 1-3-2-avoiding permutations with respect to avoidance/counting of other patterns. Such a study not only gives interesting enumerative results, but also provides a number of applications (see [4]).

To state the theorem below, we define $P_k = \sum_{n=0}^{k-1} \frac{1}{n+1}\binom{2n}{n} x^n$. That is, P_k is the sum of initial k terms in the expansion of the generating function $\frac{1-\sqrt{1-4x}}{2x}$ of the Catalan numbers.

Theorem 4.16. (Distribution of $a_1 a_2 \cdots a_k a a_{k+1} a_{k+2} \cdots a_{k+\ell}$ on the permutations $\mathcal{S}_n (2\text{-}1\text{-}3)$, [17, Theorem 24]) *Let*

$$P := P(x, y) = \sum_{n \geq 0} \sum_{\pi \in \mathcal{S}_n (2\text{-}1\text{-}3)} y^{e(\pi)} x^n$$

be the BGF of 2-1-3-avoiding permutations where $e(\pi)$ is the number of *occurrences of the SPOP* $p = a_1 a_2 \cdots a_k a a_{k+1} a_{k+2} \cdots a_{k+\ell}$ *in* π. *Then* P *is given by*

$$\frac{1 - x(1-y)(P_k + P_\ell) - \sqrt{D}}{2xy},$$

where

$$D = (x(1-y)(P_k + P_\ell) - 1)^2 - 4xy(x(y-1)P_k P_\ell + 1).$$

We now discuss several corollaries to Theorem 4.16. Note that letting $y = 1$, we obtain the GF for the Catalan numbers. Also, letting $y = 0$ in the expansion of P, we obtain the GF for the number of permutations avoiding simultaneously the patterns 2-1-3 and $a_1 a_2 \cdots a_k a a_{k+1} a_{k+2} \cdots a_{k+\ell}$.

If $k = 1$ and $\ell = 0$ in Theorem 4.16, then $P_k = 1$ and $P_\ell = 0$, and we obtain the distribution of descents in 2-1-3-avoiding permutations. This distribution gives the *triangle of Narayana numbers* (see [29, A001263]).

If $k = \ell = 1$ in Theorem 4.16, then we deal with avoiding the pattern 2-1-3 and counting occurrences of the pattern 312, since any occurrence of $a_1 a a_2$ in a legal permutation must be an occurrence of 312 and vice versa. Thus the BGF of 2-1-3-avoiding permutations with a prescribed number of occurrences of 312 is given by

$$\frac{1 - 2x(1-y) - \sqrt{4(1-y)x^2 + 1 - 4x}}{2xy}.$$

Reading off the coefficients of the terms involving only x in the expansion of the function above, we can see that the number of n-permutations avoiding simultaneously the patterns 2-1-3 and 312 is 2^{n-1}, which is known and is easy to see directly from the structure of such permutations.

Reading off the coefficients of the terms involving y to the power 1, we see that the number of n-permutations avoiding 2-1-3 and having exactly one occurrence of the pattern 312 is given by $(n-1)(n-2)2^{n-4}$. The corresponding sequence appears as [29, A001788] and it gives an interesting fact having a combinatorial proof:

Proposition 4.17. ([17, Proposition 25]) *There is a bijection between 2-dimensional faces in the $(n+1)$-dimensional hypercube and the set of 2-1-3-avoiding $(n+2)$-permutations with exactly one occurrence of the pattern 312.*

If $k = 1$ and $\ell = 2$ in Theorem 4.16, then we deal with avoiding

Fig. 6. A poset giving partial order for 1,2,3,1′, and 2′.

the pattern 2-1-3 and counting occurrences of the pattern $a_1 a a_2 a_3$. In particular, one can see that the number of permutations avoiding simultaneously 2-1-3 and $a_1 a a_2 a_3$ is given by the *Pell numbers* $p(n)$ defined as $p(n) = 2p(n-1) + p(n-2)$ for $n > 1$; $p(0) = 0$ and $p(1) = 1$. The Pell numbers appear as [29, A000129], where one can find objects related to our restricted permutations.

4.4 Non-overlapping SPOPs

This subsection deals additionally with occurrences of patterns in words. The letters $1, 2, 1′, 2′$ appearing in the examples below are ordered as in Figure 6.

Theorem 3.8 and its counterpart in the case of words [19, Theorem 4.3] and [19, Corollary 4.4], as well as Remark 3.10 applied for these results, give an interesting application of the multi-patterns in finding a certain statistic, namely the *maximum number of non-overlapping occurrences of a SPOP* in permutations and words. For instance, the maximum number of non-overlapping occurrences of the SPOP $11′2$ in the permutation 621394785 is 2, and this is given by the occurrences 213 and 478, or the occurrences 139 and 478.

Theorem 4.18 generalizes [15, Theorem 32] and [19, Theorem 5.1].

Theorem 4.18. ([16, Theorem 16]) *Let p be a SPOP and $B(x)$ (resp., $B(x; k)$) is the EGF (resp., GF) for the number of permutations (resp., words over $[k]$) avoiding p. Let $D(x, y) = \sum_\pi y^{N_p(\pi)} \frac{x^{|\pi|}}{|\pi|!}$ and $D(x, y; k) = \sum_{n \geq 0} \sum_{w \in [k]^n} y^{N(w)} x^n$ where $N_p(s)$ is the maximum number of non-overlapping occurrences of p in s. Then $D(x, y)$ and $D(x, y; k)$ are given by*

$$\frac{B(x)}{1 - y(1 + (x-1)B(x))} \quad and \quad \frac{B(x; k)}{1 - y(1 + (kx-1)B(x; k))}.$$

The following examples are corollaries to Theorem 4.18.

Example 4.19. ([16, Ex 1]) If we consider the SPOP 11′ then clearly $B(x) = 1 + x$ and $B(x; k) = 1 + kx$. Hence,

$$D(x, y) = \frac{1 + x}{1 - yx^2} = \sum_{i \geq 0}(x^{2i} + x^{2i+1})y^i,$$

and

$$D(x, y; k) = \frac{1 + kx}{1 - y(kx)^2} = \sum_{i \geq 0}((kx)^{2i} + (kx)^{2i+1})y^i.$$

Example 4.20. ([16, Ex 2]) For permutations, the distribution of the maximum number of non-overlapping occurrences of the SPOP 122′1′ is given by

$$\frac{\frac{1}{2} + \frac{1}{4}\left(\tan x\right)\left(1 + e^{2x} + 2e^x \sin x\right) + \frac{1}{2}e^x \cos x}{1 - y\left(1 + (x - 1)\left(\frac{1}{2} + \frac{1}{4}\left(\tan x\right)\left(1 + e^{2x} + 2e^x \sin x\right) + \frac{1}{2}e^x \cos x\right)\right)}.$$

4.5 q-analogues for non-overlapping SPOPs

We fix some notations. Let p be a segmented POP (SPOP) and $A_{n,k}^p$ be the number of n-permutations avoiding p and having k inversions. As usual, $[n]_q = q^0 + \cdots + q^{n-1}$, $[n]_q! = [n]_q \cdots [1]_q$, $\begin{bmatrix} n \\ i \end{bmatrix}_q = \frac{[n]_q!}{[i]_q![n-i]_q!}$, and, as above, $\text{inv}(\pi)$ denotes the number of inversions in a permutation π. We set $A_n^p(q) = \sum_{\pi \text{ avoids } p} q^{\text{inv}(\pi)}$. Moreover,

$$A_q^p(x) = \sum_{n,k} A_{n,k}^p q^k \frac{x^n}{[n]_q!} = \sum_n A_n^p(q)\frac{x^n}{[n]_q!} = \sum_{\pi \text{ avoids } p} q^{\text{inv}(\pi)}\frac{x^{|\pi|}}{[|\pi|]_q!}.$$

All the definitions above are similar in case of permutations that *quasi-avoid* p, indicated by B rather than A, namely, those permutations that have exactly one occurrence of p and this occurrence consists of the $|p|$ rightmost letters in the permutations.

Theorem 4.21. ([17, Theorem 28]; a q-analogue of [15, Theorem 28] that is valid for POPs) Let $p = p_1\text{-}\cdots\text{-}p_k$ be a multi-pattern (p_is are SPOPs, and letters of p_i and p_j are incomparable for $i \neq j$). Then

$$A_q^p(x) = \sum_{i=1}^{k} A_q^{p_i}(x)\prod_{j=1}^{i-1}B_q^{p_j}(x) = \sum_{i=1}^{k} A_q^{p_i}(x)\prod_{j=1}^{i-1}((x-1)A_q^{p_j}(x) + 1).$$

Theorem 4.22. ([17, Theorem 28]; a q-analogue of [16, Theorem 16])

If $N_p(\pi)$ denotes the maximum number of non-overlapping occurrences of a SPOP p in π, then

$$\sum_\pi y^{N_p(\pi)} q^{inv(\pi)} \frac{x^{|\pi|}}{|\pi|!} = \frac{A_q^p(x)}{1 - yB_q^p(x)} = \frac{A_q^p(x)}{1 - y((x-1)A_q^p(x) + 1)}.$$

5 POPs in compositions

Compositions are objects closely related to words, and some of the results on POPs in compositions can be viewed as generalizations of certain results on words. In this subsection we review some of the results in [11] and [20].

5.1 Avoiding POPs in compositions

Let $\mathbb{N}$ be the set of all positive integers, and let A be any ordered finite set of positive integers, say $A = \{a_1, a_2, \ldots, a_k\}$, where $a_1 < a_2 < a_3 < \cdots < a_k$. A composition $\sigma = \sigma_1 \sigma_2 \ldots \sigma_m$ of $n \in \mathbb{N}$ is an ordered collection of one or more positive integers whose sum is n. The number of *summands*, or *parts*, namely m, is called the number of *parts* of the composition. For any ordered set $A = \{a_1, a_2, \ldots, a_k\} \subseteq \mathbb{N}$, we denote the set of all compositions of n with parts in A (resp. with m parts in A) by C_n^A (resp. $C_{n;m}^A$). Occurrences of patterns, in particular, POPs in compositions are defined similarly to that in permutations and words.

Theorem 5.1. [11, Theorem 3.3] Let $A = \{a_1, a_2, \ldots, a_k\} \subseteq \mathbb{N}$.

(i) Let ϕ be a *shuffle pattern* τ-ℓ-ν, that is, ℓ is the largest element in the pattern while each letter in τ is incomparable to any letter in ν. Then for all $k \geq \ell$,

$$C_\phi^A(x, y) = \frac{C_\phi^{A-\{a_k\}}(x, y) - x^{a_k} y C_\tau^{A-\{a_k\}}(x, y) C_\nu^{A-\{a_k\}}(x, y)}{(1 - x^{a_k} y C_\tau^{A-\{a_k\}}(x, y))(1 - x^{a_k} y C_\nu^{A-\{a_k\}}(x, y))}.$$

(ii) Let ψ be a POP τ-1-ν, where 1 is the smallest element in the pattern while each letter in τ is incomparable to any letter in ν. Then for all $k \geq \ell$,

$$C_\psi^A(x, y) = \frac{C_\psi^{A-\{a_1\}}(x, y) - x^{a_1} y C_\tau^{A-\{a_1\}}(x, y) C_\nu^{A-\{a_1\}}(x, y)}{(1 - x^{a_1} y C_\tau^{A-\{a_1\}}(x, y))(1 - x^{a_1} y C_\nu^{A-\{a_1\}}(x, y))}.$$

Theorem 5.2. [11, Theorem 3.7] *Let $A \subseteq \mathbb{N}$ and let $\tau = \tau_1 \text{-} \tau_2 \text{-} \cdots \text{-} \tau_s$ be a multi-pattern (see Subsection 3.4). Then*

$$C_\tau^A(x,y) = \sum_{j=1}^{s} C_{\tau_j}^A(x,y) \prod_{i=1}^{j-1} \left[\left(y \sum_{a \in A} x^a - 1 \right) C_{\tau_i}^A(x,y) + 1 \right].$$

Theorem 5.3. [11, Theorem 4.1] *Let A be any ordered set of positive integers and let τ be a consecutive pattern. Then*

$$\sum_{n,m \geq 0} \sum_{\sigma \in C_{n;m}^A} t^{N_\tau(\sigma)} x^n y^m = \frac{C_\tau^A(x,y)}{1 - t \left[\left(y \sum_{a \in A} x^a - 1 \right) C_\tau^A(x,y) + 1 \right]},$$

where $N_\tau(\sigma)$ is the maximum number of non-overlapping occurrences of τ in σ.

5.2 Counting POPs in compositions

While dealing with counting patterns in some objects, say, permutations, we typically solve the following problem: "find the number of permutations containing certain number of occurrences of a given pattern." In [20] another problem related to counting patterns was considered: "given a POP, how many times it occurs among all compositions?" Such studies generalize some of results in the literature, for example, those in [12] (see [20, Introd.]). To state results in this direction, we need some definitions.

Given a SPOP $w = w_1 w_2 \cdots w_m$ with m parts, let $c_w(n, \ell, s)$ be the number of occurrences of w among compositions of n with $\ell + m$ parts such that the sum of the parts preceding the occurrence is s. Let $\Omega_w(x, y, z)$ be the generating function for $c_w(n, \ell, s)$:

$$\Omega_w(x,y,z) = \sum_{n,\ell,s \in \mathbb{N}} c_w(n,\ell,s) x^n y^\ell z^s.$$

Given a segmented pattern v and $n \in \mathbb{N}$, let $P_v(n)$ denote the number of compositions of n that are order isomorphic to v. If j is the largest letter of v, then $P_v(n)$ is the number of integral solutions $t_1, \ldots, t_j$ to the system

$$\mu_1 t_1 + \cdots + \mu_j t_j = n, \qquad 0 < t_1 < \cdots < t_j, \qquad (3)$$

where μ_k is the number of k's in v. We call $\mu = (\mu_1, \ldots, \mu_j)$ the *content vector* of v. By expanding terms into geometric series, one can see that the number of integral solutions to (3) is the coefficient of x^n in

$$\mathcal{P}_v(x) = \prod_{k=1}^{j} \frac{x^{m_k}}{1 - x^{m_k}}, \tag{4}$$

where $m_k = \mu_{j-k+1} + \cdots + \mu_j$ for $1 \le k \le j$.

Theorem 5.4. *Let w be a SPOP. Then*

$$\Omega_w(x,y,z) = \frac{(1-x)(1-xz)}{(1-x-xy)(1-xz-xyz)} \sum_v \mathcal{P}_v(x) \tag{5}$$

where the sum is over all linear extensions v of w.

6 Concluding remarks

The study of POPs, being a natural generalization of considering generalized patterns in permutations and words, is not only dealing with challenging enumerative problems, but also with ways to discover new connections between restricted permutations/words/compositions and other combinatorial objects. There are infinitely many partially ordered sets and patterns, which provides many opportunities for further research on POPs. Some open problems on POPs can be found in [17, Sec. 5]. We expect that POPs will play a major role in research on (permutation) patterns in the future.

References

[1] E. Babson and E. Steingrímsson. Generalized permutation patterns and a classification of the Mahonian statistics. *Sém. Lothar. Combin.*, 44:Article B44b, 18 pp., 2000.

[2] A. Björner and M. L. Wachs. Permutation statistics and linear extensions of posets. *J. Combin. Theory Ser. A*, 58(1):85–114, 1991.

[3] M. Bóna. *Combinatorics of permutations*. Discrete Mathematics and its Applications (Boca Raton). Chapman & Hall/CRC, Boca Raton, FL, 2004.

[4] P. Brändén, A. Claesson, and E. Steingrímsson. Catalan continued fractions and increasing subsequences in permutations. *Discrete Math.*, 258(1-3):275–287, 2002.

[5] A. Burstein and S. Kitaev. Partially ordered generalized patterns and their combinatorial interpretation. *The Third International Conference on Permutation Patterns*, University of Florida, Gainesville, Florida, March 7–11, 2005.

[6] A. Claesson. Generalized pattern avoidance. *European J. Combin.*, 22(7):961–971, 2001.

[7] R. C. Entringer. Enumeration of permutations of $(1, \ldots, n)$ by number of maxima. *Duke Math. J.*, 36:575–579, 1969.

[8] G. Firro and T. Mansour. Restricted permutations and polygons. *The Third International Conference on Permutation Patterns*, University of Florida, Gainesville, Florida, March 7–11, 2005.

[9] I. P. Goulden and D. M. Jackson. *Combinatorial enumeration*. Dover Publications Inc., Mineola, NY, 2004.

[10] M. T. Hardarson. Avoidance of partially ordered generalized patterns of the form k-σ-k. arXiv:0805.1872v1 [math.CO].

[11] S. Heubach, S. Kitaev, and T. Mansour. Avoidance of partially ordered patterns in compositions. *Pure Math. Appl. (PU.M.A.)*, 17(1-2):123–134, 2006.

[12] S. Heubach and T. Mansour. Counting rises, levels, and drops in compositions. *Integers*, 5(1):A11, 24 pp., 2005.

[13] Q.-H. Hou and T. Mansour. Horse paths, restricted 132-avoiding permutations, continued fractions, and chebyshev polynomials. *Discrete Appl. Math.*, 154(8):1183–1197, 2006.

[14] S. Kitaev. Multi-avoidance of generalised patterns. *Discrete Math.*, 260(1-3):89–100, 2003.

[15] S. Kitaev. Partially ordered generalized patterns. *Discrete Math.*, 298(1-3):212–229, 2005.

[16] S. Kitaev. Segmental partially ordered generalized patterns. *Theoret. Comput. Sci.*, 349(3):420–428, 2005.

[17] S. Kitaev. Introduction to partially ordered patterns. *Discrete Appl. Math.*, 155(8):929–944, 2007.

[18] S. Kitaev and T. Mansour. A survey on certain pattern problems. Available online at http://www.math.haifa.ac.il/toufik/preprint.html.

[19] S. Kitaev and T. Mansour. Partially ordered generalized patterns and k-ary words. *Ann. Comb.*, 7(2):191–200, 2003.

[20] S. Kitaev, T. B. McAllister, and T. K. Petersen. Enumerating segmented patterns in compositions and encoding by restricted permutations. *Integers*, 6:A34, 16 pp., 2006.

[21] S. Kitaev and A. Pyatkin. On avoidance of v- and λ-patterns in permutations. *Ars Combin.*, to appear.

[22] D. Knuth. *The Art of Computer Programming, Volume 4, Fascicle 3: Generating All Combinations and Partitions*. Addison-Wesley Professional, 1 edition, 2005.

[23] T. Mansour. Restricted 1-3-2 permutations and generalized patterns. *Ann. Comb.*, 6(1):65–76, 2002.

[24] T. Mansour and A. Vainshtein. Restricted 132-avoiding permutations. *Adv. in Appl. Math.*, 26(3):258–269, 2001.

[25] A. Mendes and J. B. Remmel. Generating functions via symmetric functions. In preparation.

[26] A. Mendes, J. B. Remmel, and A. Riehl. A generalization of the generating function for descent statistic. *The Fifth International Conference on Permutation Patterns*, St Andrews, Scotland, UK, June 11–15, 2007.

[27] R. Parviainen. Cycles and patterns in permutations. arXiv:math/0610616v3 [math.CO].

[28] R. G. Rieper and M. Zeleke. Valleyless sequences. In *Proceedings of the Thirty-first Southeastern International Conference on Combinatorics, Graph Theory and Computing (Boca Raton, FL, 2000)*, volume 145, pages 33–42, 2000.

[29] N. J. A. Sloane. The On-line Encyclopedia of Integer Sequences. Available online at http://www.research.att.com/~njas/sequences/.

[30] D. Warren and E. Seneta. Peaks and Eulerian numbers in a random sequence. *J. Appl. Probab.*, 33(1):101–114, 1996.

Generalized permutation patterns — a short survey

Einar Steingrímsson

The Mathematics Institute
Reykjavík University
IS-103 Reykjavík, Iceland

Abstract

An occurrence of a classical pattern p in a permutation π is a subsequence of π whose letters are in the same relative order (of size) as those in p. In an occurrence of a *generalized pattern* some letters of that subsequence may be required to be adjacent in the permutation. Subsets of permutations characterized by the avoidance—or the prescribed number of occurrences—of generalized patterns exhibit connections to an enormous variety of other combinatorial structures, some of them apparently deep. We give a short overview of the state of the art for generalized patterns.

1 Introduction

Patterns in permutations have been studied sporadically, often implicitly, for over a century, but in the last two decades this area has grown explosively, with several hundred published papers. As seems to be the case with most things in enumerative combinatorics, some instances of permutation patterns can be found already in MacMahon's classical book from 1915, *Combinatory Analysis* [45]. In the seminal paper *Restricted permutations* of Simion and Schmidt [52] from 1985 the systematic study of permutation patterns was launched, and it now seems clear that this field will continue growing for a long time to come, due to its plethora of problems that range from the easy to the seemingly impossible, with a rich middle ground of challenging but solvable problems. Most important, perhaps, for the future growth of the subject, is the wealth of connections to other branches of combinatorics, other fields of mathematics, and to other disciplines such as computer science and physics.

Whereas an occurrence of a classical pattern p in a permutation π is simply a subsequence of π whose letters are in the same relative order (of size) as those in p, in an occurrence of a *generalized pattern*, some letters of that subsequence may be required to be adjacent in the permutation. For example, the classical pattern 1-2-3-4 simply corresponds to an increasing subsequence of length four, whereas an occurrence of the generalized pattern 1-23-4 would require the middle two letters of that sequence to be adjacent in π, due to the absence of a dash between 2 and 3. Thus, the permutation 23145 contains 1-2-3-4 but not 1-23-4. Note that for the classical patterns, our notation differs from the usual one, since the dashes we have between every pair of adjacent letters in a classical pattern are usually omitted when only classical patterns are being considered.

It is well known that the number of permutations of length n avoiding any one classical pattern of length 3 is the n-th Catalan number, which counts a myriad different combinatorial objects. There are many other results in this direction, relating pattern avoiding permutations to various other combinatorial structures, either via bijections, or by such classes of permutations being equinumerous to the structures in question without there being a known bijection. Counting permutations according to the number of occurrences of generalized patterns one comes up with a vast array of known sequences, such as the Euler numbers, Stirling numbers of both kinds, Motzkin numbers, Entringer numbers, Schröder numbers, Fibonacci numbers, Pell numbers and many more. Also, one often finds lesser known sequences that are nevertheless related to known structures, such as directed animals, planar maps, permutation tableaux, various kinds of trees and involutions in S_n, to name a few. Thus, generalized patterns provide a significant addition to the already sizable flora of classical patterns and their connections to other combinatorial structures.

In fact, due to their great diversity, the non-classical generalized patterns are likely to provide richer connections to other combinatorial structures than the classical ones do. Supporting this is the fact that the recently proved Stanley-Wilf conjecture—which gives a strong bound for the growth rate of the number of permutations of length n avoiding a classical pattern—does not hold for some generalized patterns.

This paper is organized as follows: In Section 2 we introduce definitions and in Section 3 we mention implicit appearances of generalized patterns in the literature. In Sections 4 and 5 we survey what is known about the avoidance of generalized patterns of length three and four, respectively. In Section 6 we give some examples where generalized patterns have shown up in very different contexts, establishing connections to various other combinatorial structures, some of which seem quite deep. Section 7 lists several instances of so-called barred patterns that turn out to be equivalent to generalized patterns and Section 8 deals with asymptotics for avoidance of generalized patterns. Finally, in Section 9, we mention some further generalizations of the generalized patterns.

2 Some definitions

If a permutation $\pi = a_1 a_2 \ldots a_n$ contains the pattern 1-23 then clearly the *reverse* of π, that is $a_n a_{n-1} \ldots a_1$, contains the reverse of 1-23, which is the pattern 32-1. Since taking the reverse of a permutation is a bijection on the set of permutations of length n, the number of permutations avoiding a pattern p equals the number of permutations avoiding the reverse of p. More generally, the distribution—on the set of permutations of length n—of the number of occurrences of a pattern p equals the distribution for the reverse of p. The same is true of the bijection sending a permutation $\pi = a_1 a_2 \ldots a_n$ to its *complement* $\pi^c = b_1 b_2 \ldots b_n$, where $b_i = n + 1 - a_i$. (When we take the complement of a pattern we leave the dashes in place, so the complement of 1-342 is 4-213.) These two transformations, together with their compositions, generate a group of order 4 on the set of patterns, and we say that two patterns belong to the same *symmetry class* if one is transformed into the other by an element of this group. As an example, the patterns $2\text{-}31, 2\text{-}13, 13\text{-}2$ and 31-2 form an entire symmetry class.†

Clearly, two patterns in the same symmetry class have the same properties with respect to the number of permutations avoiding them, and more generally when it comes to the number of permutations with k occurrences of a pattern, for any k. However, it does happen that two patterns not belonging to the same symmetry class have the same avoidance. Thus, two patterns are said to belong to the same *Wilf class* if they have the same avoidance, that is, if the number of permutations of length n that avoid one is the same as that number for the other.

---

† In the case of classical patterns, taking the inverse of a pattern is well defined and preserves avoidance, so patterns that are each other's inverses belong to the same symmetry class. This is not the case for generalized patterns.

Clearly, a Wilf class is a union of symmetry classes (if we define both as equivalence classes).

For example, although the classical patterns of length 3 belong to two symmetry classes (represented by 1-2-3 and 1-3-2, respectively), they all belong to the same Wilf class, since their avoidance is given by the Catalan numbers. For classical patterns, much is known about Wilf classes for patterns of length up to 7, but a general solution seems distant. For the generalized patterns, much less is known. The best reference to date is probably [19]. As an example, the patterns 1-23 and 1-32 belong to the same Wilf class (the avoidance counted by the Bell numbers in both cases) but not the same symmetry class.

3 Generalized patterns in the literature

Generalized patterns have shown up implicitly in the literature in various places, and subsets of these have been studied in some generality. Namely, Simion and Stanton [53] essentially studied the patterns 2-31, 2-13, 31-2, and 13-2 and their relation to a set of orthogonal polynomials generalizing the Laguerre polynomials, and one of these patterns also played a crucial role in the proof by Foata and Zeilberger [30] that Denert's statistic is Mahonian. A permutation statistic is Mahonian if it has the same distribution—on the set of permutations of length n for each n—as the number of inversions.

Goulden and Jackson give an exponential generating function (EGF) for the number of permutations avoiding the pattern 123 (no dashes), in the book *Combinatorial Enumeration* [35, Exercise 5.2.17a, p. 310]. The formula is

$$\left(\sum_{n \geq 0} \frac{x^{3n}}{(3n)!} - \sum_{n \geq 0} \frac{x^{3n+1}}{(3n+1)!} \right)^{-1} \tag{1}$$

Although this does not seem to be mentioned in [35], the obvious generalization holds. Namely, the EGF for the number of permutations avoiding the dashless pattern $12 \cdots k$ is obtained by replacing 3 by k in 1. It is pointed out in [42, Section 3] that this general result can be obtained through an inclusion/exclusion argument similar to one given in [42].

The dashless patterns of length 3 also appeared earlier, implicitly, as the *valleys* (213 and 312), the *peaks* (132 and 231), the *double ascents* 123 and the *double descents* 321 in a permutation, the study of which was pioneered by Françon and Viennot [32], and which is intimately related to Flajolet's [28] generation of Motzkin paths by means of certain continued fractions. This will be mentioned later, in Section 6, in connection with related, recent developments.

Also, the alternating permutations, which have been studied for a long time [2, 3], are permutations that avoid the patterns 123 and 321, (with the additional restriction that the first two letters of the permutation be in decreasing order; otherwise they are *reverse alternating*. In fact, this extra restriction is equivalent to the avoidance of the pattern [12] as defined in [4]).

Generalized patterns were first defined explicitly, in full generality, in the paper *Generalized permutation patterns and a classification of the Mahonian statistics* [4], where it was shown that almost all Mahonian permutation statistics in the literature at that time (up to a certain bijective correspondence translating "excedance based" statistics to "descent based" statistics; see [4, 20]) could be written as linear combinations of generalized patterns. All but one of these statistics could be expressed as combinations of patterns of length at most 3. The odd one out was a statistic defined by Haglund [36], which, after translation by the bijection mentioned above, could be expressed as a combination of patterns of length 4 or less (see [4]). Although all the possible Mahonian statistics based on generalized patterns of length at most 3 were listed in [4], proofs were not given that all of them were indeed Mahonian. These proofs were later supplied by Foata and Zeilberger [31], who solved some of the conjectures with bijections, but others using the "Umbral Transfer Matrix Method" of Zeilberger [58]. The remaining conjectures in [4], concerning a slight generalization of generalized patterns, were proved bijectively by Foata and Randrianarivony [29].

4 Avoidance (and occurrences) of generalized patterns of length 3

The study of *avoidance* of generalized patterns—along the lines of the work of Simion and Schmidt [52] for the classical patterns—was initiated by Claesson in the paper *Generalized pattern avoidance* [16], where the enumeration was done for avoidance of any single pattern of length 3 with exactly one dash. These patterns fall into three equivalence classes with respect to avoidance. One of these classes, consisting of the patterns 2-31, 2-13, 31-2 and 13-2, has the same avoidance as both of the classes of classical patterns of length 3, and thus has avoidance enumerated by the Catalan numbers. In fact, avoiding 2-31 is *equivalent* to avoiding 2-3-1, as shown in [16], and likewise for the other three patterns. This is obviously true also for the patterns 21 and 2-1, since the only permutation with no descent and no inversion is the increasing permutation $123 \ldots n$. It is shown by Hardarson [38] that this can not occur for patterns of length greater than three, that is, two different patterns of length more than three cannot be avoided by the same permutations. The other two classes of one-dash patterns of length three, consisting of the patterns 1-23 and 1-32 and their respective sets of equivalent patterns, are enumerated by the Bell numbers, counting partitions of sets. This was proved bijectively in each case in [16].

The fact that the Bell numbers count permutations avoiding 1-23 (and some other generalized patterns) has interesting implications. Namely, this shows that the Stanley-Wilf conjecture (proved by Marcus and Tardos [46] in 2004) does not hold for some generalized patterns. This ex-conjecture says that for any classical pattern p the number of permutations of length n avoiding p is bounded by C^n for some constant C. This is easily seen to fail for the Bell numbers, whose exponential generating function is $e^{(e^x-1)}$. Apart from this, nothing seems to be known about growth rates of the number of permutations avoiding generalized patterns.

That the Bell numbers count the avoidance of 1-23 also implies the falseness for generalized patterns of another conjecture, by Noonan and Zeilberger [49] (but first mentioned by Gessel [33]), that the number of permutations avoiding a classical pattern is polynomially recursive, that is, satisfies a recursion

$$P_0(n)f(n) = P_1(n)f(n-1) + P_2(n)f(n-2) + \cdots + P_k(n)f(n-k)$$

where k is a constant and each P_i is a polynomial. This conjecture, however, is largely believed to be false even for classical patterns, although a counterexample is still missing.

Claesson [16] also enumerated the avoidance of three classes of pairs of generalized patterns of length 3 with one dash each. These turned out to be equinumerous with non-overlapping set partitions (counted by the Bessel numbers), involutions and Motzkin paths, respectively. Enumerative equivalences among the class of patterns corresponding to non-overlapping partitions, together with the connection to set partitions, naturally led to the definition of *monotone* partitions in [16].

In [19], Claesson and Mansour then completed the enumeration of permutations avoiding any pair of generalized patterns with one dash each. They also conjectured enumerative results for avoidance of any set of three or more such patterns. These conjectures were proved for sets of size three by Bernini, Ferrari and Pinzani [6], and by Bernini and Pergola [7] for the sizes 4, 5 and 6, the remaining sizes being rather trivial.

Elizalde and Noy [27] treated the *dashless patterns* (which they call "consecutive patterns"), that is, patterns with no dashes, and gave generating functions enumerating permutations according to the number k of occurrences of a pattern. This is a much stronger result than enumerating permutations *avoiding* a pattern, which is just the special case $k = 0$. In particular, they enumerated the avoidance of both Wilf classes of dashless patterns of length 3, and gave differential equations satisfied by the generating functions for three of the seven Wilf classes of patterns of length 4 (see the next section). As mentioned before, their result in the special case of avoidance of the pattern 123, was obtained already in the book *Combinatorial Enumeration* by Goulden and Jackson [35, Exercise 5.2.17a].

For subsets of two or more dashless patterns of length 3, Kitaev [41] and Kitaev and Mansour [43, 44] gave direct formulas for almost every case, and recursive formulas for the few remaining ones. Examples of the formulas thus obtained are $C_k + C_{k+1}$ where C_k is the k-th Catalan number, the central binomial coefficients $\binom{2n}{n}$, and the Entringer numbers, which also count certain permutations starting with a decreasing sequence and then alternating between ascents and descents.

Thus, the avoidance of any set of generalized patterns of length 3 has been understood, in most cases in the sense of explicit formulas or generating functions, or at least in terms of recursively defined functions.

In [18], Claesson and Mansour found the number of permutations with exactly one, two and three occurrences, respectively, of the pattern 2-31. They used the connection between continued fractions and Motzkin paths (see Section 6) that had been used in [20] to give a continued fraction for a generating function counting occurrences of 2-31, among other things. Later, Parviainen [51] showed how to use bicolored Motzkin paths to give a continued fraction counting permutations of length n according to the number of occurrences of 2-31. He gave an algorithm for finding an explicit formula, and gave this explicit formula (always a rational function in n times a binomial coefficient of the form $\binom{2n}{n-a}$) for each $n \leq 8$. Finally, Corteel and Nadeau [22] found a bijective proof of the fact, first proved in [18], that the number of permutations of length n with exactly one occurrence of 2-31 is $\binom{2n}{n-3}$. They exploited the connection between generalized patterns and *permutation tableaux*. For more about that connection, see Section 6.

5 Patterns of length 4

The classical patterns of length 4 fall into three Wilf classes, represented by the patterns 1-2-3-4, 1-3-4-2 and 1-3-2-4. The avoidance of the first two has been solved (the first by Gessel [33], the second by Bóna [8]), but 1-3-2-4 still remains to be understood, although some noteworthy progress was recently made by Albert et al. in [1].

For generalized patterns of length 4 (other than the classical ones), the situation is much more complicated than for those of length 3, as is to be expected. There are 48 symmetry classes, and computer experiments show that there are at least 24 Wilf classes, but their exact number does not seem to have been determined yet.

As mentioned above, for the dashless patterns of length 4, Elizalde and Noy [27] gave differential equations satisfied by the generating functions for the number of occurrences of three out of seven Wilf classes, namely the classes containing 1234, 1243 and 1342, respectively. The remaining classes are represented by 2413, 2143, 1324 and 1423. Note that a Wilf class is a class of patterns with the same *avoidance*, whereas Elizalde and Noy proved that in each Wilf class of dashless patterns of length four, all patterns have the same *distribution*. That is, the number of permutations of length n with k occurrences of a pattern is the same for two patterns in the same Wilf class in this case, but not in general.

Kitaev [42, Theorem 13] found an expression for the exponential generating function (EGF) for the avoidance of σ-k, where σ is any dashless pattern and k is larger than all the letters in σ, in terms of the EGFs for avoidance of the pattern σ. In particular, if σ is any dashless pattern of length 3, this, together with the results of Elizalde and Noy [27], yields explicit formulas for the EGFs for the avoidance of σ-4, where σ is any dashless pattern of length 3 (although one of these formulas involves the integral $\int_0^x e^{-t^2}\,dt$). Since there are precisely two Wilf classes for dashless patterns of length 3, this gives the avoidance of two Wilf classes of patterns of length 4.

Also, Callan [14] has given two recursive formulas for the number of permutations avoiding 31-4-2.

These seem to be all the enumerative results so far for patterns of length 4. For the non-classical patterns we thus have formulas for the avoidance of six Wilf classes out of (at least) 24, whereas for the classical patterns of length 4 there are formulas for two out of three classes. However, there is no reason to assume that the non-classical patterns should be harder to deal with than the classical ones, so explicit results for the avoidance of more such patterns should not be considered out of reach.

6 Generalized patterns appearing in other contexts

In Section 1 we mentioned the connection between dashless patterns as valleys, peaks, etc., in permutations, and Flajolet's [28] generation of Motzkin paths by means of continued fractions. Using results from Flajolet's paper [28], Clarke, Steingrímsson and Zeng [20, Corollary 11] found a continued fraction capturing, among other things, the distribution of permutations according to the number of occurrences of 2-31. This was made explicit in [18, Corollary 23]. A polynomial formula for the joint distribution of descents and 2-31 was conjectured by Steingrímsson and Williams (unpublished), after Williams [56, Corollary 5.3] had shown that formula to count permutations according to *weak excedances* and *alignments* which Williams was studying in connection with so called *permutation tableaux* (for definitions see [54, 56]). This conjecture (part of a much larger conjecture later proved in [54]) was first proved by Corteel [21]. The formula is as follows (see Corollary 30 in [54]):

The number of permutations of length n with $k - 1$ descents and m occurrences of the pattern 2-31 is equal to the coefficient of q^m in

$$q^{-k^2} \sum_{i=0}^{k-1} (-1)^i [k-i]^n q^{ki} \left(\binom{n}{i} q^{k-i} + \binom{n}{i-1} \right). \tag{2}$$

Here, $[k - i]$ is the q-bracket defined by $[m] = (1 + q + \cdots + q^{m-1})$. This is the only known polynomial formula for the entire distribution of a pattern of length greater than 2. The two cases of length 2 correspond to the Eulerian numbers, counting descents, and the coefficients of the q-factorial

$$[n]! = (1 + q)(1 + q + q^2) \cdots (1 + q + q^2 + \cdots + q^{n-1}),$$

which count permutations according to the number of inversions. A descent is an occurrence of the pattern 21 and an inversion is an occurrence of 2-1.

Moreover, this connection between patterns and permutation tableaux also led to the discovery, by Corteel [21] (see also [24, 23]), of a connection between the permutation tableaux and the partially asymmetric exclusion process (PASEP), an important model in statistical mechanics. In particular, the distribution of permutations of length n according to number of descents and number of occurrences of the pattern 2-31 equals a probability distribution studied for the PASEP.

In [17], a bijection is given between the permutations of length n avoiding both 2-41-3 and 3-1-4-2 on one hand, and so called $\beta(1,0)$-trees on the other. The $\beta(1,0)$-trees are rooted plane trees with certain labels on their vertices. These trees were defined by Jacquard and Schaeffer [39], who described a bijection from rooted nonseparable planar maps to a set of labeled plane trees including the $\beta(1,0)$-trees. These trees represent, in a rather transparent way, the recursive structure found by Brown and Tutte [10] on planar maps. As it turns out, the bijection given in [17] simultaneously translates seven different statistics on the permutations to corresponding statistics on the $\beta(1,0)$-trees. In fact, the permutations avoiding 2-41-3 and 3-1-4-2 seem to be more closely related structurally to the $\beta(1,0)$-trees—and thus to the planar maps involved—than the *two-stack sortable permutations* that had previously been shown to be in bijection with the planar maps in question (see Section 7). Earlier, Dulucq, Gire and West [25] constructed a generating tree for the permutations avoiding 2-4-1-3 and 3-14-2 that they showed to be isomorphic to a generating tree for rooted nonseparable planar maps. (Clearly, permutations avoiding these two patterns are equinumerous with the permutations avoiding 3-1-4-2 and 2-41-3, treated in [17].) However, instead of the pattern 3-14-2 they used a so-called barred pattern, which we treat in Section 7, and they only showed their bijection, which is different from the one in [17], to preserve two different statistics, rather than the seven statistics in [17].

7 Generalized patterns in disguise

As mentioned above, generalized patterns have occurred implicitly in several places in the literature, even before the systematic study of classical permutation patterns. However, they have also appeared as so-called *barred* patterns, which sometimes, but not always, turn out to be equivalent (in terms of avoidance) to some generalized patterns. An example of a barred pattern is 4-$\bar{2}$-1-3. A permutation π is said to avoid this pattern if it avoids the pattern 3-1-2 (corresponding to the unbarred elements 4-1-3) *except* where that pattern is part of the pattern 4-2-1-3.

Gire [34] showed that the *Baxter permutations*, originally defined in a very different way [15], are those avoiding the two barred patterns 4-1-$\bar{3}$-5-2 and 2-5-$\bar{3}$-1-4. It is easy to show that avoiding 4-1-$\bar{3}$-5-2 is equivalent to avoiding 3-14-2 and avoiding 2-5-$\bar{3}$-1-4 is equivalent to avoiding 2-41-3. Thus, the Baxter permutations are precisely those that avoid both 3-14-2 and 2-41-3. In fact, this was pointed out in Erik Ouchterlony's thesis [50, p. 5].

The barred pattern 4-1-$\bar{3}$-5-2 also shows up in [25], where Dulucq, Gire and West treated so called *nonseparable permutations* (bijectively related to rooted nonseparable planar maps), which they characterized by the avoidance of 2-4-1-3 and 4-1-$\bar{3}$-5-2, the latter one being equivalent to 3-14-2 as mentioned in the previous section.

Also, in [9], Bousquet-Mélou and Butler deal with the barred pattern 2-1-$\bar{3}$-5-4, avoiding which is easily shown to be equivalent to avoiding 2-14-3. Permutations avoiding that pattern and 1-3-2-4 are called forest-like permutations in [9]. It is also mentioned there that avoiding 2-1-$\bar{3}$-5-4 (and thus 2-14-3) is equivalent to avoiding 2-1-4-3 *with Bruhat condition* $(1 \leftrightarrow 4)$ in the terminology of Woo and Yong, who conjectured [57] that a Schubert variety is locally factorial if and only if its associated permutation avoids these two patterns. That conjecture was proved in [9].

Not all barred patterns can be expressed in terms of generalized patterns, however. For example, West [55] showed that two-stack sortable permutations are characterized by the simultaneous avoidance of 2-3-4-1 and the barred pattern 2-$\bar{5}$-3-4-1, and it is easy to show that no generalized pattern is avoided by the same permutations as those avoiding 2-$\bar{5}$-3-4-1. It is also easy to show that there is no pair of generalized patterns of length 4 that is avoided by the same permutations as those that avoid 2-3-4-1 and 2-$\bar{5}$-3-4-1. A consequence of this is that two-stack sortability of a permutation cannot be characterized by the avoidance of a set of generalized patterns. This is because there are precisely two permutations of length four that are not two-stack sortable, namely 2341 and 3241, so such a set would have to contain two generalized patterns whose underlying permutations (obtained by disregarding the dashes in the patterns) were 2341 and 3241. It is easy to check, by computer, that no such pair will do the job.

Also Callan [13] has shown that the number of permutations avoiding 31-4-2 is the same as the number avoiding the barred pattern 3-$\bar{5}$-2-4-1, although these are not the same permutations. (As mentioned above, Callan [14] has given two recursive formulas for this number.)

An obvious open problem here is to determine when avoiding a barred pattern is equivalent to avoiding a generalized pattern.

8 Asymptotics

As mentioned before, the Stanley-Wilf conjecture, proved by Marcus and Tardos [46], says that the number of permutations of length n avoiding any given classical pattern p is bounded by C^n for some constant C depending only on p.

In [26], Elizalde studies asymptotics for the number of permutations avoiding some generalized patterns and concludes that, in contrast to the classical patterns, there probably is "a big range of possible asymptotic behaviors." Although much work remains to be done here, and although it is not clear how varied this behavior can be, the extremes are already known. Namely, whereas the number of permutations of length n avoiding a classical pattern is bounded by c^n for a constant c, Elizalde [26, Theorem 4.1] shows that the number $\alpha_n(\sigma)$ of permutations of length n avoiding a dashless (or *consecutive*) pattern σ of length at least three satisfies $c^n n! < \alpha_n(\sigma) < d^n n!$ for some constants c and d (where, clearly, $0 < c, d < 1$).

An interesting open question related to this is which generalized patterns, apart from the classical ones, satisfy the Stanley-Wilf conjecture. It has been pointed out by Hardarson [37] that a pattern containing a block with at least two letters a and b, with $a < b$, and a letter x in some other block, with $x < a$ or $b < x$, can not satisfy the Stanley-Wilf conjecture, because there will be at least as many permutations avoiding it as there are permutations that avoid 1-23, and the number of such permutations is known not to satisfy the Stanley-Wilf conjecture (these are the Bell numbers). Thus, essentially the only open cases left are the patterns 2-3-41 and 2-41-3.

9 Further generalizations

Kitaev, in [42], introduced a further generalization of generalized patterns (GPs), namely the *partially ordered generalized patterns* or POGPs. These are GPs where some letters may be incomparable in size. An example of such a pattern is 3-12-3, an occurrence of which consists of four letters, the middle two adjacent (and in increasing order) and the first and the last both greater than the middle two, with no condition on the relative sizes of the first and the last letter. Avoiding the pattern 3-12-3 is equivalent to avoiding both 3-12-4 and 4-12-3. Indeed, an *occurrence* of 3-12-3 is equivalent to an occurerence of either 3-12-4 or 4-12-3. In general, a POGP is equal, as a function counting occurrences, to a sum of GPs.

In [38], Hardarson finds EGFs for the avoidance of k-σ-k, where σ is any dashless partially ordered pattern and k is larger than any letter in σ, in terms of the EGF for the avoidance of σ. In the special case where $\sigma = 12$ he gives a bijection between permutations avoiding 3-12-3 and bicolored set partitions, that is, all partitions of a set where each block in the partition has one of two possible colors. Also, for $\sigma = 121$, he gives a bijection between permutations of length $n + 1$ avoiding 3-121-3 and the Dowling lattice on $\{1, 2, \ldots, n\}$.

One interesting, and curious, result arising from Kitaev's study of POGPs in [42] is that knowing the EGF for the avoidance of a dashless pattern p is enough to find the EGF for the entire distribution of the maximum number of *non-overlapping occurrences* of p. Two occurrences of p in a permutation π are non-overlapping if they have no letter of π in common. For example, the permutation 4321 has three descents (occurrences of 21), but only two non-overlapping descents. Namely, Kitaev [42] proves the following theorem:

Theorem 9.1 (Kitaev [42], Theorem 32). *Let p be a dashless pattern. Let $A(x)$ be the EGF for the number of permutations that avoid p and let $N(\pi)$ be the maximum number of non-overlapping occurrences of p in π. Then*

$$\sum_{\pi} y^{N(\pi)} \frac{x^{|\pi|}}{|\pi|!} = \frac{A(x)}{1 - y((x-1)A(x) + 1)},$$

where the sum is over all permutations of lengths $0, 1, 2, \ldots$.

Alternative proofs, and some extensions, of Theorem 9.1 were given by Mendes [47] and by Mendes and Remmel [48], using the theory of symmetric functions.

A generalization in a different direction is the study of generalized patterns on words, that is, on permutations of multisets. Research in this area has only recently taken off, even in the case of classical patterns. For the generalized patterns, see [5, 12, 40].

Finally, Burstein and Lankham have considered barred generalized patterns, in relation to patience sorting problems (see [11], which also has further references).

References

[1] M. H. Albert, M. Elder, A. Rechnitzer, P. Westcott, and M. Zabrocki. On the Wilf-Stanley limit of 4231-avoiding permutations and a conjecture of Arratia. *Adv. in Appl. Math.*, 36(2):95–105, 2006.

[2] D. André. Développements de sec x et de tang x. *C. R. Math. Acad. Sci. Paris*, 88:965–967, 1879.

[3] D. André. Sur les permutations alternées. *J. Math. Pures Appl.*, 7:167–184, 1881.

[4] E. Babson and E. Steingrímsson. Generalized permutation patterns and a classification of the Mahonian statistics. *Sém. Lothar. Combin.*, 44:Article B44b, 18 pp., 2000.

[5] A. Bernini, L. Ferrari, and R. Pinzani. Enumeration of some classes of words avoiding two generalized patterns of length three. arXiv:0711.3387v1 [math.CO].

[6] A. Bernini, L. Ferrari, and R. Pinzani. Enumerating permutations avoiding three Babson-Steingrímsson patterns. *Ann. Comb.*, 9(2):137–162, 2005.

[7] A. Bernini and E. Pergola. Enumerating permutations avoiding more than three Babson-Steingrímsson patterns. *J. Integer Seq.*, 10(6):Article 07.6.4, 21 pp., 2007.

[8] M. Bóna. Exact enumeration of 1342-avoiding permutations: a close link with labeled trees and planar maps. *J. Combin. Theory Ser. A*, 80(2):257–272, 1997.

[9] M. Bousquet-Mélou and S. Butler. Forest-like permutations. *Ann. Comb.*, 11(3–4):335–354, 2007.

[10] W. G. Brown and W. T. Tutte. On the enumeration of rooted non-separable planar maps. *Canad. J. Math.*, 16:572–577, 1964.

[11] A. Burstein and I. Lankham. Restricted patience sorting and barred pattern avoidance. In *this volume*, 233–257.

[12] A. Burstein and T. Mansour. Words restricted by 3-letter generalized multipermutation patterns. *Ann. Comb.*, 7(1):1–14, 2003.

[13] D. Callan. A Wilf equivalence related to two stack sortable permutations. arXiv:math/0510211v1 [math.CO].

[14] D. Callan. A combinatorial interpretation of the eigensequence for composition. *J. Integer Seq.*, 9(1):Article 06.1.4, 12 pp., 2006.

[15] F. R. K. Chung, R. L. Graham, V. E. Hoggatt, Jr., and M. Kleiman. The number of Baxter permutations. *J. Combin. Theory Ser. A*, 24(3):382–394, 1978.

[16] A. Claesson. Generalized pattern avoidance. *European J. Combin.*, 22(7):961–971, 2001.

[17] A. Claesson, S. Kitaev, and E. Steingrímsson. Stack sorting, trees, and pattern avoidance. arXiv:0801.4037v1 [math.CO].

[18] A. Claesson and T. Mansour. Counting occurrences of a pattern of type $(1, 2)$ or $(2, 1)$ in permutations. *Adv. in Appl. Math.*, 29(2):293–310, 2002.

[19] A. Claesson and T. Mansour. Enumerating permutations avoiding a pair of Babson-Steingrímsson patterns. *Ars Combin.*, 77:17–31, 2005.

[20] R. J. Clarke, E. Steingrímsson, and J. Zeng. New Euler-Mahonian statistics on permutations and words. *Adv. in Appl. Math.*, 18(3):237–270, 1997.

[21] S. Corteel. Crossings and alignments of permutations. *Adv. in Appl. Math.*, 38(2):149–163, 2007.

[22] S. Corteel and P. Nadeau. Bijections for permutation tableaux. *European J. Combin.*, 30(1):295–310, 2009.

[23] S. Corteel and L. K. Williams. A Markov chain on permutations which projects to the PASEP. *Int. Math. Res. Not. IMRN*, 2007(17):Art. ID rnm055, 27, 2007.

[24] S. Corteel and L. K. Williams. Tableaux combinatorics for the asymmetric exclusion process. *Adv. in Appl. Math.*, 39(3):293–310, 2007.

[25] S. Dulucq, S. Gire, and J. West. Permutations with forbidden subsequences and nonseparable planar maps. *Discrete Math.*, 153(1-3):85–103, 1996.

[26] S. Elizalde. Asymptotic enumeration of permutations avoiding generalized patterns. *Adv. in Appl. Math.*, 36(2):138–155, 2006.

[27] S. Elizalde and M. Noy. Consecutive patterns in permutations. *Adv. in Appl. Math.*, 30(1-2):110–125, 2003.

[28] P. Flajolet. Combinatorial aspects of continued fractions. *Discrete Math.*, 32(2):125–161, 1980.

[29] D. Foata and A. Randrianarivony. Two oiseau decompositions of permutations and their application to Eulerian calculus. *European J. Combin.*, 27(3):342–363, 2006.

[30] D. Foata and D. Zeilberger. Denert's permutation statistic is indeed Euler-Mahonian. *Stud. Appl. Math.*, 83(1):31–59, 1990.

[31] D. Foata and D. Zeilberger. Babson–Steingrímsson statistics are indeed Mahonian (and sometimes even Euler–Mahonian). *Adv. in Appl. Math.*, 27(2-3):390–404, 2001.

[32] J. Françon and G. Viennot. Permutations selon leurs pics, creux, doubles montées et double descentes, nombres d'Euler et nombres de Genocchi. *Discrete Math.*, 28(1):21–35, 1979.

[33] I. M. Gessel. Symmetric functions and *P*-recursiveness. *J. Combin. Theory Ser. A*, 53(2):257–285, 1990.

[34] S. Gire. *Arbres, permutations à motifs exclus et cartes planaires: quelques problèmes algorithmique et combinatoire*. PhD thesis, Université Bordeaux I, 1993.

[35] I. P. Goulden and D. M. Jackson. *Combinatorial enumeration*. Dover Publications Inc., Mineola, NY, 2004.

[36] J. Haglund. *q*-rook polynomials and matrices over finite fields. *Adv. in Appl. Math.*, 20(4):450–487, 1998.

[37] M. T. Hardarson. personal communication. 2008.

[38] M. T. Hardarson. Avoidance of partially ordered generalized patterns of the form k-σ-k. arXiv:0805.1872v1 [math.CO].

[39] B. Jacquard and G. Schaeffer. A bijective census of nonseparable planar maps. *J. Combin. Theory Ser. A*, 83(1):1–20, 1998.

[40] S. Kitaev. Generalized pattern avoidance with additional restrictions. *Sém. Lothar. Combin.*, 48:Article B48e, 19 pp., 2002.

[41] S. Kitaev. Multi-avoidance of generalised patterns. *Discrete Math.*, 260(1-3):89–100, 2003.

[42] S. Kitaev. Partially ordered generalized patterns. *Discrete Math.*, 298(1-3):212–229, 2005.

[43] S. Kitaev and T. Mansour. On multi-avoidance of generalized patterns. *Ars Combin.*, 76:321–350, 2005.

[44] S. Kitaev and T. Mansour. Simultaneous avoidance of generalized patterns. *Ars Combin.*, 75:267–288, 2005.

[45] P. A. MacMahon. *Combinatory Analysis*. Cambridge University Press, London, 1915/16.

[46] A. Marcus and G. Tardos. Excluded permutation matrices and the Stanley-Wilf conjecture. *J. Combin. Theory Ser. A*, 107(1):153–160, 2004.

[47] A. Mendes. *Building generating functions brick by brick*. PhD thesis, University of California, San Diego, 2004.

[48] A. Mendes and J. B. Remmel. Permutations and words counted by consecutive patterns. *Adv. in Appl. Math.*, 37(4):443–480, 2006.

[49] J. Noonan and D. Zeilberger. The enumeration of permutations with a prescribed number of "forbidden" patterns. *Adv. in Appl. Math.*, 17(4):381–407, 1996.

[50] E. Ouchterlony. *On Young tableaux involutions and patterns in permutations*.

PhD thesis, Matematiska institutionen Linköpings Universitet, Linköping, Sweden, 2005.

[51] R. Parviainen. Lattice path enumeration of permutations with k occurrences of the pattern 2-13. *J. Integer Seq.*, 9(3):Article 06.3.2, 8 pp., 2006.

[52] R. Simion and F. W. Schmidt. Restricted permutations. *European J. Combin.*, 6(4):383–406, 1985.

[53] R. Simion and D. Stanton. Octabasic Laguerre polynomials and permutation statistics. *J. Comput. Appl. Math.*, 68(1-2):297–329, 1996.

[54] E. Steingrímsson and L. K. Williams. Permutation tableaux and permutation patterns. *J. Combin. Theory Ser. A*, 114(2):211–234, 2007.

[55] J. West. *Permutations with forbidden subsequences and stack-sortable permutations.* PhD thesis, M.I.T., 1990.

[56] L. K. Williams. Enumeration of totally positive Grassmann cells. *Adv. Math.*, 190(2):319–342, 2005.

[57] A. Woo and A. Yong. When is a Schubert variety Gorenstein? *Adv. Math.*, 207(1):205–220, 2006.

[58] D. Zeilberger. The umbral transfer-matrix method. I. Foundations. *J. Combin. Theory Ser. A*, 91(1-2):451–463, 2000.

An introduction to structural methods in permutation patterns

Michael Albert

Department of Computer Science
University of Otago
Dunedin New Zealand

Abstract

Structural methods as applied to the study of classical permutation pattern avoidance are introduced and described. These methods allow for more detailed study of pattern classes, answering questions beyond basic enumeration. Additionally, they frequently can be applied wholesale, producing results valid for a wide collection of pattern classes, rather than simply *ad hoc* application to individual classes.

1 Introduction

In the study of permutation patterns, the important aspects of permutations of $[n] = \{1, 2, \ldots, n\}$ are considered to be the relative order of both the argument and the value. Specifically, we study a partial order, denoted $\preceq$ and called *involvement*, on the set of such permutations where $\pi \in \mathcal{S}_k$ is involved in $\sigma \in \mathcal{S}_n$, i.e. $\pi \preceq \sigma$ if, for some increasing function $f : [k] \to [n]$ and all $1 \leq i < j \leq k$, $\sigma(i) < \sigma(j)$ if and only if $\pi(f(i)) < \pi(f(j))$. This dry and uninformative definition is necessary to get us started, but the reader should certainly be aware that another definition of involvement is that some of the points in the graph of π can be erased so that what remains is the graph of σ (possibly with a non-uniform scale on both axes) – in other words the pattern of σ (its graph) occurs as part of the pattern of π.

Arguably the first result in the area of permutation patterns is the famous Erdős-Szekeres theorem [24] which, in this language, states: if $n > pq$, then every permutation in $\mathcal{S}_n$ involves either $1\, 2 \cdots p\,(p+1)$ or

$(q + 1) q \cdots 2\, 1$. However, further research in the area concentrated on permutation classes (defined formally below), and perhaps even more on principal permutation classes – those sets of permutations which do not involve some specific basis permutation β. A startling number of enumerative coincidences among these classes led to the desire to find methods of enumerating them in general – a desire which is as yet unfulfilled, with exact formulas unknown even for $\beta = 4231$ (although, for all other permutations of length at most four, except symmetric cases to this one, formulas are known.) Additional impetus was provided by the long standing Stanley-Wilf conjecture, that the growth of a proper permutation class was bounded by some exponential function. This conjecture was resolved positively by Marcus and Tardos ([35]). However, their elegant proof does not provide good bounds for the actual enumeration of permutation classes. The apparent intractability of the enumeration of the principal permutation class with basis 4231 and of similar problems for most permutations of greater length suggests an approach to the study of permutation classes where we focus on classes with more tractable behaviour.

As the title suggests, the purpose of this article is to introduce a family of methods that can be applied to problems in the area of permutation patterns. It is not intended to be a comprehensive survey of the application of such methods, rather as propaganda for their application. The idea is to discover some form of structure in permutation classes with a goal to proving general results about classes in which structure can be found, for instance, that their generating functions have good properties. For reasons of space, we confine our attention here to classical pattern avoidance problems, that is those based around the relation $\preceq$. However, the structural method is central to the entire canon of *symbolic combinatorics* [25, 26], and the techniques discussed below can certainly be applied in non-classical settings, and also to other pattern type problems, for example patterns in matchings.

It is unfortunately necessary, at this point, to present (and in some cases repeat) some terminology and notation. For a positive integer n, $[n] = \{1, 2, \ldots, n\}$. All permutations, generally denoted by greek letters, will be thought of as permutations of $[n]$ for some n, and written in *one line notation* as sequences. That is, the permutation π of $[n]$ is represented by the sequence $\pi(1)\pi(2)\cdots\pi(n)$, and in fact we will also just write π_i in place of $\pi(i)$. The *length*, $|\pi|$ of a permutation π is simply the size of its domain.

A permutation is determined by the relative order between the terms of the sequence that represents it. So we frequently consider a subsequence of a permutation as a permutation in its own right, without explicit recoding. This defines, for us, the *involvement* relation, $\sigma \preceq \pi$ if some subsequence of π satisfies the same order relationships as σ (for the sake of clarity, an example: $231 \preceq 31524$ because the subsequence 352 has the pattern middle-high-low which is the same as that of 231.) A permutation π *avoids* σ if it does not involve σ. This relation is also called *containment* by some other authors.

Sets of permutations are denoted by calligraphic letters ($\mathcal{A}$, $\mathcal{B}$, $\mathcal{C}$, etc.) A *permutation class* or simply *class* is a set of permutations closed downwards under involvement. If $\mathcal{X}$ is a set of permutations, then the set $\mathrm{Av}(\mathcal{X})$ of all permutations avoiding each permutation in $\mathcal{X}$ is a class. If additionally, $\mathcal{X}$ is an antichain, then $\mathcal{X}$ is called the *basis* of that class. So, every class has a basis which consists just of the $\preceq$-minimal permutations not belonging to the class. We also use $\mathcal{C}(\pi)$ or $\mathcal{C}(\mathcal{X})$ for $\mathcal{C} \cap \mathrm{Av}(\pi)$ and $\mathcal{C} \cap \mathrm{Av}(\mathcal{X})$ respectively.

The formal variable for generating functions will always be t, and the generating function of a set of permutations will be denoted by the upper case Roman letter corresponding to its name, and defined by:

$$A = \sum_{\pi \in \mathcal{A}} t^{|\pi|}.$$

If we say that a generating function is *rational* we mean that it belongs to $\mathcal{Q}(t)$. If we say that it is *algebraic* we mean that it is algebraic over $\mathcal{Q}(t)$.

The next section is intended as a very elementary introduction to structural techniques in the study of permutation patterns, while subsequent sections deal with slightly more advanced or technical material.

2 A case study

For the purpose of illustration, we will begin with a consideration of the class $\mathcal{A} = \mathrm{Av}(231)$ and various subclasses of it. This class was considered by Knuth [31] and consists of all those permutations which can be sorted by a single pass through a stack. What do the elements of $\mathcal{A}$ look like? Consider the maximum element of such a permutation. This element could play the role of a 3 in potential 231 pattern (and can play no other role). Since there are no such patterns, we know that all the elements preceding it must be smaller than all the elements which follow it. So, the permutation has the general form illustrated in Figure

1. What of the elements preceding the maximum? Certainly these must also form a permutation avoiding 231, and the elements following the maximum must likewise form a sequence avoiding 231. Moreover, whenever these two conditions are met then the permutation as a whole avoids 231, since an occurrence of 231 cannot be split into a small part, followed by a larger part. Thus, Figure 1 completely describes $\mathcal{A}$ in

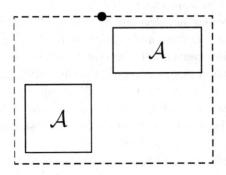

Fig. 1. The structural representation of an element of $\mathcal{A} = \mathrm{Av}(231)$.

a recursive fashion. Perhaps there is a minor issue concerning empty structures†. Suppose that we wish to compute the generating function A for $\mathcal{A}$, allowing empty permutations. Figure 1 describes only the non-empty permutations in the class, but as the collection of empty ones is not of any great complexity we immediately obtain:

$$A = 1 + AtA,$$

where we have deliberately written the second term in a non-standard fashion in order to illustrate its connection with the figure. We remark also that Figure 1 also immediately provides a bijection (or, at least, establishes that one exists) between the permutations of length n in $\mathcal{A}$ and plane binary trees with n vertices.

Can we use the structural characterization of elements of $\mathcal{A}$ in order to investigate other classes? Certainly! Consider first the class $\mathcal{B} = \mathrm{Av}(231, 132)$ and look back at Figure 1. A permutation as drawn in that figure will certainly contain a copy of 132 unless either the left

---

† In application of the symbolic method in combinatorics it is *usually* convenient to allow empty structures. But, flexibility is preferable to a dogmatic adherence to one particular scheme.

or the right box is empty. But again, provided that one of the boxes is empty, and the other contains an element of $\mathcal{B}$, then no copy of 132 will be created. In other words, the elements of $\mathcal{B}$ are those permutations formed from 1 by successively prepending or appending a new maximum element. The graph of such a permutation is V-shaped, being the juxtaposition of a decreasing and an increasing sequence. As plane trees, they correspond to those in which no proper branching occurs. It follows immediately that $b_n = 2^{n-1}$, and $B = t/(1 - 2t)$ (not including the empty permutation this time.)

Finally, let us consider the class $\mathcal{C} = \mathrm{Av}(231, 1243)$. Again, we first consult Figure 1. An obvious 1243 would occur if the lower left block contained an increasing pair, and the upper right block were non-empty. So, let us concentrate on the following cases:

- both blocks empty;
- the left hand block empty, but the right hand block non empty;
- the left hand block non empty, but the right hand block empty;
- both blocks non empty.

Only the permutation of length 1 is of the first type. In the second case, the right hand block may contain any element of $\mathcal{C}$, likewise in the third case the left hand block may contain any element of $\mathcal{C}$. In the final case, the left hand block must not contain an increasing pair (so it must be an element of $\mathcal{D}$, the class of decreasing permutations), while the right hand block must (because of the presence of an element below it and to its left) actually represent a sequence in $\mathcal{B}$ (for if it contained a 132, then those elements together with an element of the left block would form a 1243.) The latter three possibilities are illustrated in Figure 2.

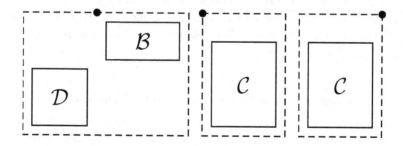

Fig. 2. The possible structural forms of an element of $\mathcal{C} = \mathrm{Av}(231, 1243)$.

It follows immediately that the generating function C, again with zero constant term, satisfies:

$$C = t + DtB + tC + Ct$$

which, using $D = t/(1-t)$ and $B = t/(1-2t)$ readily yields a formula for C. Alternatively, we could also directly derive the recurrence:

$$
\begin{aligned}
c_n &= 2c_{n-1} + \sum_{k=1}^{n-2} b_k \\
&= 2c_{n-1} + \sum_{k=1}^{n-2} 2^{k-1} \\
&= 2c_{n-1} + 2^{n-1} - 1.
\end{aligned}
$$

Even with this informal presentation, it is not hard to imagine that the techniques we have applied can be extended to all classes of the form $\mathrm{Av}(231, \pi)$ for a single permutation π. This was done by Mansour and Vainshtein [34] (modulo one of the standard symmetries), and further extended by Albert and Atkinson [2]. Perhaps the most striking consequence of these analyses is the following theorem:

Theorem 2.1. *Every proper subclass of* $\mathrm{Av}(231)$ *has a rational generating function.*

In this extended example we have concentrated on the decomposition of permutations in various classes into simpler, or at least smaller, permutations from the same or other classes. We continue and amplify this discussion in the next section. However, particularly with reference to the class $\mathcal{B}$, we also hinted at another structural viewpoint – that of permutations evolving, or developing, by the successive addition of elements (so, that we form a V-shaped permutation by beginning with a single element and then adding a new maximum to either the left or right hand end at each step.) We will return to *that* idea in Section 4 and interpret it in various ways.

3 Block decompositions and simple permutations

A *block decomposition* of a permutation π of length n is determined by two equivalence relations P (for position), and V (for value), on $[n]$ having the following properties:

• each equivalence class of P and of V is an interval;

- $P(i, j)$ if and only if $V(\pi_i, \pi_j)$.

The intuitive, and much more easily understandable viewpoint, is that the graph of π can be covered by rectangles whose projections onto the coordinate axes are pairwise disjoint as in Figure 3. When a permutation

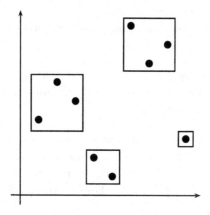

Fig. 3. A block decomposition, showing that $465219783 = 3142[132, 21, 312, 1]$

π is decomposed into blocks, then the blocks themselves form the graph of a permutation, and the elements of π within each block do likewise. So, we define for a permutation σ of length k and any permutations θ_1, θ_2, ..., θ_k the *inflation* $\sigma[\theta_1, \theta_2, \ldots, \theta_k]$ of σ by θ_1, θ_2, ..., θ_k to be the unique permutation π that has a decomposition into k blocks with the properties that the permutation induced on the blocks is σ, and the permutations within each block (read from left to right according to position) are θ_1, θ_2, ..., θ_k. This notation is also illustrated in Figure 3. For obvious reasons we will call σ the *quotient* of this decomposition, and θ_1, θ_2, ..., θ_k the *blocks*. Observe that both the quotient and the blocks of any block decomposition are involved in their inflation.

Inflation is a localization of the *wreath product* construction for permutations introduced by Atkinson and Stitt ([11]). Formally, the wreath product $\mathcal{A} \wr \mathcal{B}$ of two sets (usually permutation classes) of permutations $\mathcal{A}$ and $\mathcal{B}$ is just the set of all inflations with quotient in $\mathcal{A}$ and blocks from $\mathcal{B}$. Further investigation of properties of the wreath product of two classes, particularly with respect to finite basis considerations has been carried out by Brignall ([17]).

Every permutation has two trivial block decompositions: one in which every block is a singleton ($\pi = \pi[1, 1, \ldots, 1]$); and one in which there is a single block ($\pi = 1[\pi]$). The permutations that have no non trivial block decompositions are called *simple permutations*. They are relatively common, as asymptotically a proportion of $1/e^2$ of all permutations are simple [3]. However, among the permutations of length at most 4, only 1, 12, 21, 3142 and 2413 are simple.

There is almost a (strong) Jordan-Hölder theorem for permutations ([2]):

Theorem 3.1. *Every permutation π has a unique simple quotient σ. Moreover, unless this quotient is 12 or 21, the inflation of σ that produces π is also uniquely determined.*

The annoying qualification in the second sentence of this theorem is required because of the fact that a sequence of three or more blocks arranged in (for example) increasing order can be represented as an inflation of 12 in multiple ways. For instance: $1234 = 12[1, 123] = 12[12, 12] = 12[123, 1]$. In order to make use of block decompositions in an enumerative context, it is most convenient to privilege one of the decompositions in this case. One possible choice (followed in [2]) is to require the first block in such a decomposition *not* to have 12 as a quotient (i.e. use the first of the decompositions in the preceding example.)

If a class contains only finitely many simple permutations, then it turns out that it must be finitely based (see [2]). Since the types of block decompositions that can occur in such a class are limited by the simple permutations belonging to the class, the basic scheme of the arguments we used in the preceding section can be applied to yield:

Theorem 3.2. *If C is a class of permutations that contains only finitely many simple permutations, then the generating function of C is algebraic.*

Furthermore, the proof is essentially constructive (given knowledge of the simple permutations in the class and its basis). Together with results of Brignall, Ruškuc and Vatter [20], this implies that there is an algorithm which, given a finite basis for a permutation class, determines whether or not it has only finitely many simple permutations, and if so, determines a system of algebraic equations from which its generating function can be computed. Some example computations of this type can be found in [2], and also (in a more general setting) [19]. While it is easiest to apply these methods to solve concrete enumeration problems, in some contexts they also provide more general results. For instance:

Proposition 3.3. *Let k be a positive integer, and suppose that C is a subclass of $\mathrm{Av}(k(k-1)(k-2)\cdots 321)$ which has only finitely many simple permutations. Then, the generating function of C is rational.*

Brignall, Huczynska and Vatter in [18] and [19] have greatly extended the use and utility of block decompositions using a notion which they call *query-complete sets of properties*. To provide just a taste of what can be shown using their methods, we mention:

Theorem 3.4. *If C is a class of permutations that contains only finitely many simple permutations, then the generating functions of all of the following sets are algebraic: C itself, the alternating permutations in C, the even permutations in C, the Dumont permutations in C, the involutions in C.*

The moral of this particular story is intended to be: if a permutation class, C, contains only finitely many simple permutations, then we are justified in presuming the structure of C to be known. In particular, we should be highly confident that it is possible to resolve any *particular* question concerning C with relative ease. However, that opens up the possibility of asking more interesting *general* questions about such classes. For example:

Question 3.5. *The* separable *permutations, $\mathcal{SEP}$, are the wreath closure of 1, 12 and 21. Given a finite subset $\mathcal{X} \subseteq \mathcal{SEP}$, is there an efficient method to determine the degree of the generating function for $\mathcal{SEP}(\mathcal{X})$ over the field of rational functions?*

Of course, and perhaps fortunately, most interesting permutation classes contain infinitely many simple permutations ($\mathcal{SEP}$ being the most noteworthy exception.)

4 Encoding

The second group of structural methods that we will consider in the study of permutation patterns is connected, either explicitly or implicitly, with encodings of permutations. Roughly speaking, all of these methods obtain leverage from the following observation:

If C is a permutation class, and $\pi \in C$, then there will generally only be

*a restricted number of positions at which a new maximum element can be
inserted into π that result in a permutation $\pi' \in C$.*

Let us return, just for a moment, to the example $\mathcal{A} = \mathrm{Av}(231)$ which
we considered earlier. Let $\pi \in \mathcal{A}$ be given. Then, for some $b \geq 1$, π can
be decomposed into a rising sequence of b blocks, each of which begins
with its maximum element (and so cannot be further decomposed in this
way.) In order to maintain this property (and thus avoid the occurrence
of any 231), if a new element is to be inserted it must be immediately
before one of these blocks, or at the very end of the permutation. There
are thus only $b+1$ *active sites* at which a new maximum can be inserted.
Furthermore, for each of these we can determine the number of blocks
in the resulting permutation – which turn out to be the numbers from 1
through $b + 1$ inclusive. Thus, we get a bijection between permutations
of length n in C and those sequences of length n of positive integers
beginning with 1 that have the property that each successive symbol is
not more than one greater than its predecessor.

This string based discussion does some violence to the historical record,
where the example above (and a variety of others) were considered in
terms of a tree structure, the *generating tree*, in which there is one node
for each element of the class C, and the children of a permutation π are
those permutations in C obtained from π by the insertion of a new maximum element. Many permutation classes have been enumerated using
generating trees by a variety of authors ([16, 22, 32, 33, 38, 39, 40]). Additionally, more or less general methods have been developed for passing
from generating trees to enumerations [13]. Further classes have also
been enumerated using the closely related *ECO method* ([12, 14, 15, 23,
27]). As one might well expect, both these methods have also found
widespread application outside of the realm of permutation classes.

But, let us return to our consideration of strings. We seem to have
uncovered an idea that we might well be able to represent permutations
that we are interested in by some other *encoding* than their standard
one line representation. If the resulting language (the strings that actually represent encodings of permutations) has reasonable properties,
and if the involvement relation behaves well (whatever that means!)
with respect to the encoding, then we might hope to be able to analyze
either individual permutation classes, or whole collections of them, via
their encodings. This method was perhaps first applied in [9], though
of course it is implicit in many, perhaps most, bijective correspondences
between permutation classes and other combinatorial objects.

The most robust and well-behaved families of languages are the *regular languages*. A regular language is a subset of the strings over some finite alphabet Σ, which is recognized by a *finite state automaton*. We are about to embark on an informal description of such (and related) machines, but the careful reader might wish to consult Hopcroft and Ullman [28], or any other text on formal languages and automata theory.

A finite state automaton is a machine, $\mathbb{M}$, that has finitely many states. It consumes an input string over some finite alphabet Σ, one symbol at a time, and each symbol causes a transition between states either deterministically (that is, if the machine is in state A and receives symbol s then the subsequent state is uniquely determined by A and s), or non-deterministically. One of the states is designated as an initial state, and some of the states are designated as accepting states. The *language recognized* by $\mathbb{M}$ is the set of those strings over Σ which, when processed, leave $\mathbb{M}$ in an accepting state. In the non-deterministic case this should be taken to mean that there is *some* choice of allowed transitions that leaves $\mathbb{M}$ in an accepting state. Perhaps surprisingly, there is no difference between the languages recognized by deterministic finite state machines, and those recognized by non-deterministic ones. This is useful, in that it is frequently easier to describe non-deterministic automata that recognize a language. From a practical standpoint it should be noted however, that transforming a non-deterministic automaton into an equivalent deterministic one may result in an exponential blow up in the number of states†.

For our purposes, the important aspects of regular languages are that they are robust, and have good enumerative properties. Specifically:

Theorem 4.1. *The collection of regular languages over Σ is closed under complement, union, intersection, concatenation and repetition. Moreover, every regular language has a rational generating function.*

A related type of machine that we need to consider is called a *finite state transducer*. Like a finite state automaton, a finite state transducer, $\mathbb{T}$, has an underlying finite set of states (including an initial state and some accepting states), and processes strings over a finite alphabet Σ. However, rather than simply accepting or rejecting strings, $\mathbb{T}$, transforms them – each time it consumes an input symbol it may also produce output. The transducer thus defines a relation between pairs of strings

---

† The essential idea is that the states of the deterministic automaton represent sets of states in the non-deterministic one.

over Σ, where s is related to s' if some accepting computation of $\mathbb{T}$ that consumes s produces s'. By carrying out a certain amount of abstract engineering, the reader should be able to convince herself that if L is a regular language over Σ and L' is the set of words related according to $\mathbb{T}$ to some element of L, then L' is also a regular language.

To apply this to permutation classes, we begin with some set of permutations $\mathcal{U}$ (usually itself a class), for which we have an encoding over some finite alphabet Σ and which is represented by a regular set in this encoding. We think of $\mathcal{U}$ as the *universe* of permutations for the moment, and deal with subclasses of $\mathcal{U}$. For instance $\mathcal{U}$ might be the class of *k-bounded permutations*, those in which each symbol in the one line representation is among the k smallest remaining symbols (see the discussion of the *rank encoding* below), or, $\mathcal{U}$ might be the class of permutations having at most k descents. We suppose, though it is not *strictly* necessary, that the encoding preserves length, and that each symbol of the encoding corresponds to some particular symbol in the one line representation of a permutation. For convenience, we henceforth blur the distinction between permutations in $\mathcal{U}$ and their encodings. The key ingredient that we require is a transducer $\mathbb{D}$, the *deletion transducer* which has the property that two permutations π and π' (from $\mathcal{U}$) are related by $\mathbb{D}$ if and only if $\pi' \preceq \pi$.

Subject to making all of these requirements precise, the following theorem is implicit in [4, 7]:

Theorem 4.2. *Supposing that $\mathcal{U}$ and $\mathbb{D}$ have the properties defined above, then any permutation class contained in $\mathcal{U}$ whose basis relative to $\mathcal{U}$ is a regular set, in particular any finitely based class, is itself a regular set and conversely, if a permutation class contained in $\mathcal{U}$ is a regular set, then its basis relative to $\mathcal{U}$ is also regular.*

Furthermore, there are well known algorithms for transforming automata, and producing generating functions from automata, which mean that in any practically interesting case it will be possible to produce from the basis of say a finitely based class within $\mathcal{U}$, the generating function of the class itself.

Two encodings are of particular note. The first is the *rank encoding*. Recall that we represent a permutation in one line notation by its sequence of values. We can further modify this representation by replacing each value in that sequence by its rank among the remaining ones (itself, and the ones that follow it). Specifically, we transform the sequence $\pi_1 \pi_2 \cdots \pi_n$ to the sequence $r_1 r_2 \cdots r_n$ where, for each i between 1 and n

inclusive:

$$r_i = |\{j \,:\, i \le j \text{ and } \pi_1 \ge \pi_j\}|\,.$$

So, for instance, the permutation 35142 is encoded as 34121. A permutation is called *k-bounded* if its rank encoding uses only symbols in $[k]$. The k-bounded permutations form a permutation class which can be taken as $\mathcal{U}$ above. The deletion transducer is not entirely straightforward (it is described generally in [4]) but its existence is not surprising†.

The second is a more general encoding called the *insertion encoding*. This encoding is motivated by the previously discussed generating tree approach. Permutations are thought of as developing by repeated insertion of a new maximum element into available *slots*, denoted by ◇'s. This development begins with a single empty slot. However, instead of keeping track of active sites, where an element *might* be inserted, the insertion encoding records only those slots where a maximum element *will* be inserted. There are four possible types of insertion into any slot:

- the slot might be *filled*, meaning that no larger element will subsequently be placed between the newly added element and its immediate neighbours;

- the insertion might be at the *left* or *right* end of the slot, meaning, in the first case, that no larger element will subsequently be placed between the newly added element and its left hand neighbour, but one (or more) will be placed between it and its right hand neighbour;

- the insertion might be in the *middle* of the slot, meaning that larger elements will be placed between the newly added element and both of its immediate neighbours.

If slots are indexed from left to right, then this gives a means of encoding an arbitrary permutation over an infinite alphabet. For example, with the obvious conventions, the permutation 3742615 would be encoded as:

$$M(1)M(1)L(1)R(1)F(3)F(2)F(1).$$

† In order to recode an element, you need to know how many following elements that are smaller than it will be deleted. Only the k smallest elements at any point need to be considered, so we can use 2^k states based on a bit vector specifying which elements will be kept and which deleted. Then, when processing a symbol j, if it is to be deleted we simply output nothing. If j is to be kept, we output j minus the number of subsequent smaller elements that will be deleted. In either case, we then delete the corresponding element of the bit vector and, non-deterministically, add a new maximum element – either specifying that it will be kept or deleted.

In detail, the development of this permutation proceeds as follows:

$$\diamond \quad \overset{M(1)}{\longrightarrow} \quad \diamond\,1\,\diamond$$
$$\overset{M(1)}{\longrightarrow} \quad \diamond\,2\,\diamond\,1\,\diamond$$
$$\overset{L(1)}{\longrightarrow} \quad 3\,\diamond\,2\,\diamond\,1\,\diamond$$
$$\overset{R(1)}{\longrightarrow} \quad 3\,\diamond\,42\,\diamond\,1\,\diamond$$
$$\overset{F(3)}{\longrightarrow} \quad 3\,\diamond\,42\,\diamond\,15$$
$$\overset{F(2)}{\longrightarrow} \quad 3\,\diamond\,42615$$
$$\overset{F(1)}{\longrightarrow} \quad 3742615$$

As in the rank bounded case, we can reduce to a finite alphabet by limiting the number of slots that are allowed. It turns out that there is also a deletion transducer for this encoding (though we shall not attempt to describe it here – see [7].) The inspiration for the development of the insertion encoding was a paper of Vatter, [38], which, in particular, classified permutation classes with finitely labelled generating trees. The following result from [7] is a generalization of the main result from [38]:

Proposition 4.3. *Suppose that C is an arbitrary finitely based permutation class, and that for some $n > 0$, none of the permutations*

$$1\,(n+1)\,2\,(n+2)\,\cdots(n-1)\,(2n-1)\,n\,(2n),$$
$$1\,(2n)\,2\,(2n-1)\,\cdots(n-1)\,(n+2)\,n\,(n+1),$$
$$n\,(n+1)\,(n-1)\,(n+2)\,\cdots2\,(2n-1)\,1\,(2n),$$
$$n\,(2n)\,(n-1)\,(2n-1)\,\cdots2\,(n+2)\,1\,(n+1)$$

belong to C. Then C has a rational generating function.

In fact, as shown in [7], this proposition is really just a consequence of the application of Theorem 4.2 where $\mathcal{U}$ is the class of permutations whose insertion encodings use a bounded number of slots. Many further applications of the insertion encoding can be found in [7].

A somewhat wider collection of languages that have reasonable properties, at least with regard to enumeration, are context free languages that have an unambiguous grammar ([21]). While some results are known for classes defined by such languages, particularly with respect to the insertion encoding, it is an obvious direction in which more work is required, especially given the relationship to permutation classes defined by simple machines as in Knuth's original work ([31]).

5 Growth rates

Since the resolution by Marcus and Tardos [35] of the Stanley-Wilf conjecture it has been known that every permutation class $\mathcal{C}$ has a finite *growth rate* (or *speed*) defined as $\limsup_{n \to \infty} |\mathcal{C}_n|^{1/n}$. Naturally enough, this has further focused attention on questions related to the growth rates of permutation classes. The outstanding open question is:

Is it the case that, for every permutation class $\mathcal{C}$, $\lim_{n \to \infty} |\mathcal{C}_n|^{1/n}$ exists?

The answer is affirmative for principal classes as these have super-multiplicative growth as shown by Arratia ([8]). And, when a class has an easily described generating function, it is usually possible to apply asymptotic methods to verify that it has a limiting growth rate.

Of course, another obvious question is: which growth rates can occur? And, how "complex" must a class be if its growth rate is known to exceed some given value? The simplest version of this latter question takes us back to the Erdős-Szekeres theorem which can be recast as: a permutation class is infinite (that is, has growth rate greater than 0) if and only if it contains either every increasing permutation or every decreasing permutation.

Progress on these questions (which are obviously closely linked) was made by Kaiser and Klazar [30] who determined all possible growth rates up to 2, and more recently by Vatter [37] who has extended this analysis significantly up to the level where infinite antichains begin to appear within permutation classes. Albert and Linton [6] showed that the set of possible growth rates becomes very rich (precisely, they showed that it contains a perfect set), while Vatter [36] has shown that the set of possible growth rates contains every real number greater than an algebraic number which is approximately 2.48188.

There are various ways in which structural methods can be applied in this area. For example, one might begin with a class with good structure and consider local versions of the questions, specifically, what growth rates can occur among its subclasses. For the class of separable permutations, some progress in this direction is reported in [1]. Alternatively, it seems natural to consider classes, $\mathcal{C}$, which cannot be approximated by their proper subclasses in the sense that the supremum of the growth rates of the proper subclasses of $\mathcal{C}$ is not equal to the growth rate of $\mathcal{C}$. Such classes exist – for instance the class of permutations having at most one descent has growth rate 2 and every proper subclass has polynomial growth rate†, but beyond this little is known about them. By contrast, a conjecture of Arratia providing an upper bound of 9 on the growth rate of Av(4231) was refuted in [5] essentially by constructing subclasses of Av(4231) with good structure (regular for the insertion encoding) whose growth rates could be computed (exactly in principle, approximately in practice.)

6 Conclusions

This article has focused on the use of structure in permutation classes to obtain combinatorial information about them. One type of structure that we have not mentioned in any detail is *partial well order*. There is a good case to be made that any class which is partially well ordered should be considered as having reasonable structure, and there has been some progress, initiated in [10], in pursuing that idea.

In model theory, a dichotomy, or series of dichotomies between structure and complexity is used to understand and classify theories†. A corresponding programme in permutation patterns would presumably focus on the presence of certain families of patterns as an indication of complexity, and their absence as an indication of structure. Indeed, the results of Kaiser and Klazar on classes of small growth rate can be viewed in precisely this way. It seems likely that, taken literally, such a programme would simply throw almost all permutation classes into a sack marked "unstructured". However, we have tried to point out above, that there are several places where one might adopt a more local viewpoint of this programme with, perhaps, more satisfying results.

To us, one of the exciting aspects of pursuing structural methods in studying problems about permutation patterns, is that there seems to be no single characterization of structure that can be applied globally. In addition to the methods mentioned above concerning block structure, generating trees or encodings, and partial well order, we fully expect that new structural motifs will be discovered and used effectively to solve previously difficult problems. What is significant, in our eyes, is that these methods have the scope to be applied, if not universally, at least to a wide variety of problems, and to provide general theorems of

† This is related to the results of [29]. Even this simple example illustrates the importance of the underlying structural ideas. A proper subclass must avoid one of the permutations $24 \cdots 2n\, 1\, 3 \cdots (2n-1)$ as well as all of the basis elements of the full class, but this implies polynomial growth.

† For instance, in first order logic, the definability of an infinite linear order immediately leads to a non-structured theory

the form "If such and such a type of structure exists in a permutation class, then additionally the following nice things happen ..." (possibly including converse results as well, as in the Erdős Szekeres theorem) rather than dealing with classes on a one by one *ad hoc* basis.

While the resolution of the Stanley-Wilf conjecture solves what had been regarded as the major open problem in the area of permutation patterns, we believe that there is still a wide variety of interesting and important problems concerning permutation patterns and permutation classes that remain to be solved, and that structural methods will be the key to solving many of them.

Acknowledgment I would like to thank an anonymous referee for some very helpful comments.

References

[1] M. H. Albert. Aspects of separability. Abstract for Permutation Patterns 2007.

[2] M. H. Albert and M. D. Atkinson. Simple permutations and pattern restricted permutations. *Discrete Math.*, 300(1-3):1–15, 2005.

[3] M. H. Albert, M. D. Atkinson, and M. Klazar. The enumeration of simple permutations. *J. Integer Seq.*, 6(4):Article 03.4.4, 18 pp., 2003.

[4] M. H. Albert, M. D. Atkinson, and N. Ruškuc. Regular closed sets of permutations. *Theoret. Comput. Sci.*, 306(1-3):85–100, 2003.

[5] M. H. Albert, M. Elder, A. Rechnitzer, P. Westcott, and M. Zabrocki. On the Wilf-Stanley limit of 4231-avoiding permutations and a conjecture of Arratia. *Adv. in Appl. Math.*, 36(2):95–105, 2006.

[6] M. H. Albert and S. Linton. Growing at a perfect speed. *Combin. Probab. Comput.*, 18:301–308, 2009.

[7] M. H. Albert, S. Linton, and N. Ruškuc. The insertion encoding of permutations. *Electron. J. Combin.*, 12(1):Research paper 47, 31 pp., 2005.

[8] R. Arratia. On the Stanley-Wilf conjecture for the number of permutations avoiding a given pattern. *Electron. J. Combin.*, 6:Note, N1, 4 pp., 1999.

[9] M. D. Atkinson, M. J. Livesey, and D. Tulley. Permutations generated by token passing in graphs. *Theoret. Comput. Sci.*, 178(1-2):103–118, 1997.

[10] M. D. Atkinson, M. M. Murphy, and N. Ruškuc. Partially well-ordered closed sets of permutations. *Order*, 19(2):101–113, 2002.

[11] M. D. Atkinson and T. Stitt. Restricted permutations and the wreath product. *Discrete Math.*, 259(1-3):19–36, 2002.

[12] S. Bacchelli, E. Barcucci, E. Grazzini, and E. Pergola. Exhaustive generation of combinatorial objects by ECO. *Acta Inform.*, 40(8):585–602, 2004.

[13] C. Banderier, M. Bousquet-Mélou, A. Denise, P. Flajolet, D. Gardy, and D. Gouyou-Beauchamps. Generating functions for generating trees. *Discrete Math.*, 246(1-3):29–55, 2002.

[14] E. Barcucci, A. Del Lungo, E. Pergola, and R. Pinzani. From Motzkin to Catalan permutations. *Discrete Math.*, 217(1-3):33–49, 2000.

[15] E. Barcucci, A. Del Lungo, E. Pergola, and R. Pinzani. Permutations avoiding an increasing number of length-increasing forbidden subsequences. *Discrete Math. Theor. Comput. Sci.*, 4(1):31–44, 2000.

[16] M. Bousquet-Mélou. Four classes of pattern-avoiding permutations under one roof: generating trees with two labels. *Electron. J. Combin.*, 9(2):Research paper 19, 31 pp., 2003.

[17] R. Brignall. Wreath products of permutation classes. *Electron. J. Combin.*, 14(1):Research paper 46, 15 pp., 2007.

[18] R. Brignall, S. Huczynska, and V. Vatter. Decomposing simple permutations, with enumerative consequences. *Combinatorica*, 28:385–400, 2008.

[19] R. Brignall, S. Huczynska, and V. Vatter. Simple permutations and algebraic generating functions. *J. Combin. Theory Ser. A*, 115(3):423–441, 2008.

[20] R. Brignall, N. Ruškuc, and V. Vatter. Simple permutations: decidability and unavoidable substructures. *Theoret. Comput. Sci.*, 391(1–2):150–163, 2008.

[21] N. Chomsky and M. P. Schützenberger. The algebraic theory of context-free languages. In *Computer programming and formal systems*, pages 118–161. North-Holland, Amsterdam, 1963.

[22] T. Chow and J. West. Forbidden subsequences and Chebyshev polynomials. *Discrete Math.*, 204(1-3):119–128, 1999.

[23] E. Duchi, J.-M. Fedou, and S. Rinaldi. From object grammars to ECO systems. *Theoret. Comput. Sci.*, 314(1-2):57–95, 2004.

[24] P. Erdős and G. Szekeres. A combinatorial problem in geometry. *Compos. Math.*, 2:463–470, 1935.

[25] P. Flajolet and R. Sedgewick. *Analytic combinatorics*. Cambridge University Press, Cambridge, 2009.

[26] I. P. Goulden and D. M. Jackson. *Combinatorial enumeration*. Dover Publications Inc., Mineola, NY, 2004.

[27] O. Guibert, E. Pergola, and R. Pinzani. Vexillary involutions are enumerated by Motzkin numbers. *Ann. Comb.*, 5(2):153–147, 2001.

[28] J. E. Hopcroft and J. D. Ullman. *Introduction to automata theory, languages, and computation*. Addison-Wesley Publishing Co., Reading, Mass., 1979. Addison-Wesley Series in Computer Science.

[29] S. Huczynska and V. Vatter. Grid classes and the Fibonacci dichotomy for restricted permutations. *Electron. J. Combin.*, 13:Research paper 54, 14 pp., 2006.

[30] T. Kaiser and M. Klazar. On growth rates of closed permutation classes. *Electron. J. Combin.*, 9(2):Research paper 10, 20 pp., 2003.

[31] D. E. Knuth. *The art of computer programming. Vol. 1: Fundamental algorithms*. Addison-Wesley Publishing Co., Reading, Mass., 1969.

[32] D. Kremer and W. C. Shiu. Finite transition matrices for permutations avoiding pairs of length four patterns. *Discrete Math.*, 268(1-3):171–183, 2003.

[33] M. Lipson. Completion of the Wilf-classification of 3-5 pairs using generating trees. *Electron. J. Combin.*, 13(1):Research paper 31, 19 pp., 2006.

[34] T. Mansour and A. Vainshtein. Restricted 132-avoiding permutations. *Adv. in Appl. Math.*, 26(3):258–269, 2001.

[35] A. Marcus and G. Tardos. Excluded permutation matrices and the Stanley-Wilf conjecture. *J. Combin. Theory Ser. A*, 107(1):153–160, 2004.

[36] V. Vatter. Permutation classes of every growth rate above 2.48188. *Mathematika*, 56:182–192, 2010.

[37] V. Vatter. Small permutation classes. arXiv:0712.4006v2 [math.CO].

[38] V. Vatter. Finitely labeled generating trees and restricted permutations. *J. Symbolic Comput.*, 41(5):559–572, 2006.

[39] J. West. Generating trees and the Catalan and Schröder numbers. *Discrete Math.*, 146(1-3):247–262, 1995.

[40] J. West. Generating trees and forbidden subsequences. *Discrete Math.*, 157(1-3):363–374, 1996.

Combinatorial properties of permutation tableaux

Alexander Burstein†

Department of Mathematics
Howard University
Washington, DC 20059 USA

Niklas Eriksen

Department of Mathematical Sciences
Göteborg University and Chalmers University of Technology
SE-412 96 Göteborg, Sweden

Abstract

We give another construction of a permutation tableau from its corresponding permutation and construct a permutation-preserving bijection between 1-hinge and 0-hinge tableaux. We also consider certain alignment and crossing statistics on permutation tableaux that are known to be equidistributed via a complicated map to permutations that translates those to occurrences of certain patterns. We give two direct maps on tableaux that proves the equidistribution of those statistics by exchanging some statistics and preserving the rest. Finally, we enumerate some sets of permutations that are restricted both by pattern avoidance and by certain parameters of their associated permutation tableaux.

1 Introduction

Permutation tableaux are combinatorial objects that are in bijection with permutations. They originally turned up in the enumeration of totally positive Grassmannian cells [8, 11]. Permutation tableaux have then been studied either in their own right [1, 10] to produce enumeration results for permutations, or in connection with the PASEP model in statistical mechanics [3, 5, 6].

† Research supported in part by NSA Young Investigator Grant H98230-06-1-0037.

0	1	0	1
1	1	0	1
1	1	1	2
0	1	1	2

0	1	0	1
1	1	0	1
1	1	1	
0	1	1	

Fig. 1. Permutation tableaux (1-hinge) of $\pi = 36187425$. To the right is illustrated $\pi(1) = 3$, $\pi(2) = 6$ and $\pi(6) = 4$

A **permutation tableaux** $\mathcal{T}$ is usually defined as a $k \times (n - k)$ array filled with zeroes, ones and twos such that the cells filled with zeroes and ones form a Young tableau Y_λ of an integer partition λ with $n - k = \lambda_1 \geq \lambda_2 \geq \ldots \geq \lambda_k \geq 0$ (note that zero parts are allowed), and such that these rules are obeyed:

(column): Each column in the tableau contains at least one 1.
(1-hinge): Each cell in the Young tableau Y_λ with a 1 above in the same column and to its left in the same row must contain a 1.

Equivalently, one forgets about the twos and considers only the Young tableau of zeroes and ones, which are sometimes encoded as blanks and bullets, respectively (see Figure 1). The **shape** of $\mathcal{T}$ is the integer partition $\text{sh}(\mathcal{T}) = \lambda$ and its **length** is the number of parts in λ, $\ell(\mathcal{T}) = k$.

The second rule above can, however, take several forms. An alternative, presented in [1], is this:

(0-hinge): Each cell in the Young tableau Y_λ with a 1 above in the same column and to its left in the same row must contain a 0.

In this paper, we generalize these hinge rules:

(0/1-hinge): Given any partition $\mu \leq \lambda$ (i.e. $\mu_i \leq \lambda_i$ for all $i \geq 0$), each cell in the Young tableau Y_λ with a 1 above in the same column and to its left in the same row must contain a 1 if the cell is in μ and a 0 otherwise.

The 0/1-hinge rule specializes to the 0-hinge rule for $\mu = \emptyset$ an the 1-hinge rule for $\mu = \lambda$. We let $\mathfrak{T}_n$ denote the set of 1-hinge permutation tableaux $\mathcal{T}$ such that, with $\text{sh}(\mathcal{T}) = \lambda$, $\lambda_1 + \ell(\mathcal{T}) = n$.

Permutation tableaux have their name from a natural bijection $\Phi : \mathfrak{T}_n \to \mathfrak{S}_n$ between (1-hinge) permutations tableaux and permutations [10]. Let $n = \lambda_1 + k$. Label the south-east boundary of the tableau with 1 to n, starting in the north-east corner, and extend these labels to the rows and columns they belong to. Then, starting at the top of a column (or the left of a row) labeled i, we follow the zig-zag path obtained by bouncing right or down every time we hit a 1. If the label of the exit is j, we put $\pi(i) = j$. This is illustrated to the far right in Figure 1.

There are several important statistics of $\pi = \Phi(\mathcal{T})$, typically related to the order relation between a position i and its letter $\pi(i)$ which are easily deduced from $\mathcal{T}$. The **weak excedances** (positions i such that $\pi(i) \geq i$) are given by the row labels and the **deficiencies** (the opposites of weak excedances, that is positions i such that $\pi(i) < i$) are given by the column labels. Fixed points correspond to empty rows. We let wex(π) denote the number of weak excedances in π and conclude that wex(π) $= k$. We also note that a bijection between permutations tableaux and permutations with $n - k$ **descents** (elements $\pi(i)$ such that $\pi(i) > \pi(i+1)$) has been proposed [4], which amongst other things reproves that descents and deficiencies are equidistributed.

The inverse $\Gamma = \Phi^{-1} : \mathfrak{S}_n \to \mathfrak{T}_n$ of the presented bijection is less natural, and several algorithms describing Γ have been proposed [1, 10]. We continue this tradition with another algorithm which we consider simpler than the previous ones. It is presented in Section 2 together with the presentation of 0/1-hinge tableaux.

Further, we study two bijections on tableaux. Alignment and crossing statistics on permutations have been proposed by Corteel [3], and they have natural interpretations on tableaux. First we give in Section 3 a direct description on tableaux for the bijection between tableaux that belong to permutation π and irc(π), where irc is the inverse of the reversal of the complement. This exchanges A_{EN} and A_{NE} statistics and preserves statistics wex, A_{EE}, A_{NN}, C_{EE}, C_{NN}. Then we give a simple bijection on tableaux that preserves the A_{EN} and A_{NE} statistics, and exchange the $A_{\mathrm{NN}} + A_{\mathrm{EE}}$ and $C_{\mathrm{NN}} + C_{\mathrm{EE}}$ statistics. Definitions and algorithm are presented in Section 4.

Pattern avoidance (or pattern statistics in general) and tableau restrictions do not always combine easily and naturally. Patterns deal with comparisons of different letters in a permutation, in particular, they are better suited for considering descents and inversions. Permutation tableaux, on the other hand, naturally emphasize excedances, fixed points and deficiencies, i.e. comparisons of letters with their positions. Thus, permutation tableaux are useful in considering pattern-restricted sets when the information about descents and inversions can be translated into information about weak excedances and deficiencies. A good example of such a situation is [2] where alternating permutations (descent-related objects) with the maximum number of fixed points (excedance-related property) are considered.

In a 1-hinge tableau $\mathcal{T}$, those ones that are not forced by the 1-hinge rule are called **essential**. We conclude this article in Section 5 by giving some enumeration results on sets of permutations avoiding pairs of patterns of length 3 and whose associated tableaux (via maps Φ and Γ) have the maximum number of essential ones. The eventual goal of this undertaking is to refine pattern-occurrence statistics with respect to the number of essential ones in the associated tableaux.

2 Combining 1-hinge and 0-hinge

An important property of 1-hinge permutation tableaux is that every zero has a clear view (only zeroes) to its left or above it. We will use this property in two ways. First, we give a new algorithm for computing the 1-hinge tableaux of a permutation, and then we show that the 1-hinge and 0-hinge tableaux of a permutation are connected via a series of mixed hinge tableaux.

In a tableau $\mathcal{T}$, two paths are said to **meet** at a cell if each of these paths enters the cell. If the cell contains a zero, the paths will **cross**, and if it contains a one they will **bounce**. By the 1-hinge rule, two paths can only cross at their first meet.

Definition 2.1. In a tableau $\mathcal{T}$ of π, we have columns $c_i(\mathcal{T}), 1 \leq i \leq \lambda_1$ and rows $r_i(\mathcal{T}), 1 \leq i \leq n - \lambda_1$. When no confusion can arise, we write $c_i = c_i(\mathcal{T})$ and $r_i = r_i(\mathcal{T})$. Further, let $\mathrm{ent}(c_i)$ be the label of column c_i and let $\mathrm{ext}(c_i) = \pi(\mathrm{ent}(c_i))$. We call these **entry** and **exit** labels of column c_i. Similar definitions apply for the rows. The row (resp. column) number with exit label j is denoted $\mathrm{row}(j)$ (resp. $\mathrm{col}(j)$). In other words, $\mathrm{col}(j) = i \Leftrightarrow \mathrm{ext}(c_i) = j$.

The **initial zeroes** in a column (or row) are the zeroes in that column (row) that have no ones above (to the left). The number of initial zeroes in a column or row is denoted $z(c_i)$ and $z(r_i)$, respectively, and similarly $t(c_i)$ and $t(r_i)$ for the number of twos in the column or row.

The exit labels of rows are of course the weak excedance letters and the exit labels of columns are the deficiency letters.

We note that given the shape of a tableau $\mathcal{T} \in \mathfrak{T}_n$, information on the number of initial zeroes in all rows and columns completely determines $\mathcal{T}$, since each zero is an initial zero of some row or column. The initial zeroes also determines the essential ones, and thus any 0/1-hinge tableau. Thus, to compute $\Gamma(\pi)$ we need only determine the initial zeroes. To accomplish this, only the relative order of the exit labels of rows and columns are important.

Definition 2.2. For a row r_i in $\mathcal{T}$, we say that row r_i has m **inversions** if there are m rows above with higher exit labels. Likewise, we say that column c_i has m **inversions** if there are m columns to its left with lower exit labels. The number of inversions are denoted $\text{inv}(r_j, \mathcal{T})$ and $\text{inv}(c_j, \mathcal{T})$, respectively, or $\text{inv}(r_j)$ and $\text{inv}(c_j)$ for short.

As for usual permutation inversions, we need only to know the inversion numbers of all row (column) exit labels to compute their order.

Lemma 2.3. *Consider a permutation tableau $\mathcal{T}$ with $\text{sh}(\mathcal{T}) = \lambda$. Let r contain the indices of the rows r_i such that $\text{ext}(r_i) > \text{ext}(c_j)$ in increasing order. Then, $z(c_j) = r_{\text{inv}(c_j)+1} - 1$. Similarly, letting c contain the columns c_i such that $\text{ext}(c_i) < \text{ext}(r_j)$ in increasing order, with $\lambda_j + 1$ appended at the end, we have $z(r_j) = c_{\text{inv}(r_j)+1} - 1$.*

Example 2.4. The lemma is best appreciated after an example. Consider the permutation $\pi = 463785912$. Its tableau has shape $\lambda = (3^5, 2)$. The column exit labels are 215 (deficiency letters read from right to left) and the row exit labels are 463789 (weak excedance letters read from left to right).

Now, for the third column we get $r = (2, 4, 5, 6)$ and hence $z(c_3) = r_{\text{inv}(c_3)+1} - 1 = r_{2+1} - 1 = 5 - 1 = 4$. For c_2 and c_1, we find no inversions, and the first entry in r is 1, yielding no initial zeroes.

For the rows, the only row with positive inversion number is r_3, with $\text{inv}(r_3) = 2$. Computing $c = (1, 2, \lambda_3 + 1)$, we get $z(r_3) = c_{2+1} - 1 = \lambda_3 + 1 - 1 = 3$.

$$\begin{array}{c|ccc}
 & 2 & 1 & 5 \\
\hline
4 & \bullet & \bullet & \\
6 & \bullet & \bullet & \\
3 & & & \\
7 & \bullet & \bullet & \bullet \\
8 & \bullet & \bullet & \\
9 & \bullet & \bullet & \\
\end{array}$$

We are now ready to prove Lemma 2.3.

Proof. Assume that $z(c_j) = m$ and consider the array from $\mathcal{T}$ of the intersection of rows r_1 to r_m and columns c_1 to c_{j-1}. The paths that enter the array must also exit the array. Those that exit horizontally will cross the path from c_j and those that exit vertically will not, since paths starting at columns must cross at an initial zero of c_j, and paths starting at rows r_i must cross at an initial zero of either c_j or r_i, neither of which is possible with a vertical exit.

Thus, with $j - 1$ vertical exits and $j - 1 - \text{inv}(c_j)$ taken by columns, there must be $j - 1 - (j - 1 - \text{inv}(c_j)) = \text{inv}(c_j)$ rows with higher exit labels in the array. Since r_{m+1} does not cross c_j we find that $r_{\text{inv}(c_j)+1} = m + 1$, which proves the lemma for columns. For rows, the situation is completely analogous, although it should be noted that rows without ones get the same number of initial zeroes as the length of the row. □

We now turn to 0/1-hinge tableaux. We will prove that for any $\mu < \lambda = \text{sh}(\Gamma(\pi))$, the permutation π has a unique tableau such that the 1-hinge property is fulfilled on Y_μ and the 0-hinge property is fulfilled on $Y_{\lambda/\mu}$. To prove this, we need a few definitions and a lemma. We also take the liberty of extending the function $\Phi : \mathfrak{T}_n \to \mathfrak{S}_n$ to allow any permutation tableau, in particular, a 0/1-hinge tableaux.

Definition 2.5. Let $\mathfrak{S}_\lambda$ denote the set of permutations π such that $\text{sh}(\Gamma(\pi)) = \lambda$. Also, let $\mathfrak{T}_\lambda^\mu$ denote the set of tableaux T such that T fulfills the 1-hinge property on Y_μ and the 0-hinge property on $Y_{\lambda/\mu}$.

Lemma 2.6. *Let $\pi \in \mathfrak{S}_\lambda$, and consider the integer partitions $\mu < \nu \leq \lambda$ such that $Y_{\nu/\mu}$ has exactly one cell. Further let $T \in \mathfrak{T}_\lambda^\nu$ be a tableau such that $\Phi(T) = \pi$. Then, switching ones to zeroes and zeroes to ones on the first and last meetings in ν of the two paths that meet at ν/μ, we obtain a tableau $T' \in \mathfrak{T}_\lambda^\mu$ such that $\Phi(T') = \pi$.*

Proof. Let $i \to \pi(i)$ and $j \to \pi(j)$ be the paths that meet at ν/μ. It is obvious that $\Phi(T') = \pi$ since each of the changes correspond to the transposition $(i\ j)$. Thus, what remains is to show that $T' \in \mathfrak{T}_\lambda^\mu$.

The paths can only cross at their first meet. Thus, if $T(\nu/\mu) = 0$, the first and last meetings coincide and no changes are made. The 0-hinge property is trivially fulfilled on ν/μ, since only 1s can violate it.

If $T(\nu/\mu) = 1$, the 0-hinge property on λ/ν is still trivially fulfilled. On ν/μ, it is trivial if the value changes to 0, and otherwise the 1 must be essential and cannot violate the 0-hinge property. What remains is to show that changing the first meeting of paths does not violate the 1-hinge property. But changing the value at the first meeting to 0 is not a problem, since then the 1 was essential, and changing the value to 1 is legal too, by the following argument.

The new 1 must be essential, so assume without loss of generality that it is essential in its row. Then, any 0s immediately to its right must have 0s below in the row of the second meeting. These 0s have a 1 to their left and hence can have no 1 above, which shows that the 0s immediately to the right of the new 1 do not violate the 1-hinge property.   □

Theorem 2.7. *Let* $\pi \in \mathfrak{S}_\lambda$ *and let* $\mu \leq \lambda$. *Then, there is a unique tableau* $T \in \mathfrak{T}^\mu_\lambda$ *such that* $\Phi(T) = \pi$.

Proof. Given a tableau $T \in \mathfrak{T}^\lambda_\lambda$, which is known to be unique from [10], we can use the algorithm of Lemma 2.6 to reduce the 1-hinge part to μ cell by cell. Thus, it is clear that there is at least one $T \in \mathfrak{T}^\mu_\lambda$ such that $\Phi(T) = \pi$.

At any given moment during the reduction there are usually several cells that can be moved from the 1-hinge area to the 0-hinge area. We need to show that regardless how we choose the order of them, we still end up with the same tableau.

It is fairly easy to realize that for any two cells that can be chosen at the same time, the order of these is insignificant. The cells of their last meetings are on different rows and columns, so all paths through these cells are distinct. Thus, any changes induced by one of the cells will not affect the paths through another one of these cells, and hence will not affect the changes induced by that other cell.

By the strong convergence theorem [7], it suffices to show that any two moves that are valid at the same time commute and that the sequence of moves is finite for a game to have a unique end result. We have shown that any two moves commute and since the number of moves is $|\lambda/\mu|$, the uniqueness is proved.   □

We let $\Gamma^\mu_\lambda : \mathfrak{S}_\lambda \to \mathfrak{T}^\mu_\lambda$ map any permutation whose weak excedance pattern matches λ to the tableau with the 1-hinge property on $\mu \leq \lambda$ and the 0-hinge property on λ/μ. Of course, its inverse is Φ restricted to $\mathfrak{T}^\mu_\lambda$.

3 The *irc* map

In [3], Sylvie Corteel defined the permutation statistics

$$A_{EE}(\pi) = |\{(i,j) \mid j < i \le \pi(i) < \pi(j)\}|,$$
$$A_{NN}(\pi) = |\{(i,j) \mid \pi(j) < \pi(i) < i < j\}|,$$
$$A_{EN}(\pi) = |\{(i,j) \mid j \le \pi(j) < \pi(i) < i\}|,$$
$$A_{NE}(\pi) = |\{(i,j) \mid \pi(i) < i < j \le \pi(j)\}|,$$
$$C_{EE}(\pi) = |\{(i,j) \mid j < i \le \pi(j) < \pi(i)\}|,$$
$$C_{NN}(\pi) = |\{(i,j) \mid \pi(i) < \pi(j) < i < j\}|.$$

They are related to the 1-hinge permutation tableaux $T = \Gamma(\pi)$ in the following way. Label the 0-cells in T with EN if the paths that cross there originated from one column and one row, NN if both these paths originated from columns and EE if both paths originated from rows. The 2-cells are labeled NE. Let $EN(T)$ be the number of cells labeled EN in this labeling, and use similar notation for the other three labels. From [10] and [1] we know that

$$A_{EE}(\pi) = EE(T),$$
$$A_{NN}(\pi) = NN(T),$$
$$A_{EN}(\pi) = EN(T),$$
$$A_{NE}(\pi) = NE(T),$$

and

$$C_{EE}(\pi) + C_{NN}(\pi) = \#\text{nontop 1s}(T).$$

The map irc $= i \circ r \circ c$ (inverse of reversal of complement) is known to preserve all alignment and crossing statistics on permutation except for exchanging statistics A_{NE} and A_{EN}. We now show that the tableau of irc(π) can be easily computed from the tableau of π. The irc map on tableaux is also named irc(T) $= \Gamma(\text{irc}(\Phi(T)))$.

To simplify matters, we start by showing this bijection for quite restricted tableaux and successively remove the restriction until the general case is reached. The condition $A_{NE}(\pi) = 0$ simply means that the shape of $\Gamma(\pi)$ is a rectangle, and the condition $A_{EN}(\pi) = 0$ implies that the main diagonal of $\Gamma(\pi)$ (starting at the northwest corner) contains only ones. It is easy to see that if the main diagonal contains only ones then no path can cross it and hence $A_{EN}(\pi) = EN(\Gamma(\pi)) = 0$, and conversely, if a row (or column) has initial zeroes reaching the main diagonal, there are not enough rows (columns) to account for all these crossings, and hence $A_{\text{EN}}(\pi) > 0$.

Proposition 3.1. *Let $\pi \in \mathfrak{S}_n$ and let $T = \Gamma(\pi) \in \mathfrak{T}_n$ and assume $A_{\text{EN}}(\pi) = A_{\text{NE}}(\pi) = 0$. To compute $T' = \text{irc}(T)$, associate the number of initial zeroes to each row and column, and then order the columns by increasing $\text{ext}(c_i)$ and the rows by decreasing $\text{ext}(r_i)$.*

Proof. If $\pi(i) = j$, then $\text{irc}(\pi)(n+1-j) = n+1-i$. Thus, if i is a weak excedance in π, then $n+1-j$ will be a weak excedance in $\text{irc}(\pi)$. Assuming $A_{\text{EN}}(\pi) = A_{\text{NE}}(\pi) = 0$, all these weak excedances map one of the $k = \text{wex}(\pi)$ lowest elements on one of the k highest elements. Since $i \leq k$ implies $n+1-i \geq n+1-k$ and vice versa, this holds for $\text{irc}(\pi)$ as well, and hence $\text{sh}(T') = \text{sh}(T)$. Further, for rectangular tableaux, we have $i = n + 1 - \text{ent}(c_i)$, and for tableaux with only ones on the main diagonal, we have $z(c_i) = \text{inv}(c_i)$.

Consider column $c_i(T)$. For each inversion $\pi(k) < \pi(i) < i < k$ we get an inversion in T', since then $\text{irc}(\pi)(n+1-\pi(k)) < \text{irc}(\pi)(n+1-\pi(i)) < n+1-\pi(i) < n+1-\pi(k)$. All these inversions contribute to initial zeroes in the same column c in T'. Since $\text{ent}(c) = n + 1 - \pi(i) = n + 1 - \text{ext}(c_i(T))$, we get $c = c_{n+1-\text{ent}(c)} = c_{\text{ext}(c_i(T))}$, which is the statement of the proposition. Rows are handled in a similar fashion. $\square$

Example 3.2. Consider the permutation $\pi = 76485132$. Below we have its 1-hinge tableau T with rows and columns labeled with their exit labels $\text{ext}(r_i)$ and $\text{ext}(c_i)$, as well as the tableau T' obtained by reordering columns and rows according to the proposition above. It is easy to check that $\Phi(T') = \text{irc}(\pi) = 58746213$. Note that the number of initial 0s for each exit label remains the same after reordering rows and columns.

We now continue with permutations π such that $A_{\text{EN}}(\pi) > 0$. The idea is to remove all EN cells, use the transformation for permutations with $A_{\text{EN}}(\pi) = A_{\text{NE}}(\pi) = 0$ and then put them back as NE cells, which contain 2s.

Proposition 3.3. *Let $\pi \in \mathfrak{S}_n$ and let $T = \Gamma(\pi) \in \mathfrak{T}_n$. Assuming $A_{NE}(\pi) = 0$, then the following algorithm will give $T'' = \mathrm{irc}(T)$. Let T' have the same shape as T and let $z(c_i(T')) = \mathrm{inv}(c_i(T))$ for all columns (and similarly for the rows). Letting the column exit labels of T be $i_1 < i_2 < \ldots < i_{n-\mathrm{wex}(\pi)}$, compute $T'' = \mathrm{irc}(T')$ and let $t(c_j(T'')) = i_j - j$ for all j.*

Proof. The tableau T' will have column and row exit labels in the same relative order as T, since only EN cells are changed, but the column exit labels will be 1 to $n - \mathrm{wex}(\pi)$ and the row exit labels will be $n - \mathrm{wex}(\pi) + 1$ to n. Let $k = \mathrm{wex}(\pi)$ and let the column exit labels in T be $i_1 < i_2 < \ldots < i_k$. Then, each column exit label is reduced by $a(j) = \mathrm{ext}(c_j(T)) - \mathrm{ext}(c_j(T')) = i_j - j$. We define permutations $\sigma_j = ((i_j - a(j)) \; (i_j - a(j) + 1) \; \ldots \; i_j)$ and obtain $\Phi(T') = \sigma_k \ldots \sigma_1 \pi$, since each σ_j replaces the exit label i_j with j, while maintaining the relative order of all other exit labels.

Further, $\mathrm{irc}(T')$ is the tableau of

$$\mathrm{irc}(\sigma_k \ldots \sigma_1 \pi) = \mathrm{irc}(\pi)\mathrm{irc}(\sigma_1) \ldots \mathrm{irc}(\sigma_k),$$

where $\mathrm{irc}(\sigma_j) = ((n + 1 - i_j) \; (n + 1 - i_j + 1) \ldots (n + 1 - i_j + a(j)))$. Multiplying from the right with $((n + 1 - i_k) \; (n + 1 - i_k + 1) \ldots (n + 1 - i_k + a(k)))$ in a permutation circularly changes the positions of the letters in positions $n + 1 - i_k, n + 1 - i_k + 1, \ldots, n + 1 - i_k + a(k)$, which is equivalent with introducing $a(k)$ twos in the rightmost column. Similarly, removing the other cycles introduce $a(j)$ in the $(k - j + 1)$th column from the right. $\square$

The irc bijection when $A_{NE}(\pi) = 0$ is clearly bijective. Thus, its inverse gives $\mathrm{irc}(\Gamma(\pi))$ when $A_{EN}(\pi) = 0$, and we state the proposition without proof.

Proposition 3.4. *Let $\pi \in \mathfrak{S}_n$ and let $T = \Gamma(\pi) \in \mathfrak{T}_n$. Assuming $A_{EN}(\pi) = 0$, then the following algorithm will give $T'' = \mathrm{irc}(T)$. Given $\lambda = \mathrm{sh}(T)$ and $k = \mathrm{wex}(\pi)$, let T' have shape λ_1^k and let $z(c_i(T')) = z(c_i(T))$ for all columns (and similarly for the rows). Then, let $T'' = \mathrm{irc}(T')$ and increase column labels j by $t(c_j(T))$.*

Example 3.5. The permutation $\pi = 38652417$ has $A_{EN}(\pi) = 4$ and $A_{NE}(\pi) = 0$. Its tableau is below to the right. Removing the EN zeroes gives the second tableau, where the column exit label 7 has been reduced by 3 and 4 has been reduced by 1. Applying irc gives the third tableau, and inserting 3 twos in the forth column and 1 in the third gives the final tableau of irc(π) = 71653842 to the right. To compute irc of the last tableau, just follow the tableaux below from right to left.

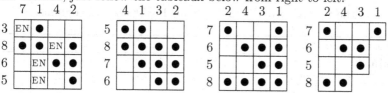

As it turns out, the processes of turning EN zeroes into NE twos and vice versa are independent processes. Combining the last two Propositions thus gives us an algorithm for computing irc(T) for any 1-hinge permutation tableau T.

Theorem 3.6. *Let $\pi \in \mathfrak{S}_n$ and let $T = \Gamma(\pi) \in \mathfrak{T}_n$, with $\lambda = \mathrm{sh}(T)$, $k = \mathrm{wex}(\pi)$ and column exit labels $i_1 < i_2 < \ldots < i_k$. To compute $T'' = \mathrm{irc}(T)$, let T' have shape λ_1^k and let $z(c_i(T')) = \mathrm{inv}(c_i(T))$ for all columns (and similarly for the rows). Then, let $T'' = \mathrm{irc}(T')$, increase column labels j by $t(c_j(T))$ and $t(c_j(T'')) = i_j - j$ for all j.*

Proof. Removing the twos of T corresponds to multiplying π by cycles on the right as in the proof of Proposition 3.3. Removing the EN zeroes corresponds to multiplying π by cycles on the left. These processes are commutative. We can then compute irc and interpret the cycles moving from left to right as NE twos and the other cycles as EN zeroes. $\square$

Example 3.7. The permutation $\pi = 38265417$ has $A_{\mathrm{EN}}(\pi) = 4$ and $A_{\mathrm{NE}}(\pi) = 2$. Its tableau is below to the right. Removing the EN zeroes and the twos gives the second tableau, which is the same as the second tableau in the previous example. Applying irc gives the third tableau, and inserting twos as before, as well as two EN zeroes by increasing column exit label 4 to 6 gives the final tableau of irc(π) = 71543862 to the right.

7	1	4	2			4	1	3	2			2	4	3	1			2	4	3	1

3 EN ● · ·   5 ● ● · ·   7 ● · · ●   7 ● · · ●
8 ● ● EN ●   8 ● ● ● ●   6 · ● ● ●   6 · EN ●
6 · EN ● ·   7 · ● ● ●   5 · · ● ●   5 · EN
5 · EN · ·   6 · · ● ●   8 ● ● ● ●   8 ● ●

4 The $A \leftrightarrow C$ bijection

Steingrímsson and Williams showed that $A_{\mathrm{NN}} + A_{\mathrm{EE}}$ is equidistributed with $C_{\mathrm{NN}} + C_{\mathrm{EE}}$. We would like to show this with a simple bijection, which exchanges these statistics while preserving A_{NE}, A_{EN} and wex. We name this bijection $\psi : \Gamma(\mathfrak{S}_n) \to \Gamma(\mathfrak{S}_n)$. To find ψ, we need to keep track of the relative order of the exit labels of rows and columns.

Algorithm 1 The bijection ψ

Data: T
Result: $T' = \psi(T)$.
$T' \leftarrow T$
$S_c \leftarrow$ exit labels of the columns of T, sorted in descending order
$S_r \leftarrow$ exit labels of the rows of T, sorted in ascending order
for $i \leftarrow 1$ **to** λ_1 **do**
$\quad \text{inv}_{S_c(i)}(T') \leftarrow n - i - S_c(i) - |\{j : \text{ent}(r_j) > S_c(i)\}| - \text{inv}_{S_c(i)}(T)$
end
for $i \leftarrow 1$ **to** $n - \lambda_1$ **do**
$\quad \text{inv}_{S_r(i)}(T') \leftarrow S_r(i) - i - |\{j : \text{ent}(c_j) \le S_r(i)\}| - \text{inv}_{S_r(i)}(T)$
end

Theorem 4.1. *Algorithm 1 is an involution on permutation tableaux T such that $T' = \psi(T)$ fulfills*

- $A_{\text{NN}}(\Phi(T')) + A_{\text{EE}}(\Phi(T')) = C_{\text{NN}}(\Phi(T)) + C_{\text{EE}}(\Phi(T))$;
- $C_{\text{NN}}(\Phi(T')) + C_{\text{EE}}(\Phi(T')) = A_{\text{NN}}(\Phi(T)) + A_{\text{EE}}(\Phi(T))$;
- $A_{\text{EN}}(\Phi(T')) = A_{\text{EN}}(\Phi(T))$;
- $A_{\text{NE}}(\Phi(T')) = A_{\text{NE}}(\Phi(T))$;
- $\text{sh}(T') = \text{sh}(T)$.

Proof. It is clear that the proposed map is an involution, but there are three additional items for us to prove: that each tableau T is a valid input, that the obtained tableau T' is valid and finally that the statistics are transformed as stated. We take these matters in order.

In the rest of the proof, we will consider rows only, but the columns are treated analogously. Now, the maximal number of rows between the entry and the exit of the path with exit label $S_r(i)$ is $S_r(i) - i - |\{j : \text{ent}(c_j) \le S_r(i)\}|$. Since $\text{inv}_{S_r(i)}$ is bounded from above by this number and from below by zero for both T and T', it is clear that each tableau T is a valid input, and that $\psi(T)$ is a valid tableau.

Since we do not change the exit labels of the rows, except for their order, the value of A_{EN} stays constant. By the argument that $\text{inv}_{S_r(i)}(T')$ stays within the given bounds, it follows that the shape and A_{NE} does not change either. If $A_{\text{EN}} = A_{\text{NE}} = 0$, the total number of inversions for T and T' clearly is $(k-1)(n-k)$, which shows that for this case, $A_{EE} + A_{NN}$ and $C_{EE} + C_{NN}$ must exchange. But since each increase in A_{NE} or A_{EN} decreases the number of inversions equally, the exchange must still hold. $\qquad\square$

If $\Phi(T)$ avoids the patterns 231 and 321 (in classical notation), then the exit labels of the columns in T will be sorted in descending order (no inversions) and $i + S_c(i) + |\{j : \mathrm{ent}(r_j) > S_c(i)\}| = n$. Thus, we get $\mathrm{inv}_{S_c(i)}(\psi(T)) = \mathrm{inv}_{S_c(i)}(T) = 0$. Similarly, $\mathrm{inv}_{S_r(i)}(\psi(T)) = \mathrm{inv}_{S_r(i)}(T) = 0$. Hence, $\psi(T) = T$.

Let the **extended diagonal** be the usual main diagonal followed to the right by the remainder of the lowest row extending past the main diagonal. It is not too hard to deduce the following special cases of the ψ involution.

Corollary 4.2. *For T such that $A_{EN}(T) = A_{NE}(T) = 0$, we get*

$$
\begin{aligned}
z(\mathrm{col}(i, \psi(T))) + z(\mathrm{col}(i, T)) &= \mathrm{inv}(\mathrm{col}(i, \psi(T))) + \mathrm{inv}(\mathrm{col}(i, T)) \\
&= \min(i, \mathrm{wex}(\Phi(T))) - 1,
\end{aligned}
$$

where $\min(i, \mathrm{wex}(\Phi(T))) - 1$ is the number of cells above the extended diagonal in column c_i, and

$$
\begin{aligned}
z(\mathrm{row}(i, \psi(T))) + z(\mathrm{row}(i, T)) &= \mathrm{inv}(\mathrm{row}(i, \psi(T))) + \mathrm{inv}(\mathrm{row}(i, T)) \\
&= n - \max(i, \mathrm{wex}(\Phi(T))),
\end{aligned}
$$

where $n - \max(i, \mathrm{wex}(\Phi(T)))$ is the number of cells to the left of the extended diagonal in row r_{n-i+1}.

Example 4.3. Consider the permutation $\pi = 76813524$ with tableau as below. The number of initial zeroes is zero for exit labels $1, 2, 4$, one for 3 and two for 5. Hence, in $\psi(T)$ we get $z(\mathrm{col}(5)) = 2 - 2 = 0$, $z(\mathrm{col}(4)) = 2 - 0 = 2$, $z(\mathrm{col}(3)) = 2 - 1 = 1$, $z(\mathrm{col}(2)) = 1 - 0 = 1$ and $z(\mathrm{col}(1)) = 0 - 0 = 0$.

$$
T = \quad
\begin{array}{c|c|c|c|c|c|}
 & 4 & 2 & 5 & 3 & 1 \\
\hline
7 & \bullet & \bullet & & & \bullet \\
\hline
6 & & \bullet & & \bullet & \bullet \\
\hline
8 & \bullet & \bullet & \bullet & \bullet & \bullet \\
\hline
\end{array}
\qquad
\psi(T) = \quad
\begin{array}{c|c|c|c|c|c|}
 & 5 & 1 & 3 & 4 & 2 \\
\hline
8 & \bullet & \bullet & & & \\
\hline
6 & & \bullet & \bullet & & \bullet \\
\hline
7 & & \bullet & \bullet & \bullet & \bullet \\
\hline
\end{array}
$$

Corollary 4.4. *Consider any tableau T such that $A_{NN}(T) = A_{EE}(T) = A_{EN}(T) = 0$. Then, the tableau of $\psi(T)$ contains only zeroes and the extended diagonal filled with ones, possibly pushed up by the south-east border.*

Example 4.5. For the permutation 671283945, the tableaux $\mathcal{T} = \Phi(\pi)$ and $\psi(\mathcal{T})$ become

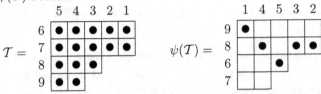

Corollary 4.6. *Consider any tableau* $\mathcal{T}$ *such that* $A_{\mathrm{NN}}(\mathcal{T}) = A_{\mathrm{EE}}(\mathcal{T}) = A_{\mathrm{NE}}(\mathcal{T}) = 0$, *and split it in a lower part consisting of the bottom row, a left part where coordinates* (i,j) *satisfy* $i - j \geq n - 2\lambda_1$ *and* $i < n - \lambda_1$ *and an upper part where* $\lambda_1 + i - j < n - \lambda_1$. *Then,* $\mathrm{inv}(c_i(\psi(\mathcal{T})))$ *equals the number of ones in column* i *in the left part and* $\mathrm{inv}(r_i(\psi(\mathcal{T})))$ *the number of ones in row* i *in the upper part of* $\mathcal{T}$.

Example 4.7. The permutation $\pi = 157923468$ fulfills the conditions of the previous corollary. We take its tableau and split it.

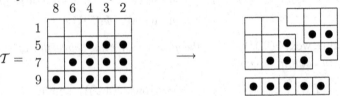

The number of ones in the left part is, from the left, 0, 1, 2 and 1, and the number of ones in the upper part is, from above, 0, 2 and 1. Thus, the permutation becomes 197543628, with two inversions on 5 and one on 7, and similarly for the columns.

5 Tableaux of restricted permutations

In this section, we will enumerate some restricted sets of permutations whose 1-hinge tableaux have the maximum number of essential 1s as a first step in determining the distribution of restricted permutations according to the number of the essential 1s of their associated tableaux. Let M_n denote the set of permutations in $\mathfrak{S}_n$ whose 1-hinge tableaux have $n - 1$ essential 1s (i.e. the maximum number). As we noted in the introduction, not every set of pattern restrictions refines well by the essential 1s statistic, so we find analyze some of the "nicer" cases here.

Definition 5.1. We say that a permutation $\sigma \in \mathfrak{S}_n$ contains a **pattern** τ if τ is a permutation order-isomorphic to a subsequence of σ. If σ does not contain the pattern τ, then we say that σ **avoids** τ. We denote the set of τ-avoiding permutations in $\mathfrak{S}_n$ by $\mathfrak{S}_n(\tau)$. Given a set of patterns T, we let $\mathfrak{S}_n(T)$ be the set of permutations avoiding all patterns in T.

Similarly, given a pattern (or a set of patterns) τ, let $M_n(\tau)$ denote the set of permutations in $\mathfrak{S}_n(\tau)$ whose 1-hinge tableaux have $n - 1$ essential 1s, i.e. a single **doubly essential** 1 (i.e. both the leftmost 1 in its row and the topmost 1 in its column). We then proceed to determine the structure of these tableaux to prove some enumerative results.

We note that, for simplicity, the patterns here are denoted the "old" way, without using the newer generalized-pattern hyphenated notation.

Remark 5.2. Note that any permutation $\pi \in M_n$ must have a 1 in upper left corner of its corresponding 1-hinge tableau. In particular, this implies that $\pi(1) > \pi(n)$. Also, the 1-hinge tableau of π must contain at least one 1 in every row, and hence π does not have fixed points.

Theorem 5.3. *The number of permutations in $M_n(132, 231)$ (resp. in $M_n(213, 312)$) whose 1-hinge tableaux have k rows (resp. k columns) is equal to*

$$2^{k-1} \binom{n - k - 1}{k - 1} - 2^{k-2} \binom{n - k - 2}{k - 2}.$$

Proof. Any permutation in $\pi \in \mathfrak{S}_n(132, 231)$ can be written as $\pi = \pi'1\pi''$, where π' is a decreasing sequence and π'' is an increasing sequence. Thus, either $n = \pi(1)$ or $n = \pi(n)$, so by Remark 5.2, we have $n = \pi(1)$ for $\pi \in M_n(132, 231)$. Therefore, the leftmost column of the 1-hinge tableau $\mathcal{T}$ of π contains a 1 only in the first row. We also conclude that $n - 1 = \pi(2)$ or $n - 1 = \pi(n)$. In addition, since π is a derangement and $n = \pi(1)$, it follows that $\pi(n - 1) < n - 1$, so $n - 1$ is a column label in π.

From this, we can conclude that the first $k < \pi^{-1}(1)$ positions are excedances, and the remaining ones are deficiencies, so the shape of $\mathcal{T}$ is a $k \times (n - k)$ rectangle. Further, $\text{ext}(r_i) > \text{ext}(r_{i+1})$, $\text{ext}(r_k) = \pi(k) > \pi(k + 1) = \text{ext}(c_{n-k})$ and $\text{ext}(c_i) > \text{ext}(c_{i+1})$ for $i \geq m$, where m is the column with exit label 1. We can thus conclude that $z(r_i) = n - \text{ext}(r_i)$, as well as $z(c_i) = \text{ext}(c_i) - 1$ for $i > m$ and zero otherwise. The 1s in the tableau are thus on a northwest-southeast directed band so that no 1 has only 0s in both its row and its column.

To determine such a tableau, we need to determine for each row the leftmost and rightmost position. The leftmost position in r_i is at least one more than in row r_{i-1}, since $z(r_i) = n - \text{ext}(r_i) > n - \text{ext}(r_{i-1}) = z(r_{i-1})$, and the rightmost position changes by at most one, since two adjacent columns cannot share the same number of zeroes to the right of m. Further, the leftmost one of the first row and the rightmost one of the last row are fixed.

Thus, we need to pick a subset of $k - 1$ columns among all columns but the first one to get the leftmost positions in rows 2 to k, and for each of the first $k - 1$ rows, we need to determine if the rightmost 1 should be to the left of the row below or not. The number of such choices is $2^{k-1}\binom{n-k-1}{k-1}$. However, we cannot leave a 1 in the southeast corner with no other 1s above it or to its left, so all such tableaux must be removed. They are counted by $2^{k-2}\binom{n-k-2}{k-2}$, and we are done. $\square$

Theorem 5.4. *If $M_n^k(123, 213)$ is the set of permutations in $M_n(123, 213)$ with k nonessential 1s, then*

$$|M_{2n}^k(123, 213)| = a(2n - 2, k) + a(2n - 3, k),$$
$$|M_{2n+1}^k(123, 213)| = 2a(2n - 2, k) + a(2n - 3, k),$$

where $a(n, k) = $ A037027(n, k), the kth entry in row n of the Fibonacci-Pascal triangle.

Proof. Let $\pi \in M_n^k(123, 213)$ and let $\mathcal{T} = \Phi^{-1}(\pi)$ be its permutation tableau. Simply avoidance of the both pattern implies the following: $n = \pi(1)$ or $n = \pi(2)$, and $\pi(n) = 1$ or $\pi(n) = 2$. Also, if $\pi(1) = n$ and $\pi(n) = 1$, then the top row (labeled 1) and leftmost column (labeled n) of $\mathcal{T}$ both have a single 1 in the top left cell (labeled $(1, n)$). Thus, the cell labeled $(2, n - 1)$ in the second row from the top and next-to-leftmost column is also a doubly essential 1, and hence, π does not have the maximum number of essential 1s. Therefore, either $\pi(2) = n$ and $\pi(n) = 1$, or $\pi(1) = n$ and $\pi(n) = 2$, or $\pi(2) = n$ and $\pi(n) = 2$.

Case 1. Let $n \geq 3$, $\pi(n) = 1$ and $\pi(2) = n$. Since $\pi(n) = 1$, the top row has a single 1 in the leftmost column. Since $\pi(2) = n$, the leftmost column has a 1 in row 2 and no 1s below it. Thus, we have

Let $\mathcal{T}'$ be the tableau obtained by removing the top row and leftmost column of $\mathcal{T}$. Then, for $\mathcal{T}$ to have the maximum number of 1s, $\mathcal{T}'$ must also have the maximum number of essential 1s, and in particular, an essential 1 in the top left corner. Moreover, the nonessential 1s of T' are exactly the nonessential 1s of T. Let $\pi' = \Phi(T')$, then $\pi(1) = \pi'(1) + 1$, $\pi(2) = n$, $\pi(i) = \pi'(i - 1) + 1$ for $3 \leq i \leq n - 1$, and $\pi(n) = 1$. In other words, π' is obtained from π by removing 1 and n and subtracting 1 from each of the remaining values. Hence, $\pi' \in M^k_{n-2}(123, 213)$.

Case 2. Let $n \geq 3$, $\pi(2) = n$ and $\pi(n) = 2$. Since the leftmost column has 1s in the rows 1 and 2, any 1 in row 1 induces a (nonessential) 1 in the cell directly below it. Since $\pi(n) \neq 1$, there is at least one 1 in the top row in addition to the 1 in the leftmost cell. Suppose the second leftmost 1 in the top row is in column labeled i. Then the path starting south at column n turns east at cell $(1, n)$, south at $(1, i)$, and east at $(2, i)$. Since $\pi(n) = 2$, that path exits the tableau at row 2, so there are no 1s in row 2 to the right of column i. Hence, there are also no 1s in row 1 to the right of column i, so $\pi(i) = 1$ and row 1 has, in fact, only two 1s, in the leftmost column and in column i. Let T' be the tableau obtained by removing the top row and leftmost column of T. Then, just as in the previous case, it is easy to see that T has exactly one more nonessential 1 than T', namely, the 1 in the cell labeled $(2, i)$. Moreover, T' also has the maximum number of essential 1s, and $\pi' = \Phi(T')$ is obtained by deleting 2 and n from π and subtracting 1 from each remaining value except 1. Therefore, π' also avoids patterns 123 and 213, and hence, $\pi' \in M^{k-1}_{n-2}(123, 213)$.

Case 3. Let $n \geq 3$, $\pi(1) = n$ and $\pi(n) = 2$. Since $\pi(1) = n$, the leftmost column of T contains a single 1 in the top row. Therefore, the leftmost column cannot be longer than the column to its right, and hence the second leftmost column has label $n - 1$. Since T has the maximum number of essential 1s and 0 in cell $(2, n)$, either $n = 3$ and $\pi = 312$ or there must be a 1 in the cell $(2, n - 1)$. Thus, the path starting south at column n, turns east at cell $(1, n)$, then again south at $(1, n - 1)$, and so passes through cell $(2, n - 1)$. Since $\pi(n) = 2$, this means that T has a 1 in the cell $(2, n - 1)$ and only 0s to its right. Hence, this is the single 1 in row 2, so either $n = 4$ and $\pi = 4312$ or $n \geq 5$ and T must have a 1 in the cell $(3, n - 1)$. Also, the fact that row 2 has no 1s to the right of row $n - 1$ means that row 1 has no 1s to the right of column $n - 1$, so $\pi(n - 1) = 1$. Finally, since $\pi(1) = n$ and π avoids patterns 123 and 213, we must have $n - 1 = \pi(2)$ or $n - 1 = \pi(3)$. Since column n has no 1s other than in top row, $\pi(i) = n - 1$ implies that i is the lowest row with a 1 in column $n - 1$. Since there is a 1 in cell $(3, n - 1)$, we must have $n - 1 = \pi(3)$, so there are no 1s in column $n - 1$ below row

3. Therefore, either $n = 5$ and $\pi = 53412$ or $n \geq 6$ and there is a 1 in cell $(3, n - 2)$.

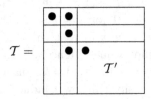

$$\mathcal{T} = \qquad \mathcal{T}'$$

Let $\mathcal{T}'$ be the tableau obtained by deleting the top two rows and leftmost two columns of $\mathcal{T}$, and let $\pi' = \Phi(\mathcal{T}')$. Then $\mathcal{T}'$ has the maximum number of essential 1s and the same number of nonessential 1s as $\mathcal{T}$, and $\pi(1) = n$, $\pi(2) = \pi'(1) + 2$, $\pi(3) = n - 1$, $\pi(i) = \pi(i - 2) + 2$ for $4 \leq i \leq n - 2$, and $\pi(n - 1) = 1$, $\pi(n) = 2$. In other words, π' is obtained from π by deleting $n, n - 1, 1, 2$ at positions $1, 3, n - 1, n$, respectively, and subtracting 2 from each of the remaining values. Therefore, π' also avoids 123 and 213, so $\pi' \in M_{n-4}^k(123, 213)$.

Thus, it is easy to see that $f(n, k) = |M_n^k(123, 213)|$ satisfies the recurrence relation

$$f(n, k) = f(n - 2, k) + f(n - 2, k - 1) + f(n - 4, k) \qquad (1)$$

with initial values $f(0, 0) = 1$, $f(1, 0) = 1$, $f(2, 0) = 1$, $f(3, 0) = 2$, $f(3, 1) = 0$ $f(4, 0) = 2$, $f(4, 1) = 1$, and $f(n, k) = 0$ if $k < 0$ or $k > (n - 2)/2$ for $n \geq 2$. Routine application of generating functions now yields the theorem. $\qquad \square$

Corollary 5.5. $|M_n(123, 213)| = A002965(n + 1)$ *for* $n \geq 0$ *where A002965 is a sequence number from [9].*

Proof. Let $g(n) = |M_n(123, 213)|$. Then $g(n) = \displaystyle\sum_{k=0}^{\lfloor \frac{n-2}{2} \rfloor} f(n, k)$, so summing Equation (1) over k, we get

$$g(n) = 2g(n - 2) + g(n - 4), n \geq 4,$$

and $g(0) = 1$, $g(1) = 1$, $g(2) = 1$, $g(3) = 2$, which implies the first equality in the theorem. $\qquad \square$

We can refine this result differently as follows.

Theorem 5.6. *Let* $M_n(123, 213; k)$ *be the set of permutations in the* set $M_n(123, 213)$ *that start with the letter* k. *Then, using the sequence numbering from [9], for* $1 \leq k \leq n \leq 2k + 3$ *we have*

$$|M_n(123, 213; k)| = 2^{n-k} A002965(2k - n - 3)$$

when $n \geq 4$ *and* $\lfloor n/2 \rfloor + 2 \leq k \leq n$, *while*

$$|M_n(123, 213; k)| = 1$$

when $(n, k) \in \{(1, 1), (2, 2), (3, 3), (3, 2)\}$.

Proof. Let $g(n, k) = |M_n(123, 213; k)|$, and let $g_1(n, k)$, $g_2(n, k)$ and $g_3(n, k)$ be the number of permutations in $M_n(123, 213; k)$ that fall, respectively, into Case 1, 2 or 3 of the proof of Theorem 5.4, so that $g(n, k) = g_1(n, k) + g_2(n, k) + g_3(n, k)$. Then $g_3(n, k) = 0$ for $k < n$, $g_1(n, k) = 0$ for $k = 1, 2, n$, and $g_2(n, k) = 0$ for $k = 1, 2, n$. Moreover, for $k \geq 3$, there is a bijection between the permutations in Case 1 and Case 2, and it is simply the transposition of values 1 and 2. Hence, $g_2(n, k) = g_1(n, k)$ for $k \geq 3$. (Recall also that either 1 or 2 is the rightmost value in Cases 1 and 2.) As in the proof of Theorem 5.4, given a permutation $\pi \in M_n(123, 213; k)$ for $k \geq 3$, we can delete values n and $\pi(n)$ (recall that $\pi(n) = 1$ or $\pi(n) = 2$) and subtract 1 from each value except 1 to obtain a permutation in $M_{n-2}(123, 213; k - 1)$. Hence,

$$g(n, k) = 2g(n - 2, k - 1), \quad 3 \leq k < n.$$

Likewise, if $k = n$, then $g(n, n) = g_3(n, n)$, and there is a bijection $M_n(123, 213; n) \rightarrow M_{n-4}(123, 213)$ defined by deleting values n, $n - 1$, 1 and 2 and subtracting 2 from the remaining values. Since for $\pi \in M_n(123, 213; n)$ we have $n = \pi(1)$, $n - 1 = \pi(3)$, $1 = \pi(n - 1)$ and $2 = \pi(n)$, it follows that $3 \leq \pi(2) \leq n - 2$, so

$$g(n, n) = g_3(n, n) = \sum_{k=3}^{n-2} g(n - 4, k - 2) = \sum_{k=1}^{n-4} g(n - 4, k) = g(n - 4)$$

for $n \geq 4$. Note also that $g(n) = 1$ for $n = 0, 1, 2$, and $g(3) = 2$. Therefore,

$$
\begin{aligned}
g(n, k) &= 2^{n-k} g(n - 2(n - k), k - (n - k)) \\
&= 2^{n-k} g(2k - n, 2k - n) \\
&= 2^{n-k} g(2k - n - 4).
\end{aligned}
$$

Note that $g(n, k) > 0$ if and only if $2k - n - 4 \geq 0$, i.e. if and only if $k \geq \lfloor n/2 \rfloor + 2$. This implies the second equality in the theorem. $\square$

Theorem 5.7. *For any $n \geq 1$,*

$$|M_n(132, 213)| = \begin{cases} 3^m - 1 & \text{if } n = 2m + 1, \ m \geq 0, \\ 2 \cdot 3^m - 1 & \text{if } n = 2m + 2, \ m \geq 0. \end{cases}$$

In fact, if $M_n(132, 213; k)$ is the set of permutations in $M_n(132, 213)$ that start with k, then

$$|M_n(132, 213; k)| = 2^{k-2}$$

when $n = 2m$ or $n = 2m + 1$, $m \geq 0$, $2 \leq k \leq m + 1$, while

$$|M_n(132, 213; m + 2 + k)| = 3^k \cdot 2^{m-1-k}$$

when $n = 2m$ or $n = 2m + 1$, $m \geq 0$, $0 \leq k \leq n - m - 2$.

Proof. Permutations avoiding patterns 132 and 213 are exactly decreasing sequences of increasing blocks, e.g. $\overset{\frown}{567}\overset{\frown}{34}\overset{\frown}{21}$. Such permutations are called *reversed layered permutations* and their blocks (denoted by hats in our example) are called *reversed layers*. It is easy to see that in any such permutation all excedances precede all deficiencies, so the tableau of any $\pi \in S_n(132, 213)$ has no 2s (i.e. its shape is a rectangle). Note also that since $\pi \in M_n(132, 213)$ cannot have any fixed points, it must have at least two reversed layers.

Now we claim that if $\pi \in S_n(132)$ and the tableau of π has rectangular shape, then the lowest 1 in its leftmost column is either in the top row or in the same row as the lowest 1 in the second column. Indeed, if not, then $n - 1$ is to the right of n, which in turn is to the right of $\pi(1)$, so $\pi(1) \neq n - 1$, and hence $(\pi(1), n, n - 1)$ is an occurrence of pattern 132 in π. Moreover, all 1s in the top row (i.e. row 1) of the tableau of π are to the left of all 0s. (Indeed, if not, then there are columns labeled $i < j < k$, such that columns i and k contain two consecutive 1s in the top row, and column j contains a 0 in the top row. But then $\pi(i) < \pi(k) < \pi(j)$, so $(\pi(i), \pi(j), \pi(k))$ is an instance of pattern 132 in π.)

Likewise, if $\pi \in S_n(213)$ and the tableau of π has rectangular shape, then the rightmost 1 in row 1 is either in the leftmost column or in the same column as the rightmost 1 of the second row. (Otherwise, π would contain a subsequence $(2, 1, \pi(n))$ with $\pi(n) \geq 3$.) Similarly, all 1s in the leftmost column of the tableau of π must be above all 0s. (Again, if not, then there are rows labeled $i < j < k$, such that rows i and k contain two consecutive 1s in the leftmost column, and column j contains a 0 in the leftmost column. But then $\pi(j) < \pi(i) < \pi(k)$, so $(\pi(i), \pi(j), \pi(k))$ is an instance of pattern 213 in π.)

Now suppose $\pi \in M_n(132, 213)$. Then we cannot have a single 1 both in the top row and the leftmost column, or else there would be another doubly essential 1 in row 2 and column labeled $n-1$ (i.e. second column from the left).

Furthermore, note that deleting the row 1 and column n (the leftmost column) from the tableau $\Phi(\pi)$ of π is equivalent to deleting values n and 1 from π and subtracting 1 from the remaining entries. It is easy to see that doing so to a reversed layered permutation $\pi \in S_n(132, 213)$ yields another reversed layered permutation $\pi' \in S_{n-2}(132, 213)$. Moreover, it follows from the discussion above that π and π' have the same number of essential 1s unless the 1 in the top left cell is the only one both in its row and in its column. Therefore, $\pi \in M_n(132, 213)$ if and only if $\pi' \in M_{n-2}(132, 213)$ or $\pi' = id_{n-2} = 12\widehat{\ldots n} - 2$ (recall that $\pi' \in M_n$ cannot have a single reversed layer).

Conversely, given a permutation $\pi' \in M_{n-2}(132, 213)$, one can obtain π from π' as follows: add 1 to each value of π' and call the resulting string $\pi' + 1$, then insert 1 either just before the last reversed layer of $\pi' + 1$ (thus extending the last reversed layer) or at the end (thus creating a new last reversed layer), then insert n either at the beginning of $\pi'+1$ (thus creating a new first reversed layer) or at the end of the first reversed layer of $\pi' + 1$ (thus extending the first reversed layer). This yields 3 ways of producing π from $\pi' \neq id_{n-2}$. In two of those cases, we have $\pi(1) = \pi'(1) + 1$, and in the remaining third case, $\pi(1) = n$.

Similarly, if $\pi = id_{n-2}$ then there are only 2 ways to produce $\pi \in M_n$: we must insert n at the beginning and 1 at position 2 to get $\pi = \widehat{n}\, 12\ldots n - 1$, or insert n at position $n - 1$ and 1 at position n to get $\pi = 23\widehat{\ldots n} - 1\widehat{1}$.

Now let $h(n) = |M_n(132, 213)|$ and $h(n, k) = |M_n(132, 213; k)|$. Then $h(n, 1) = 0$ for $n \geq 2$, and for $n \geq 3$

$$\begin{cases} h(n) & = 3h(n-2) + 2, \\ h(n, 2) & = 1, \\ h(n, k) & = 2h(n-2, k-1), \quad 2 < k < n, \\ h(n, n) & = h(n-2) + 1. \end{cases}$$

This, together with the initial conditions $h(2) = 1$, $h(3) = 2$, yields the desired result. $\qquad \square$

References

[1] A. Burstein. On some properties of permutation tableaux. *Ann. Comb.*, 11(3-4):355–368, 2007.

[2] R. Chapman and L. K. Williams. A conjecture of Stanley on alternating permutations. *Electron. J. Combin.*, 14(1):Note 16, 7 pp., 2007.

[3] S. Corteel. Crossings and alignments of permutations. *Adv. in Appl. Math.*, 38(2):149–163, 2007.

[4] S. Corteel and P. Nadeau. Bijections for permutation tableaux. *European J. Combin.*, 30(1):295–310, 2009.

[5] S. Corteel and L. K. Williams. A Markov chain on permutations which projects to the PASEP. *Int. Math. Res. Not. IMRN*, 2007(17):Art. ID rnm055, 27, 2007.

[6] S. Corteel and L. K. Williams. Tableaux combinatorics for the asymmetric exclusion process. *Adv. in Appl. Math.*, 39(3):293–310, 2007.

[7] K. Eriksson. *Strongly convergent games and Coxeter groups.* PhD thesis, KTH Royal Institute of Technology, 1993.

[8] A. Postnikov. Webs in totally positive Grassmann cells. Manuscript, 2001.

[9] N. J. A. Sloane. The On-line Encyclopedia of Integer Sequences. Available online at http://www.research.att.com/~njas/sequences/.

[10] E. Steingrímsson and L. K. Williams. Permutation tableaux and permutation patterns. *J. Combin. Theory Ser. A*, 114(2):211–234, 2007.

[11] L. K. Williams. Enumeration of totally positive Grassmann cells. *Adv. Math.*, 190(2):319–342, 2005.

Enumeration schemes for words avoiding permutations

Lara Pudwell

Department of Mathematics and Computer Science
Valparaiso University
Valparaiso, IN 46383 USA

Abstract

The enumeration of permutation classes has been accomplished with a variety of techniques. One wide-reaching method is that of enumeration schemes, introduced by Zeilberger and extended by Vatter. In this paper we further extend the method of enumeration schemes to words avoiding permutation patterns. The process of finding enumeration schemes is programmable and allows for the automatic enumeration of many classes of pattern-avoiding words.

1 Background

The enumeration of permutation classes has been accomplished by many beautiful techniques. One natural extension of permutation classes is pattern-avoiding words. Our concern in this paper is not attractive methods for counting individual classes, but rather developing a systematic technique for enumerating many classes of words. Four main techniques with wide success exist for the systematic enumeration of permutation classes. These are generating trees, insertion encoding, substitution decomposition, and enumeration schemes. In this paper we adapt the method of enumeration schemes, first introduced for permutations by Zeilberger [9] and extended by Vatter [8] to the case of enumerating pattern-restricted words.

Definition 1.1. Let $[k]^n$ denote the set of words of length n in the alphabet $\{1, \ldots, k\}$, and let $w \in [k]^n, w = w_1 \cdots w_n$. The *reduction* of w, denoted by $red(w)$, is the unique word of length n obtained by replacing the ith smallest entries of w with i, for each i.

For example, the reduction of $w = 2674423$ is $r = red(w) = 1453312$. Notice that r uses *every* letter from $\min(r)=1$ to $\max(r)$, and furthermore $r_i \leq r_j \iff w_i \leq w_j$. We also say that w_1 and w_2 are *order-isomorphic* if $red(w_1) = red(w_2)$.

Definition 1.2. Let $w \in [k]^n, w = w_1 \cdots w_n$ as above, and let $q \in [k]^m, q = q_1 \cdots q_m$. We say that w *contains* q if there exist $1 \leq i_1 < i_2 < \cdots < i_m \leq n$ so that $w_{i_1} \cdots w_{i_m}$ is order-isomorphic to q. Otherwise w *avoids* q.

It is an easy exercise to fix a word w and to list all patterns of length m that w contains. However, it is a much more difficult question to fix a pattern (or set of patterns) q and enumerate all words in $[k]^n$ that avoid q for symbolic n. Our main object of study is:

Definition 1.3. A *frequency vector* is a vector $\mathbf{a} = [a_1, \ldots, a_k]$ such that $k \geq 1$ and $a_i \geq 0$ for $1 \leq i \leq k$. Let $\|\mathbf{a}\| := \sum_{i=1}^{k} a_i$. Then, given a frequency vector $\mathbf{a}$ and a set of reduced words Q in $[k]^m$ for some $m > 0$, we define $A_{\mathbf{a},Q}$ as

$$\{w \in [k]^{\|\mathbf{a}\|} \mid w \text{ avoids } q \text{ for every } q \in Q, w \text{ has } a_i \text{ } i\text{'s for } 1 \leq i \leq k\}.$$

Notice that if $a_1 = \cdots = a_k = 1$, we reduce to the case of counting pattern-avoiding permutations. Also note that if $a_i = 0$ for some i, then we have $A_{\mathbf{a},Q} = A_{\mathbf{a}',Q}$, where $A_{\mathbf{a}'} = [a_1, \ldots, a_{i-1}, a_{i+1}, \ldots, a_k]$. Thus, we may assume that $a_i > 0$ for $1 \leq i \leq k$. When the set of patterns Q is clear from context, we may simply write $A_{\mathbf{a}}$.

2 Previous Work

The study of pattern avoidance in permutations (where neither w nor q have repeated letters) has been well-studied, but less is known for the more general case of words. The ground-breaking work in this area was done by Burstein in his 1998 Ph.D. thesis [3], where he discusses words avoiding sets of permutations and uses generating function techniques to prove his results. Three years later, sets of words avoiding a single pattern of length three were completely classified by Albert, Aldred, Atkinson, Handley, and Holton [1].

Brändén and Mansour [2] were the first to study pattern-avoiding words in a more general context, using finite automata to aid in the enumeration of these objects. Given a fixed alphabet size k, and a forbidden pattern q, they construct a finite automata to enumerate the elements of $A_{\mathbf{a},Q}$. Then, they use the transfer matrix method to find the asymptotics of $A_{\mathbf{a},Q}$. as $\|\mathbf{a}\|$ goes to infinity. This work gave more straightforward proofs of some results that were previously known, as well as a number of new results.

The enumeration techniques discussed thus far are dependent on what patterns are being avoided, or in the case of the automata of Brändén and Mansour, dependent on the size of the alphabet. However, one would hope for a more universal approach that enumerates pattern-avoiding strings independent of the pattern and the alphabet size. In 1998, Zeilberger [9] addressed this problem for pattern-avoiding *permutations* when he introduced prefix enumeration schemes, giving a more universal framework for counting these permutation classes. Unfortunately, this method did not yet have as strong a success rate as hoped for. In 2005, Vatter [8] extended these schemes, completely automating the enumeration of many more permutation classes. Vatter's work studies a symmetry of prefix schemes to ease notation. In 2006, Zeilberger [10] reformulated Vatter's schemes in his original notation of prefix schemes, allowing even quicker enumeration of many permutation classes. In this paper, I extend the notion of Zeilberger and Vatter's prefix schemes to enumerate words avoiding sets of permutations (i.e. pattern q has no repeated letters, but w may), and detail the success rate of this method.

The current work also bears some striking similarities to the *scanning elements method* of Firro and Mansour. Indeed, as in the following sections of this paper, the scanning elements method partitions a set of pattern-avoiding permutations into disjoint subsets depending on the initial letters of the words involved, and then looks for recurrences between these subsets [5]. More explicitly, for any infinite family $P(n)$ of finite subsets of S_n, they define

$$P(n; b_1, b_2, \ldots, b_l) = \{\pi_1 \pi_2 \ldots \pi_n \in P(n) \mid \pi_1 \pi_2 \ldots \pi_l = b_1 b_2 \ldots b_l\}.$$

Then, if there exists a bijection between the set $P(n; b_1, \ldots, b_l)$ and the set $P(n - s; a_1, \ldots a_{l-s})$, the set $P(n - s; a_1, \ldots a_{l-s})$ is said to be the *reduction* of $P(n; b_1, \ldots, b_l)$. The partitioning of $P(n)$ by initial letters and this definition of reduction respectively correspond to the notions of *refinement* and *reversibly deletable elements*, which are two of the three components of an enumeration scheme. Although there is no explicit equivalent to the third component, *gap vectors*, in the scanning elements method, Firro and Mansour make use of similar logic to deduce recurrences counting words which avoid certain forbidden patterns. They go one step further and use the kernel method to solve these recurrences, yielding generating functions for these sets of words when possible [4], [6]. The method of enumeration schemes, while highly similar in structure, has one advantage worth noting: the formal definition of reversibly deletable elements and of gap vectors makes the method of enumeration schemes completely programmable so that the systems of recurrences encoded in an enumeration scheme can be discovered completely algorithmically by computer.

3 Refinement

Assume we want to enumerate a set $A(n)$. If we cannot find a closed formula for $|A(n)|$, then ideally, we want to find a recurrence involving only n. Unfortunately, this is not always possible.

If we cannot find a direct recurrence, following Zeilberger, we introduce the notion of *refinement* as follows. Decompose $A(n)$ as $A(n) = \bigcup_{i \in I} B(n, i)$, so that the $B(n, i)$ are disjoint. Then, if we can find a recurrence for each $B(n, i)$ in terms of the other $B(n, i)$ and $A(n)$, we have a recursive formula for $A(n)$ as well. If not, then refine each $B(n, i)$ as the disjoint union $B(n, i) = \bigcup_{j \in J} C(n, i, j)$, and repeat.

In the case of words, we will use *reduced prefixes* as our refinement parameter.

Definition 3.1. Let $w \in [k]^n$, $w = w_1 \cdots w_n$. The *i-prefix* of w is the word obtained by reducing $w_1 \cdots w_i$, $1 \leq i \leq n$.

For example, if $w = 152243$, the 1-prefix of w is 1, the 2-prefix of w is 12, the 3-prefix of w is 132, the 4-prefix of w is 1322, the 5-prefix of w is 14223, and the 6-prefix of w is 152243. In particular, the n-prefix of w is $red(w)$.

Now, to allow us to talk about sets, we introduce the following notation:

Definition 3.2. $A_{\mathbf{a},Q} (p_1 \cdots p_l) := \{w \in [k]^{\|\mathbf{a}\|} \mid w \text{ avoids } Q, w \text{ has } l\text{-prefix } p_1 \cdots p_l\}$ and

$$A_{\mathbf{a},Q} \begin{pmatrix} p_1 \cdots p_l \\ i_1 \cdots i_l \end{pmatrix} := \{w \in [k]^{\|\mathbf{a}\|} \mid w \text{ avoids } Q, w = i_1 \cdots i_l w_{l+1} \cdots w_n \text{ has}$$

l-prefix $p_1 \cdots p_l\}$

For example:

$$A_{[1,2,1],\{\}}\,(21)$$
$$= \{3122, 3212, 3221, 2123, 2132\},$$

$$A_{[2,1,2],\{\}}\begin{pmatrix}121\\131\end{pmatrix}$$
$$= \{13123, 13132\},$$

$$A_{[2,1,1,1,1],\{\}}\begin{pmatrix}132\\154\end{pmatrix}$$
$$= \{154123, 154132, 154312, 154321, 154213, 154231\}.$$

Thus, for any set of forbidden patterns Q, we have

$$A_{\mathbf{a},Q}(\emptyset)$$
$$= A_{\mathbf{a},Q}(1)$$
$$= A_{\mathbf{a},Q}(12) \cup A_{\mathbf{a},Q}(11) \cup A_{\mathbf{a},Q}(21)$$
$$= (A_{\mathbf{a},Q}(231) \cup A_{\mathbf{a},Q}(121) \cup A_{\mathbf{a},Q}(132) \cup A_{\mathbf{a},Q}(122) \cup A_{\mathbf{a},Q}(123))$$
$$\cup (A_{\mathbf{a},Q}(221) \cup A_{\mathbf{a},Q}(111) \cup A_{\mathbf{a},Q}(112))$$
$$\cup (A_{\mathbf{a},Q}(321) \cup A_{\mathbf{a},Q}(211) \cup A_{\mathbf{a},Q}(312) \cup A_{\mathbf{a},Q}(212) \cup A_{\mathbf{a},Q}(213))$$
$$= \dots.$$

Finally, for ease of notation, we make the following definition:

Definition 3.3. Given a prefix p of length l, the set of *refinements* of p is the set of all prefixes of length $l + 1$ whose l-prefix is p.

For example, the set of refinements of 1 is $\{11, 12, 21\}$. The set of refinements of 11 is $\{221, 111, 112\}$. The set of refinements of 12 is $\{231, 121, 132, 122, 123\}$. This simplifies our notation to the following:
$$A_{\mathbf{a},Q}(p) = \bigcup_{r \in \{\text{refinements of } p\}} A_{\mathbf{a},Q}(r).$$
Now that we have developed a way to partition $A_{\mathbf{a}}$ into disjoint subsets, we investigate methods to find recurrences between these subsets.

4 Reversibly Deletable

Following Zeilberger, we have the following:

Definition 4.1. Given a forbidden pattern q, $p = p_1 \cdots p_l$ an l-prefix, and $1 \le t \le l$, we say that p_t is *reversibly deletable* if every instance of q in a word with prefix p involving p_t implies the presence of an instance of q without p_t.

For example, let $q = 123$ and $p = 21$. Then $w = ij\cdots$ with $i > j$. p_1 is reversibly deletable since the only way for $p_1 = i$ to be involved in a 123 pattern is if $w = ij\cdots a\cdots b\cdots$ with $i < a < b$. But since $i > j$, we have $j < a < b$ as well so jab forms a 123 pattern without using position p_1.

Now, if p_t is reversibly deletable, we have the following recurrence, where $\hat{p}_t$ (resp. $\hat{i}_t$) indicates that the letter p_t (resp. i_t) has been deleted:

$$\left| A_{\mathbf{a}} \begin{pmatrix} p_1 \cdots p_l \\ i_1 \cdots i_l \end{pmatrix} \right| = \left| A_{[a_1,\ldots,a_t-1,\ldots,a_k]} \begin{pmatrix} p_1 \cdots \hat{p}_t \cdots p_l \\ i_1 \cdots \hat{i}_t \cdots i_l \end{pmatrix} \right|$$

That is, deleting and replacing p_t provides a bijection between the two sets.

It should be noted that in Zeilberger's original schemes for permutations, if positions t and s are both reversibly deletable, then

$$\left| A_{\mathbf{a}} \begin{pmatrix} p_1 \cdots p_l \\ i_1 \cdots i_l \end{pmatrix} \right| = \left| A_{[a_1,\ldots,a_t-1,\ldots,a_s-1,\ldots,a_k]} \begin{pmatrix} p_1 \cdots \hat{p}_t \cdots \hat{p}_s \cdots p_l \\ i_1 \cdots \hat{i}_t \cdots \hat{i}_s \cdots i_l \end{pmatrix} \right|$$

However, in the case of pattern-avoiding words, this is no longer true. Consider for example $q = 123$, $p = 11$. Both p_1 and p_2 are reversibly deletable independently, but not together. In a later section, we will revisit the question of when reversibly deletable letters in the same prefix can be deleted at the same time.

5 Gap Vectors

Thus far, knowing only a prefix and a forbidden pattern are enough to determine reversibly deletable positions. However, there are instances when this is not the case. Consider for example $q = 123$ and $p = 12$. With our current definition neither position of the prefix is reversibly deletable. However, observe that if $w \in [k]^n$, $w = ij\cdots$ with $i < j < k$, then k eventually appears in the word w. Thus, $w = ij\cdots k\cdots$ and ijk is a 123 pattern. Since every word with prefix 12 where the letter playing the role of 2 is less than k has a 123 pattern, we know that in any word with prefix 12, the second letter is necessarily k. This is the largest letter in the alphabet, so it cannot be involved in a 123 pattern, so $p_2 = k$ is trivially reversibly deletable.

To help determine the reversibly deletable positions in these more sophisticated cases, we introduce the following:

Definition 5.1. Given a pattern q, a prefix $p = p_1\cdots p_l$, and letters

$i_1 \cdots i_l$ comprising the prefix p, let $s_1 \leq \cdots \leq s_l$ such that $\{s_1, \ldots, s_l\} = \{i_1, \ldots, i_l\}$. We say that $\mathbf{g} = \langle g_1, \ldots, g_{l+1} \rangle$ is a gap vector for $[q, p, [k]^n]$ if there are no words $w \in [k]^n$ avoiding q, with prefix p and with $s_1 - 1 \geq g_1$, $s_j - s_{j-1} \geq g_j$ $(2 \leq j \leq l)$, and $k - s_l \geq g_{l+1}$.

For example, in the case of $q = 123$ and $p = 12$ above, $\mathbf{g} = \langle 0, 1, 1 \rangle$ is a gap vector since the set of all words in $[k]^n$ where $w = i_1 i_2 \cdots$, and $i_1 - 1 \geq 0$, $i_2 - i_1 \geq 1$, and $k - i_2 \geq 1$, (i.e. all words that begin with an increasing pair where $i_2 < k$) is empty. (Otherwise, $w = i_1 i_2 \cdots k \cdots$ contains a 123 pattern, namely $i_1 i_2 k$.)

This definition may at first seem awkward in that the entries at the beginning and end of a vector have a slightly different meaning from the interior entries. An interior 0 denotes a repeated letter in the prefix, while an interior 1 denotes two necessarily adjacent letters. In the convention of Zeilberger and Vatter, we would have used -1 and 0 respectively instead. This change gives one advantage of notation. If prefix $p = p_1 \cdots p_l$ has gap vector $\langle g_1, \ldots, g_{l+1} \rangle$, then prefix $p_1 \cdots \hat{p}_t \cdots p_l$ has gap vector $\langle g_1, \ldots, g_t + g_{t+1}, \ldots, g_{l+1} \rangle$. With the notation of 0s and -1s, further adjustment would need to be made.

Notice that as in Vatter's schemes, if $\mathbf{g}$ and $\mathbf{h}$ are vectors of length l, $g_i < h_i$ $(1 \leq i \leq l)$, and $\mathbf{g}$ is a gap vector for some $[q, p, [k]^n]$, then so is $\mathbf{h}$. In other words, the set of gap vectors for a given pattern, alphabet, and prefix form an upper ideal in the poset of vectors in $\mathbb{N}^l$. Such an ideal can be uniquely denoted by its set of minimal elements. We call this set of minimal elements the basis of the ideal. By the structure of $\mathbb{N}^l$, we are guaranteed a finite basis of gap vectors for $[q, p, [k]^n]$.

6 Enumeration Schemes for Words

We define an *abstract enumeration scheme* S to be a set of triples of the form $[p, G, R]$ where each p is a reduced prefix of length l, G is a (possibly empty) set of vectors of length $l + 1$ and R is a subset of $\{1, \ldots, l\}$. If d is the maximum length of a prefix p in S, we say that S is a scheme of depth d.

Such an enumeration scheme is said to be a *concrete enumeration scheme* if for all triples in S, either R is non-empty or all refinements of p are also in S. Once we have such an enumeration scheme, it can be considered as an encoding of a system of recurrences. The simplest example of such a scheme is

$$S = \{\ [\emptyset, \{\}, \{\}],\ [1, \{\}, \{1\}]\ \}.$$

For prefix $\emptyset$, all refinements, i.e. $\{1\}$, belong to S. For prefix 1, $R \neq \emptyset$.

In fact, this is the scheme for counting all words in $[k]^n$. First note that it is equivalent to count all words or to count all words beginning with a 1 pattern. For words beginning with a 1 pattern, the 1 is trivially reversibly deletable (there is no forbidden pattern to avoid). This gives the following recurrence:

$$
\begin{aligned}
|A_\mathbf{a}(\emptyset)| &= |A_\mathbf{a}(1)| \\
&= \sum_{i=1}^{k} \left| A_\mathbf{a}\binom{1}{i} \right| \\
&= \sum_{i=1}^{k} \left| A_{[a_1,\ldots,a_i-1,\ldots,a_k]}(\emptyset) \right|, \\
|A_{[a_1]}| &= 1.
\end{aligned}
$$

As expected, this gives the unique solution $|A_\mathbf{a}(\emptyset)| = \binom{\|\mathbf{a}\|}{a_1,\ldots,a_k}$.

We have now developed all the necessary tools to completely automatically find concrete enumeration schemes that count pattern-avoiding words in the following way:

(i) Initialize $S := \{[\emptyset, \{\}, \{\}]\}$.

(ii) Let $P = \{$refinements of all prefixes in S with no reversibly deletable elements$\}$.

(iii) For each prefix in P, find its set G_p of gap vectors.

(iv) For each pair $[p, G_p]$, find the set R_p of all reversibly deletable elements, and let and $S2 = \cup_{p \in P}\{[p, G_p, R_p]\}$.

(v) If $R_p \neq \{\}$ for all triples in $S2$, then return $S \cup S2$. Otherwise let $S = S \cup S2$, and return to step 2.

It is clear that steps 1, 2, and 5 can be done completely automatically. In the following sections, we will prove that steps 3 and 4 can be done completely rigorously and automatically as well.

It should be noted that as in the case of permutations, the operations of complement and reversal are involutions on the set of words in $[k]^n$ with some useful properties. Namely, if p is a forbidden pattern, p^c is its complement (formed by replacing $i \rightarrow k+1-i$), and p^r is its reversal, in the notation of section 1, we have:

$$A_{[a_1,\ldots,a_k],\{p\}} = A_{[a_k,\ldots,a_1],\{p^c\}},$$

$$A_{[a_1,\ldots,a_k],\{p\}} = A_{[a_1,\ldots,a_k],\{p^r\}}.$$

Let $Av(p)$ denote the set of all words avoiding p. If, we can find a scheme for $Av(p)$, then we have a system of recurrences for counting $Av(p)$, $Av(p^c)$, $Av(p^r)$, and $Av(p^{rc})$. If we are unable to automatically find a scheme for $Av(Q)$ directly, we may use these natural symmetries, or Wilf equivalences, on patterns to find an equivalent scheme.

7 Finding Gap Vectors Automatically and Rigorously

Recall that for a fixed set of patterns Q and a prefix p of length l, $\mathbf{g}$ is a gap vector if there are no words avoiding Q with prefix p and spacing given by $\mathbf{g}$. Thus, to study gap vectors we consider the following sets:

$A(Q,p,\mathbf{g}) := \{w \in [1+\|\mathbf{g}\|]^n \mid n \geq l,\ w$ avoids Q, w has prefix p, $\{s_1,\ldots,s_l\} = \{w_1,\ldots,w_l\}$ with $s_1 \leq \cdots \leq s_l$, and $s_1 = g_1 + 1$, $s_j = s_{j-1} + g_j\ (j > 1)\}$.

Thus, $A(Q,p,\mathbf{g})$ is the set of all Q-avoiding words with alphabet size $k = 1 + \|\mathbf{g}\|$ whose first l elements form a p pattern composed of the letters $g_1 + 1, g_2 + g_1 + 1, \ldots, g_l + \cdots + g_1 + 1$.

Not all pairs $(p, \mathbf{g})$ result in a non-empty set $A(Q,p,\mathbf{g})$. If the set $A(Q,p,\mathbf{g})$ is empty, then $\mathbf{g}$ is a gap vector for (Q,p).

Denote $G(p) := \{\mathbf{g} \in \mathbb{N}^{l+1} | A(Q,p,\mathbf{g}) \neq \emptyset$ for all $n \geq l\}$. The set $\mathbb{N}^{l+1} \setminus G(p)$ is the same as the set of all gap vectors that was introduced previously. We observed before that the set of gap vectors is an upper ideal in $\mathbb{N}^{l+1}$. Since $\mathbb{N}^{l+1}$ is partially well-ordered, we may define the set of gap vectors in terms of a basis by specifying the minimal elements not in $G(p)$. We are guaranteed, by the poset structure of $\mathbb{N}^{l+1}$ under product order, that this basis is finite.

Now that we are concerned with determining a *finite* set of vectors, two questions remain: (1) How can we determine all gap vectors of a particular norm?, and (2) What is the maximum norm of a gap vector in the basis?

First, following Zeilberger, we may find all gap vectors of a specific norm $k = \|\mathbf{g}\|$ in the following way. Intuitively, a gap vector $\mathbf{g}$ specifies the relative spacing of the initial entries of a word beginning with prefix p. Consider prefix $p = p_1 \ldots p_l$, sorted and reduced to be $s = s_1 \cdots s_l$ and potential gap vector $\mathbf{g} = \langle g_1, \ldots, g_{l+1} \rangle$. This means that there are

g_1 entries smaller than s_1, $max\{0, g_{i+1} - 1\}$ entries between s_i and s_{i+1} (for $2 \le i \le l - 1$), and finally g_{l+1} entries larger than s_l.

Let $\frac{1}{g_1+1}, \frac{2}{g_1+1}, \ldots, \frac{g_1}{g_1+1}$ be the g_1 elements smaller than s_1.

Let $s_i + \frac{1}{g_{i+1}+1}, s_i + \frac{2}{g_{i+1}+1}, \ldots, s_i + \frac{max\{0,g_{i+1}\}}{g_{i+1}+1}$ be the elements between s_i and s_{i+1}, $(2 \le i \le l - 1)$.

Let $s_l + \frac{1}{g_{l+1}+1}, \ldots, s_l + \frac{g_{l+1}}{g_{l+1}+1}$ be the g_{l+1} elements larger than s_l.

Extending the definition of reduction to fractional elements, we may consider all words of length $l + \|\mathbf{g}\|$ which begin with s and end with some permutation of the set of fractional letters above. There are at most $(g_1 + \cdots + g_{l+1})!$ such possibilities. If *each and every* one of these words contains an element of Q, then we know that $\mathbf{g}$ is a gap vector for prefix p since the set of words beginning with p, avoiding Q, and obeying the gap conditions imposed by $\mathbf{g}$ is the empty set.

Now, we have a rigorous way to find all gap vectors of a specific norm, but the question remains: what is the maximum norm of elements in the (finite) basis of gap vectors guaranteed above?

First, it should help to remember how gap vectors are used. The notion of gap vector was introduced to help determine when a particular letter of a word prefix is reversibly deletable. We revisit this concept more rigorously.

For any $r \in [l]$, the set $A(Q, p, \mathbf{g})$ embeds naturally (remove the entry (p_r) and reduce) into $A(Q, d_r(p), d_r(\mathbf{g}))$ where $d_r(p)$ is obtained by deleting the rth entry of p and reducing. $d_r(\mathbf{g})$ is obtained by sorting p, and finding the index i corresponding to p_r, then letting $d_r(\mathbf{g}) = \langle g_1, \ldots, g_{i-1}, g_i + g_{i+1}, g_{i+2}, \ldots, g_{l+1} \rangle$.

Sometimes this embedding of $A(Q, p, \mathbf{g})$ into $A(Q, d_r(p), d_r(\mathbf{g}))$ is a bijection. If this is true for all gap vectors $\mathbf{g}$ that obey $G(p)$, that is, this embedding is a bijection whenever the set $A(Q, p, \mathbf{g})$ is non-empty, then we say that p_r is *reversibly deletable* for p with respect to Q. Notice that this equivalent to the notion of reversibly deletable introduced previously.

Adapting notation from Vatter, we have the following proposition, which puts a bound on the number of gap conditions to check before declaring an element to be reversibly deletable.

Proposition 7.1. *The entry p_r of the prefix p is reversibly deletable if and only if*

$$|A(Q, p, \boldsymbol{g})| = |A(Q, d_r(p), d_r(\boldsymbol{g}))|$$

for all $\mathbf{g} \in G(p)$ with $\|\mathbf{g}\| \leq \|Q\|_\infty + l - 2$, $\|Q\|_\infty$ denotes the maximum length of a pattern in Q, and l is the length of p.

Proof. If p_r is reversibly deletable then the claim follows by definition. To prove the converse, suppose that p_r is not reversibly deletable. We trivially have that $|A(Q, d_r(p), d_r(\mathbf{g}))| \geq |A(Q, p, \mathbf{g})|$, and since p_r is not reversibly deletable, we now have $|A(Q, d_r(p), d_r(\mathbf{g}))| > |A(Q, p, \mathbf{g})|$ for some $\mathbf{g} \in G(p)$. Pick $\mathbf{g} \in G(p)$ and $p^* \in A(Q, d_r(p), d_r(\mathbf{g}))$ so that p^* cannot be obtained from a word in $A(Q, p, \mathbf{g})$ by removing p_r and reducing.

Now, form the Q-containing word p' by incrementing every entry of p^* that is at least p_r by 1 and inserting p_r into position r. p' is the word that would have mapped to p^*, except that p' contains a pattern $\rho \in Q$, and thus is in $A(\emptyset, p, \mathbf{g}) \setminus A(Q, p, \mathbf{g})$.

Now, pick a specific occurrence of $\rho \in Q$ that is contained in p'. Since $p^* = red(p' - p(r))$ avoids Q, this occurrence of ρ must include the entry p_r. Let p'' be the reduction of the subsequence of p' formed by all entries that are either in the chosen occurrence of ρ or in prefix p (or both). p'' is now a word of length $\leq \|Q\|_\infty + l - 1$. Since all gap vectors $\mathbf{g}$ have $\|\mathbf{g}\| = k - 1$ where k is the size of the alphabet, we have that p'' lies in $A(\emptyset, p, \mathbf{h})$ for some $\mathbf{h}$ with $\|\mathbf{h}\| \leq \|Q\|_\infty + l - 2$. On the other hand, $red(p'' - p(r))$ avoids Q, so that $|A(Q, d_r(p), \mathbf{h})| > |A(Q, d_r(p), \mathbf{h})|$, as desired. $\qquad\square$

Although not as sharp as the original bound of $\|\mathbf{g}\| \leq \|Q\|_\infty - 1$ given by Vatter for pattern-avoiding *permutations*, this still gives a bound on the depth of gap vectors that only increases linearly with the depth of the enumeration scheme. Now we have found a completely rigorous way to compute a basis for all gap vectors corresponding to a given prefix p. Finally, we turn our attention to the notion of reversibly deletable elements.

8 Finding Reversibly Deletable Elements Rigorously

Intuitively, to show that p_r is reversibly deletable, we must show that every conceivable forbidden pattern involving p_r implies the presence of another forbidden pattern not involving p_r. For example, in the case of $Q = \{1234\}$ and $p = 123$, for position p_3 we first compute that $\mathbf{g} = \langle 0, 1, 1, 1 \rangle$ is a basis for $G(p)$ with respect to p, thus $p_3 = k$, the largest letter in the word. Since the third (and largest) letter cannot be in a 1234-pattern, p_3 is trivially reversibly deletable.

As a more instructive example, consider $p = 4213$ and $Q = \{43215\}$ and check if p_3 is reversibly deletable. First, we compute that the set of gap vectors for (Q, p) is empty; that is, given any set of letters to append to prefix p, there is always a way to arrange them so that they avoid Q, so we may not rule out scenarios of forbidden patterns using gap vectors. Now, note that the only ways that $p_3 = 1$ can participate in a 43215 pattern is if we have (1) **4213abc**, where $c > 4 > 1 > a > b$, (2) **4213abc** where $c > 2 > 1 > a > b$, or (3) **4213ab** where $b > 4 > 2 > 1 > a$.

Consider the first case. If this happens, then we have 2 letters smaller than "1", and one letter larger than "4", i.e. our word has the gap condition $\langle 2, 0, 0, 0, 1 \rangle$. In this case, form the word $4213abc$ and delete the 1, to obtain $423abc$. In this case $43abc$ forms a 43215 pattern, so p_3 is ok.

Now, consider the case where $21abc$ is our 43215 pattern. Again, we have 2 letters smaller than "1", but our final letter is bigger than "2", so we may have any of the following gap vectors: $\langle 2, 0, 1, 0, 0 \rangle$ (that is, $2 < c < 3$), $\langle 2, 0, 0, 1, 0 \rangle$ (that is, $3 < c < 4$), or $\langle 2, 0, 0, 0, 1 \rangle$ (that is, $c > 4$). Again, we must test all 3 cases, to check for implied instances of 43215. For gap $\langle 2, 0, 1, 0, 0 \rangle$, we have $4213abc$ with $b < a < 1 < 2 < c < 3 < 4$, so $423abc$ reduces to 635214. There is no implied 43215 pattern, so p_3 is not reversibly deletable.

These two examples give the general idea for how to test if a position p_r is reversibly deletable:

(i) List all possible forbidden patterns involving p_r.
(ii) For each possible forbidden pattern involving p_r, list all gap spacings the pattern may have with respect to the prefix p.
(iii) If each gap spacing of the forbidden pattern implies a different instance of a forbidden pattern, then p_r is reversibly deletable. Otherwise, it is not.

Furthermore, if there are non-trivial gap vectors, we may rule out many of the above cases in our computation because the gap vectors imply that the set of all such words with no bad pattern is empty.

Now that we have shown how to completely automatically determine the set of all gap vectors, and the set of reversibly deletable entries of a given prefix, we revisit the notion of independence of reversibly deletable elements. We showed earlier that if both p_r and p_s are reversibly deletable entries of prefix p, we cannot necessarily delete both p_r and p_s. We now show an important case where elements *may* be deleted simultaneously.

Proposition 8.1. *Let S be a concrete enumeration scheme, let p be a prefix in S and let p_r, p_s be reversibly deletable elements of p. If neither $d_r(p)$ nor $d_s(p)$ is a member of S, and p_s is reversibly deletable for some i-prefix of $d_r(p)$ in S, then p_r and p_s may be deleted simultaneously.*

We will denote this embedding (deleting p_r and p_s simultaneously) as $d_{r,s}$.

Proof. Suppose that p, r, s, S are as above. Since $d_r(p)$ is not in S, there must be some i-prefix p^* of $d_r(p)$ in S with a reversibly deletable element p_j^*. Since p_j^* is reversibly deletable for p^*, it is also reversibly deletable for $d_r(p)$ (which begins with p^*), therefore this position is also reversibly deletable. $\qquad\square$

In short, this proposition shows that while we may not always be able to delete more than one prefix entry at a time, when it is necessary to obtain a prefix in scheme S, it can always be done.

We have now shown how to completely rigorously find all components of a concrete enumeration scheme for pattern-avoiding words.

9 The Maple Package mVATTER

The algorithm described above has been programmed in the Maple package mVATTER, available from the author's website. The main functions are SchemeF, MiklosA, MiklosTot, and SipurF.

SchemeF inputs a set of patterns and a maximum depth scheme to search for, and outputs a concrete enumeration scheme for words avoiding q of the specified maximum depth. SchemeF also makes use of the natural symmetries of pattern avoiding words: reversal and complement. If it cannot find a scheme for a set of patterns, it tries to find a scheme for a symmetry-equivalent pattern set and returns that scheme instead.

MiklosA inputs a scheme, a prefix, and a frequency vector and returns the number of words obeying the scheme and the vector, having that prefix. To count *all* words with a specific frequency vector avoiding a specific set of patterns, try MiklosA(SchemeF(Patterns, SchemeDepth), [], [frequency vector]).

MiklosTot inputs a scheme, and positive integers k and n, and outputs the total number of words in $[k]^n$ obeying the scheme.

SipurF takes as its input a list $[L]$, a maximum scheme depth, an integer r, and a list of length r. It outputs all information about schemes for words avoiding one word of each length in L. For example, SipurF([3],2,4,[10,8,6,6]) outputs all information about words avoiding permutation patterns of length 3. It will output the first 10 terms in the sequence of the number of permutations (1 copy of each letter) avoiding a pattern, the first 8 terms in the sequence of the number of words with exactly 2 copies of each letter, and the first 6 terms in the sequences with exactly 3 or 4 copies of each letter.

SipurF has been run on $[L]$ for various lists of the form $[3^a, 4^b]$, and the output is available from the author's website.

10 A Collection of Failures

Although this notion of enumeration schemes for words is successful for many sets of patterns, there is more to be done.

There are many cases where enumeration schemes of Vatter and Zeilberger fail. Unfortunately, the new schemes for words avoiding permutation patterns will necessarily fail whenever Zeilberger and Vatter's schemes fail for the same patterns. Namely, the chain of prefixes with no reversibly deletable elements from the permutation class enumeration scheme will still have no reversibly deletable elements for words, since there are even more possibilities for a bad pattern to occur.

Further, this paradigm only succeeds for avoiding permutations (i.e., the patterns to be avoided have precisely one copy of each letter). The key observation is that gap vectors keep track of spacing, but they do not keep track of frequency. More precisely:

Proposition 10.1. *If $Q = \{\rho\}$ where ρ has a repeated letter, then there is no finite enumeration scheme for words avoiding Q.*

Proof. To show that there is no finite enumeration scheme, we must exhibit a chain of prefixes which have no reversibly deletable elements with respect to Q. Consider the structure of ρ. We have $\rho = q_1 l q_2 l q_3$, where l is the first repeated letter of ρ. Thus, $q_1 l q_2$ is a permutation.

First we consider a simple case. Suppose that $q_1 = \emptyset$. Then, consider the chain of prefixes of the form $p_i = 1 \ldots i$. Consider an occurrence of ρ beginning with j, $1 \leq j \leq i$. Since j is the first repeated letter in forbidden pattern ρ, there is no other letter in p_i which can take its place. Thus we have an infinite chain of prefixes with no reversibly deletable element.

Now, suppose that $|q_1| \geq 1$ and the final letter of q_1 is $> l$. For

$1 \leq i \leq |q_1|$, we let $p_i = (q_1)_1 \cdots (q_1)_i$. Now, for $i > |q_1|$, let $d = i - |q_1|$, and make the following construction:

$$(q_1^*)_i = \begin{cases} (q_1)_i, & \text{if } (q_1)_i < l; \\ (q_1)_i + d & \text{if } (q_1)_i > l; \end{cases}$$

Then, $p_i = (p_i)_1 \cdots (p_i)_i$ where

$$(p_i)_j = \begin{cases} (q_1^*)_i, & \text{if } j \leq |q_1^*|; \\ l + (j - (|q_1^*| + 1)) & \text{if } j > |q_1^*|; \end{cases}$$

In essence, for large i, $p_i = q_1^* l^*$, where l has been replaced by increasing sequence l^*, and all entries of q_1 greater than l are incremented accordingly.

Viewing a prefix as a function from $\{1, \ldots, i\}$ to $\{1, \ldots, i\}$, the prefixes p_i of length $|q_1| + 1$ and $|q_1| + 4$ are displayed below as an example.

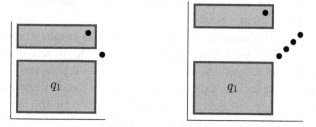

Now, consider the occurrence of forbidden pattern ρ that uses element j of the monotone run at the end of p_i as l. Since this is the first repeated letter in the pattern, no matter how ρ occurs in the word, the role of l must be played by j. Thus there are no reversibly deletable entries of p_i.

For the remaining case: $|q_1| \geq 1$ and the final letter of q_1 is smaller than l, repeat the construction above, but with a decreasing run instead of an increasing run at the end of p_i for large i. Again, for each letter in p_i, we can pick an occurrence of ρ that demonstrates that letter is not reversibly deletable.

$\square$

Proposition 10.1 raises the question whether there is yet another way to extend schemes. Recall that in this paper, we have modified Zeilberger's original schemes which use prefixes for refinement. On the other hand, Vatter took symmetries and refined by the patterns formed by the smallest entries of a permutation. In the study of restricted permutations, these two notions are equivalent, but for words, this is no longer true. Indeed, in Vatter's notation, if we refine by adding one letter at a time, the repeated letters in words cause $1 \to 11 \to 111 \to \cdots$ to often be an infinite chain in schemes for pattern-avoiding words. Ideally we would like to find schemes that do not depend on the alphabet size or on specifying frequency of letters. One way to circumvent this difficulty is to refine words by adding multiple letters at a time. Enumeration results obtained in this manner are addressed in [7].

11 Examples and Successes

Despite the holes for future progress discussed in section 9, prefix enumeration schemes for words have a reasonable success rate, especially when avoiding *sets* of permutation patterns. This method is 100% successful when avoiding sets of patterns of length 3, and enjoys fairly high success when avoiding sets of 3 or more patterns simultaneously. Some of the nicer results are displayed below.

We draw an enumeration scheme as a directed graph, where the vertices are prefixes. A solid arrow goes from a prefix to any of its refinements. A dotted arrow, denoting reversibly deletable letters, goes from a prefix to one of its i-prefixes, and is labeled by the corresponding deletion map. If there are any gap vectors for a given prefix, the basis for those gap vectors is written below that prefix.

Many of the enumeration schemes for permutations carry over to enumerating words almost directly. Some simple examples include the scheme for counting all words, and the scheme for counting words avoiding the pattern 12.

The first non-trivial examples are schemes for avoiding one pattern of length 3. These schemes are nearly identical to the permutation schemes, only with the 11 prefix now included. The symmetry of these schemes gives an alternate explanation that $Av(123)$ and $Av(132)$ are Wilf-equivalent for words as well as for permutations.

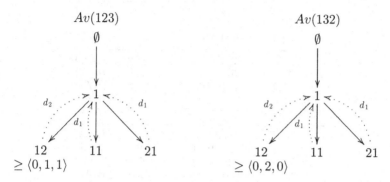

A more interesting example is:

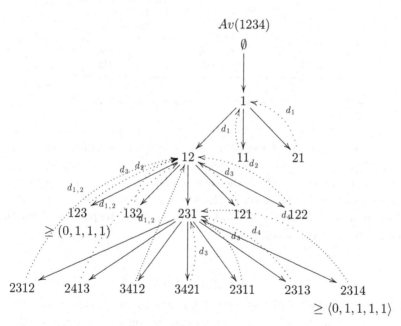

We conclude this section with Table 1, which shows statistics comparing the success rate of Vatter and Zeilberger's schemes for permutations versus the success rate of schemes for words. As discussed above, the current success of word schemes is bounded above by the success of permutation schemes. We consider success rate to be the percentage of trivial Wilf classes of patterns which can be enumerated via schemes.

Pattern Lengths to Avoid	Permutation Scheme Success	Word Scheme Success
[2]	1/1 (100%)	1/1 (100%)
[2,3]	1/1 (100%)	1/1 (100%)
[2,4]	1/1 (100%)	1/1 (100%)
[3]	2/2 (100%)	2/2 (100%)
[3,3]	5/5 (100%)	6/6 (100%)
[3,3,3]	5/5 (100%)	6/6 (100%)
[3,3,3,3]	5/5 (100%)	6/6 (100%)
[3,3,3,3,3]	2/2 (100%)	2/2 (100%)
[4]	2/7 (28.6%)	2/8 (25%)
[3,4]	17/18 (94.4%)	9/24 (37.5%)
[3,3,4]	23/23 (100%)	27/31 (87.1%)
[3,3,3,4]	16/16 (100%)	20/20 (100%)
[3,3,3,3,4]	6/6 (100%)	6/6 (100%)
[3,3,3,3,3,4]	1/1 (100%)	1/1 (100%)
[4,4]	29/56 (51.8%)	?/84 (in process)
[3,4,4]	92/92 (100%)	38/146 (26%)
[3,3,4,4]	68/68 (100%)	89/103 (86.4%)
[3,3,3,4,4]	23/23 (100%)	29/29 (100%)
[3,3,3,3,4,4]	3/3 (100%)	3/3 (100%)

Table 1. *The success rate for permutations versus that for words.*

Notice that there are fewer Wilf classes when enumerating permutations, since the operations of reverse, complement, and inverse all give trivial equivalences, while in the case of words, inverses no longer exist. Finally, we use the notation of the program SipurF, described above. For example, avoiding the list $[3, 4]$ means to avoid one pattern of length 3 and one pattern of length 4.

12 Future Work

This modification of Zeilberger and Vatter's enumeration schemes provides the beginning of a universal method for counting pattern-avoiding words. The success rate is quite good for avoiding sets of patterns, and has reasonable success for avoiding single patterns. The following problems remain open:

- Find new techniques for enumerating classes of pattern-avoiding words not counted by prefix schemes or by the scanning elements method.

- Find ways to simplify schemes to compute even more values in the sequence of the number of pattern-avoiding words for fixed k and q.
- Find ways to convert concrete enumeration schemes to closed forms or generating functions when possible. In some cases, the kernel method, as utilized in the work of Firro and Mansour, should apply.

13 Acknowledgements

Special thanks to Andrew Baxter, Eric Rowland, and Doron Zeilberger for many helpful comments during the writing of this paper. Also, thank you to two anonymous referees for a number of valuable suggestions to improve the manuscript.

References

[1] M. H. Albert, R. E. L. Aldred, M. D. Atkinson, C. Handley, and D. Holton. Permutations of a multiset avoiding permutations of length 3. *European J. Combin.*, 22(8):1021–1031, 2001.

[2] P. Brändén and T. Mansour. Finite automata and pattern avoidance in words. *J. Combin. Theory Ser. A*, 110(1):127–145, 2005.

[3] A. Burstein. *Enumeration of Words with Forbidden Patterns*. PhD thesis, University of Pennsylvania, 1998.

[4] G. Firro. *Distanced patterns*. PhD thesis, University of Haifa, 2007.

[5] G. Firro and T. Mansour. Three-letter-pattern-avoiding permutations and functional equations. *Electron. J. Combin.*, 13(1):Research Paper 51, 14 pp., 2006.

[6] G. Firro and T. Mansour. Restricted k-ary words and functional equations. *Discrete Appl. Math.*, 157(4):602–616, 2009.

[7] L. Pudwell. Enumeration schemes for words avoiding patterns with repeated letters. *Integers*, 8:A40, 19, 2008.

[8] V. Vatter. Enumeration schemes for restricted permutations. *Combin. Probab. Comput.*, 17:137–159, 2008.

[9] D. Zeilberger. Enumeration schemes and, more importantly, their automatic generation. *Ann. Comb.*, 2(2):185–195, 1998.

[10] D. Zeilberger. On Vince Vatter's brilliant extension of Doron Zeilberger's enumeration schemes for Herb Wilf's classes. *The Personal Journal of Ekhad and Zeilberger*, published electronically at http://www.math.rutgers.edu/~zeilberg/pj.html, 2006.

The lexicographic first occurrence of a I-II-III-pattern

Torey Burton, Anant P. Godbole, Brett M. Kindle

Department of Mathematics
East Tennessee State University
Johnson City, TN 37614 USA

Abstract

Consider a random permutation $\pi \in \mathcal{S}_n$, which we associate with a sequence $\{Y_i\}_{i=1}^n$ of independent and identically distributed uniform random variables. In this paper, perhaps best classified as a contribution to discrete probability distribution theory, we study the *first* occurrence $X = X_n$ of a I-II-III-pattern, where "first" is interpreted in the lexicographic order induced by the 3-subsets of $[n] = \{1, 2, \ldots, n\}$. Of course if the permutation is I-II-III-avoiding then the first I-II-III-pattern never occurs; to address this case, we also study the first occurrence of a I-II-III-pattern given an infinite sequence $\{Y_i\}_{i=1}^\infty$ of uniform random variables (or any other sequence of random variables with a non-atomic distribution).

1 Introduction

Consider a random permutation $\pi \in \mathcal{S}_n$. In this short note, perhaps best classified as a contribution to discrete probability distribution theory, we study the *first* occurrence $X = X_n$ of a I-II-III-pattern, defined as follows: Order the 3-subsets of $[n] = \{1, 2, \ldots, n\}$ in the "obvious" lexicographic fashion

$$\{1, 2, 3\} < \{1, 2, 4\} < \{1, 2, 5\} < \ldots \{1, 2, n\} < \{1, 3, 4\} < \{1, 3, 5\} < \ldots$$

$$< \{1, n-1, n\} < \{2, 3, 4\} < \ldots \{n-2, n-1, n\}.$$

We say that the first I-II-III-pattern occurs at $\{a, b, c\}$ if $\pi(a) < \pi(b) < \pi(c)$ and if $\pi(d) < \pi(e) < \pi(f)$ does not hold for any $\{d, e, f\} < \{a, b, c\}$.

For example, if $\pi = (4,3,6,1,2,5,8,7)$, then the first ascending 3-sequence is $(4,6,8)$ and thus $X = \{1,3,7\}$. The probability model we adopt to model this situation, for reasons that will become clear, is the following: Consider a sequence $\{Y_i\}_{i=1}^{\infty}$ of independent and identically distributed (i.i.d.) random variables, each with the uniform distribution on $[0,1]$. Alternately we may use any other sequence of i.i.d. random variables with a non atomic distribution. The non-atomicity of the random variables guarantees that $\mathbb{P}(Y_i = Y_j; i \neq j) = 0$. The values of the sequence $\{Y_i\}_{i=1}^{n}$ are used to determine the random permutation on $[n]$; if Y_i is the jth largest member of the sequence, we set $\pi(i) = j$. It is clear that $\mathbb{P}(\pi = \delta) = 1/n!$ for each $\delta \in S_n$, as must be the case. (For example, if $Y_2 = 0.3, Y_4 = 0.4, Y_5 = 0.6, Y_1 = 0.7$ and $Y_3 = 0.9$ are the ordered values of the Ys, then we consider the sequence $(0.7, 0.3, 0.9, 0.4, 0.6)$ which leads to $\pi = (41523)$, and thus $X = 245$. Also $\{2,4,5\}$ is the first set $\{a,b,c\}$ for which $Y_a < Y_b < Y_c$) The advantage of this model is that it allows us to additionally consider infinite sequences $\{Y_i\}_{i=1}^{\infty}$, which becomes necessary if the permutation π on $[n]$ is I-II-III-avoiding, which occurs (see [1]) with probability $\binom{2n}{n}/(n+1)!$, and in which case the "first" I-II-III-pattern never occurs. This cannot happen, as we shall see, for an infinite sequence. It is typically impossible in the infinite case, however, to recover the bijection $f : \mathbf{Z}^+ \to \mathbf{Z}^+$ that represents the ranked values of the Y_is, but we shall see that the distribution of X *may still* be derived.

In what follows, we will state our results, as convenient, both in terms of the random permutation and the underlying uniform random variables. Also, we will use the notation $X = abc$ as short for the event $\{X = \{a,b,c\}\}$, and refer to the case of an infinite sequence of random variables as the $n = \infty$ case.

2 Results

Proposition 2.1. *For each $n \leq \infty$,*

$$\mathbb{P}(X = 12r) = \frac{1}{r-1} - \frac{1}{r}.$$

Proof. Let the set $\{\pi(1), \pi(2), \ldots \pi(r)\}$ be ordered increasingly as $x_1 < x_2 < \ldots < x_r$. Then, if $\pi(2) = x_{r-1}$ and $\pi(r) = x_r$, we clearly have $X = 12r$. We claim the converse is true as well, and next provide the proof of this fact in detail since the key idea will be used in later results as well. Assume that the second largest of $\{\pi(1), \pi(2), \ldots \pi(r)\}$ is not in the

First I-II-III Pattern	Probability	Cumulative Probability
123	120/720	0.1666
124	60/720	0.2500
125	36/720	0.3000
126	24/720	0.3333
134	50/720	0.4028
135	28/720	0.4417
136	18/720	0.4667
145	26/720	0.5028
146	16/720	0.5250
156	16/720	0.5472
234	48/720	0.6139
235	22/720	0.6444
236	12/720	0.6611
245	24/720	0.6944
246	12/720	0.7111
256	14/720	0.7306
345	24/720	0.7639
346	10/720	0.7778
356	14/720	0.7972
456	14/720	0.8167
I-II-III avoiding	132/720	1.0000

Table 1. *The First Occurrence of a I-II-III Pattern when $n = 6$.*

2nd spot, then either $\pi(2) = x_r$, or $\pi(2) < x_{r-1}$. In the former case, the only way that we can have $X = 12s$ is with $s > r$. If $\pi(2) < x_{r-1}$, there are two possibilities: If $\pi(1) < \pi(2)$, then $X = 12s$ for some $s < r$, and if $\pi(1) > \pi(2)$ then $X \neq 12s$ for any s. Thus we must have $\pi(2) = x_{r-1}$. This forces $\pi(r) = x_r$, and thus $\mathbb{P}(X = 12r) = \frac{(r-2)!}{r!} = \frac{1}{r-1} - \frac{1}{r}$, as desired. □

Proposition 2.2. *If $n = \infty$, then $X = 1sr$ for some $2 \leq s < r$ with probability one.*

Proof. We have $Y_1 < 1$ with probability one. Also with probability one there are infinitely many js such that $Y_1 < Y_j < 1$; let s be the smallest such j. Finally let r be the smallest integer larger than s for which $Y_1 < Y_s < Y_r$. □

Our ultimate goal is to try to determine the entire probability distribution of X; for $n = 6$, for example, we can check that the ensemble $\{\mathbb{P}(X = abc) : 1 \leq a < b < c \leq 6\}$ is as follows:

Recall that the *median* of any random variable X is any number m such that $\mathbb{P}(X \leq m) \geq 1/2$ and $\mathbb{P}(X \geq m) \geq 1/2$. Now Propositions 1 and 2 together reveal that for $n = \infty$,

$$\mathbb{P}(X \leq 134) \geq \sum_{r=3}^{\infty} \frac{1}{r-1} - \frac{1}{r} = \frac{1}{2}$$

and

$$\mathbb{P}(X \geq 134) = 1 - \sum_{r=3}^{\infty} \frac{1}{r-1} - \frac{1}{r} = \frac{1}{2},$$

which shows that X has 134 as its unique median. For finite n, however, the median is larger – Table 1 reveals, for example, that $m = 145$ for $n = 6$.

Proposition 2.3. *For $n = \infty$,*

$$\mathbb{P}(X = 1sr) = \frac{1}{(s-1)(r-1)r} = \frac{1}{s-1}\left(\frac{1}{r-1} - \frac{1}{r}\right).$$

Proof. $X = 1sr$ iff $\{1, s, r\}$ is the first lexicographically ordered 3-subset $\{a, b, c\}$ for which $Y_a < Y_b < Y_c$. Now it can easily be proved, as in Proposition 1 and keeping in mind that $n = \infty$, that this occurs iff (i) $Y_s = x_{r-1}$ and $Y_r = x_r$ are the second largest and largest among the first r random variables $Y_1, \ldots, Y_r$; and (ii) $Y_1 = \max_{1 \leq j \leq s-1} \pi(j)$. It now follows that

$$\mathbb{P}(X = 1sr) = \frac{\binom{r-2}{s-1}(s-2)!(r-s-1)!}{r!} = \frac{1}{(s-1)(r-1)r},$$

as claimed. $\square$

Proposition 3 provides us with the entire distribution of X when $n = \infty$; note that

$$\sum_{s=2}^{\infty} \sum_{r=s+1}^{\infty} \frac{1}{s-1}\left(\frac{1}{r-1} - \frac{1}{r}\right) = \sum_{s=2}^{\infty} \frac{1}{s-1} - \frac{1}{s} = 1.$$

The probability of the first I-II-III pattern occurring at positions $12r$ is the same for all $n \leq \infty$, noting, though, that for finite n, $\sum_r \mathbb{P}(X = 12r) = \frac{1}{2} - \frac{1}{n}$. There is, however, a subtle and fundamental difference in general between $\mathbb{P}(X = 1sr), s \geq 3$, when $n = \infty$ and when n is finite. We illustrate this fact for $\mathbb{P}(X = 13r)$ when $n < \infty$. Recall from the proof of Proposition 3 that for X to equal $13r$ in the infinite case, we had to have $\pi(3) = x_{r-1}, \pi(r) = x_r$, and $\pi(2) < \pi(1)$. The above scenario *will still*, in the finite case, cause the first I-II-III pattern to occur at positions $13r$, but there is another case to consider. If $n = \pi(2) > \pi(1)$

then it is *impossible* for X to equal $12s$ for any s; in this case we must have $\pi(3) = x_{r-2}$ and $\pi(r) = x_{r-1}$. The probability of this second scenario is

$$\frac{(r-3)!}{n(r-1)!} = \frac{1}{n(r-2)(r-1)}.$$

Adding, we see that

$$\mathbb{P}(X = 13r) = \frac{1}{2}\left(\frac{1}{r-1} - \frac{1}{r}\right) + \frac{1}{n}\left(\frac{1}{r-2} - \frac{1}{r-1}\right),$$

and, in contrast to the $n = \infty$ case where the net contribution of $\mathbb{P}(X = 13r; r \geq 4)$ was $1/6$, we have

$$\sum_{r=4}^{n} \mathbb{P}(X = 13r) = \frac{1}{2}\sum_{r=4}^{n}\left(\frac{1}{r-1} - \frac{1}{r}\right) + \frac{1}{n}\sum_{r=4}^{n}\left(\frac{1}{r-2} - \frac{1}{r-1}\right)$$

$$= \frac{1}{6} - \frac{1}{n(n-1)}.$$

The above example illustrates a general fact:

Theorem 2.4. *For finite n,*

$$\mathbb{P}(X = 1sr) = \sum_{k=0}^{s-2} \frac{\binom{s-2}{k}\binom{r-k-2}{s-k-1}(s-k-2)!(r-s-1)!}{n(n-1)\ldots(n-k+1)(r-k)!}. \tag{1}$$

Proof. We may have k of the quantities $\pi(2), \pi(3), \ldots, \pi(s-1)$ being greater than $\pi(1)$, where k ranges from 0 to $s-2$. In this case, however, these πs must equal, from left to right, $(n, n-1, \ldots, n-k+1)$. This makes it impossible for $X = 1jk; 2 \leq j \leq s-1; k \geq s$. Arguing as before, we must have $\pi(s) = x_{r-1-k}$ and $\pi(r) = x_{r-k}$. We next compute the probability $\mathbb{P}(X = 1sr)$. The denominator of (1) represents the number of ways we may arrange k specific integers from $[n]$, followed by an ordering of the unspecified integers in the remaining $r - k$ positions. Consider the numerator. Here we pick k of the $s-2$ spots in $\{2, 3, \ldots, s-1\}$ to be occupied, from left to right, by $n, n-1, \ldots, n-k+1$. The largest and second largest of the remaining $r - k$ numbers, no matter what these are, must occupy positions r and s respectively. Of the remaining $r - k - 2$ numbers, we choose $s - k - 1$ to go to the left of s, place the largest of these in spot 1, and arrange the $s - k - 2$ others in $(s - k - 2)!$ ways, doing the same with the $r - s - 1$ numbers to the right of s. $\qquad\square$

Unlike the infinite case, $\sum_s \sum_r \mathbb{P}(X = 1sr) \neq 1$. So how much *is* $\mathbb{P}(X \geq 234)$, or alternatively, how close is

$$\mathbb{P}(X = 1sr) = \sum_{s \geq 2} \sum_{r \geq s+1} \sum_{k=0}^{s-2} \frac{\binom{s-2}{k}\binom{r-k-2}{s-k-1}(s-k-2)!(r-s-1)!}{n(n-1)\ldots(n-k+1)(r-k)!}$$

to unity?

We obtain the answer in closed form as follows: Conditioning on $\pi(1)$, we see that $X \geq 234$ iff for $j = n, n-1, \ldots, 1$, $\pi(1) = j$, and the integers $n, n-1, \ldots, j+1$ appear from left to right in π. Summing the corresponding probabilities $1/n$, $1/n$, $1/(2!n)$, $1/(3!n)$, etc yields

Proposition 2.5.

$$\mathbb{P}(X \geq 234) \sim \frac{e}{n}.$$

3 Open Problems

- Lexicographic ordering is not our only option; in fact it is somewhat unnatural. Consider another possibility: What is

$$\inf\{k : \text{there is a } I - II - III \text{ pattern in } (\pi(1), \ldots, \pi(k))\}?$$

This question is not too hard to answer from Stanley-Wilf theory. Since

$$\mathbb{P}((\pi(1), \ldots, \pi(k)) \text{ is } I - II - III \text{ free}) = \frac{\binom{2k}{k}}{(k+1)!},$$

the probability that k is the first integer for which $(\pi(1), \ldots, \pi(k))$ contains a I-II-III pattern is

$$\frac{\binom{2k-2}{k-1}}{k!} - \frac{\binom{2k}{k}}{(k+1)!}.$$

A more interesting question is the following: Conditional on the fact that first I-II-III pattern occurs only after the kth "spot" is revealed, what is the distribution of the first 3-subset, interpreted in the sense of this paper, that causes this to occur? For example, if we let $n = 6$, and are told that the first k for which there is a I-II-II pattern in $(\pi(1), \ldots, \pi(k))$ is 5, what is the chance that the first set that causes this to happen is $\{1, 2, 5\}, \{1, 3, 5\}, \{1, 4, 5\}, \{2, 3, 5\}, \{2, 4, 5\}$ or $\{3, 4, 5\}$?

- Can the results of this paper be readily generalized to other patterns of length 3? To patterns of length 4?

- Theorem 4 and Proposition 5 fall short of providing the exact probabilities $\mathbb{P}(X = rst)$ for $r \geq 2$. Can these admittedly small probabilities be computed exactly or to a high degree of precision?
- Does the distribution of X consist, as it does for $n = 6$, of a series of decreasing segments with the initial probability of segment $j + 1$ no smaller than the final probability of segment j?

Acknowledgments The research of the second named author was supported by NSF Grants DMS-0139286 and DMS-0552730. We thank the anonymous referee for several valuable suggestions, particularly that we model both finite and infinite permutations through a sequence of i.i.d. variables with non-atomic distribution.

References

[1] M. Bóna. *Combinatorics of permutations.* Discrete Mathematics and its Applications (Boca Raton). Chapman & Hall/CRC, Boca Raton, FL, 2004.

Enumeration of partitions by rises, levels and descents

Toufik Mansour

Department of Mathematics
Haifa University
31905 Haifa, Israel

Augustine O. Munagi

The John Knopfmacher Centre for Applicable Analysis and Number Theory
School of Mathematics
University of the Witwatersrand
Johannesburg 2050, South Africa

Abstract

A *descent* in a permutation $\alpha_1 \alpha_2 \cdots \alpha_n$ is an index i for which $\alpha_i > \alpha_{i+1}$. The number of descents in a permutation is a classical permutation statistic which was first studied by P. A. MacMahon almost a hundred years ago, and it still plays an important role in the study of permutations. Representing set partitions by equivalent canonical sequences of integers, we study this statistic among the set partitions, as well as the numbers of *rises* and *levels*. We enumerate set partitions with respect to these statistics by means of generating functions, and present some combinatorial proofs. Applications are obtained to new combinatorial results and previously-known ones.

1 Introduction

A *descent* in a permutation $\alpha = \alpha_1 \alpha_2 \cdots \alpha_n$ is an index i for which $\alpha_i > \alpha_{i+1}$. The number of descents in a permutation is a classical permutation statistic. This statistic was first studied by MacMahon [6], and it still plays an important role in the study of permutation statistics. In this paper we study the statistics of numbers of *rises*, *levels* and

221

descents among set partitions expressed as canonical sequences, defined below.

A *partition* of $[n] = \{1, 2, \ldots, n\}$ is a (finite) collection $B_1, B_2, \ldots$ of nonempty disjoint sets, called *blocks*, whose union is $[n]$. We will assume that the blocks are listed in the increasing order of their minimum elements, that is, $\min B_1 < \min B_2 < \cdots$. We represent a partition $\pi = B_1, B_2, \ldots$ in the canonical sequential form $\pi = \pi_1 \pi_2 \cdots \pi_n$ such that $j \in B_{\pi_j}, 1 \leq j \leq n$. For instance, 1231242 is the canonical sequential form of the partition $\{1, 4\}, \{2, 5, 7\}, \{3\}, \{6\}$ of $[7]$. The set of partitions of $[n]$ is denoted by P_n.

Note that a sequence π over the alphabet $[d]$ represents a partition of $[n]$ with $[d]$ blocks if and only if it has the following properties:

- Each number from the set $[d]$ appears at least once in π.
- For each i, j such that $1 \leq i < j \leq d$, the first occurrence of i precedes the first occurrence of j.

We remark that sequences satisfying these properties are also known as *restricted growth functions*, and they are often encountered in the study of set partitions [3, 8, 10] as well as other related topics, such as Davenport-Schinzel sequences [1, 4, 5, 7].

Throughout this paper, we identify a set partition with the corresponding canonical sequence, and employ this representation to define patterns among set partitions. Let $\pi = \pi_1 \pi_2 \cdots \pi_n$ be any partition represented by its canonical sequence. We say that π has a *rise, level*, and *descent* at i if $\pi_i < \pi_{i+1}$, $\pi_i = \pi_{i+1}$, and $\pi_i > \pi_{i+1}$, respectively. For example, the partition 12311242 of [8] has four rises (at $i = 1$, $i = 2$, $i = 5$ and $i = 6$), one level (at $i = 4$), and two descents (at $i = 3$ and $i = 7$).

For any partition π, we denote the number of blocks, rises, levels, and descents by blocks(π), rises(π), levels(π), and descents(π), respectively. We define the generating function for the number of partitions of $[n]$ specifying the number of blocks, rises, levels, and descents as

$$
\begin{aligned}
P(x, y) &= P(x, y; r, \ell, d) \\
&= \sum_{n \geq 0} \sum_{\pi \in P_n} x^n y^{\text{blocks}(\pi)} r^{\text{rises}(\pi)} \ell^{\text{levels}(\pi)} d^{\text{descents}(\pi)}.
\end{aligned}
$$

Our main result can be formulated as follows.

Theorem 1.1. *We have*

$$P(x, y; r, \ell, d) = 1 + \sum_{i \geq 0} \frac{x^i y^i r^{i-1} (r - d)^i}{(1 + x(d - \ell))^i \prod_{j=1}^{i} \left(r - d \left(\frac{1 + x(r - \ell)}{1 + x(d - \ell)} \right)^j \right)}.$$

Moreover, the generating function for the number of partitions of $[n]$ according to number of rises, levels, and descents with exactly $k \geq 1$ blocks is given by

$$\frac{x^k r^{k-1} (r - d)^k}{(1 + x(d - \ell))^k \prod_{j=1}^{k} \left(r - d \left(\frac{1 + x(r - \ell)}{1 + x(d - \ell)} \right)^j \right)}.$$

The following result (Proposition 2.4) is crucial to the proof of the theorem.

If a sequence f_ℓ, is defined by $f_\ell = a_\ell + b_\ell \sum_{j \geq \ell} f_j$, $\ell = 0, 1, \ldots$, then $f = \sum_{\ell \geq 0} f_\ell$ is given by

$$f = \sum_{i \geq 0} \frac{a_i}{\prod_{j=0}^{i} (1 - b_j)}.$$

We state some combinatorial applications of the main theorem in the form of corollaries, and present some direct proofs.

2 Proof of Theorem 1.1

In order to find the two generating functions, we derive recursions for partitions that start with a specific subword. Thus, we define a second set of generating functions

$$\begin{aligned} P(s_1 \cdots s_i | x, y) &= P(s_1 \cdots s_i | x, y; r, \ell, d) \\ &= \sum_{n \geq 0} \sum_{\pi} x^n y^{\text{blocks}(\pi)} r^{\text{rises}(\pi)} \ell^{\text{levels}(\pi)} d^{\text{descents}(\pi)}, \end{aligned}$$

where the sum on the right side of the equation is over all the partitions $\pi \in P_n$ such that π starts with $s_1 s_2 \cdots s_i$, that is, $\pi_1 \pi_2 \cdots \pi_i = s_1 s_2 \ldots s_i$, $i \geq 0$. We denote the generating function $P(12 \cdots i | x, y) - P(12 \cdots i + 1 | x, y)$ by $P_i(x, y)$. Clearly, $P_i(x, y)$ is a generating function for the number of partitions $\pi \in P_n$ according to number of blocks, rises, levels, and descents such that $\pi_1 \pi_2 \cdots \pi_i \pi_{i+1} = 12 \cdots ij$ where $1 \leq j \leq i$. From the definitions of the generating functions, we immediately have the following

$$P(x, y; r, \ell, d) = 1 + \sum_{k \geq 1} P_k(x, y; r, \ell, d), \tag{1}$$

where the summand 1 covers the case $n = 0$, and

$$P_i(x, y; r, \ell, d) = x^i y^i r^{i-1} + \sum_{j=1}^{i} P(12 \cdots ij | x, y; r, \ell, d), \tag{2}$$

where the summand $x^i y^i r^{i-1}$ covers the partition $12 \cdots i$. Define $P_i^*(x, y)$ to be $\sum_{j \geq i} P_j(x, y)$. The strategy is now to find expressions for the generating functions $P_i(x, y)$, which will allow us to prove the theorem.

Lemma 2.1. *For all $1 \leq j \leq i - 2$,*

$$P(12 \cdots i(j + 1) | x, y) = \frac{1 + xd - x\ell}{1 + xr - x\ell} P(12 \cdots ij | x, y), \tag{3}$$

with $P(12 \cdots ii | x, y) = x\ell P_i^*(x, y)$ where $P(12 \cdots i1 | x, y)$ is equal to

$$\frac{x^{i+1} y^i r^{i-1}(d - r) + x(xdr - xr\ell + d)P_i^*(x, y) + x(r - d)P_i(x, y)}{1 + x(r - \ell)}. \tag{4}$$

Proof. By the definitions for all $1 \leq j < i$ we have that

$$P(12 \cdots ij | x, y)$$

$$= \quad x^{i+1} y^i r^{i-1} d + \sum_{k=1}^{j-1} P(12 \cdots ijk | x, y) + P(12 \cdots ijj | x, y)$$

$$+ \sum_{k=j+1}^{i} P(12 \cdots ijk | x, y) + P(12 \cdots ij(i + 1) | x, y)$$

$$= \quad x^{i+1} y^i r^{i-1} d + xd \sum_{k=1}^{j-1} P(12 \cdots ik | x, y) + x\ell P(12 \cdots ij | x, y)$$

$$+ xr \sum_{k=j+1}^{i-1} P(12 \cdots ik | x, y) + x^2 dr P_i^*(x, y) + xd P_{i+1}^*(x, y). \tag{5}$$

Therefore,

$$P(12 \cdots i(j + 1) | x, y) - P(12 \cdots ij | x, y)$$
$$= \quad (xd - x\ell) P(12 \cdots ij | x, y) + (x\ell - xr) P(12 \cdots i(j + 1) | x, y),$$

which implies (3). Also, (5) gives

$$(1 - x\ell)P(12\cdots i1|x,y)$$
$$= x^{i+1}y^i r^{i-1}d$$
$$+ xr\sum_{k=2}^{i-1} P(12\cdots ik|x,y) + x^2 dr P_i^*(x,y) + xd P_{i+1}^*(x,y),$$

and using (2) we obtain

$$(1 + xr - x\ell)P(12\cdots i1|x,y)$$
$$= x^{i+1}y^i r^{i-1}d + xr\left(P_i(x,y) - x^i y^i r^{i-1} - P(12\cdots ii|x,y)\right)$$
$$+ x^2 dr P_i^*(x,y) + xd P_{i+1}^*(x,y),$$

which is equivalent to

$$(1 + xr - x\ell)P(12\cdots i1|x,y)$$
$$= x^{i+1}y^i r^{i-1}d + xr\left(P_i(x,y) - x^i y^i r^{i-1} - x\ell P_i^*(x,y)\right)$$
$$+ x^2 dr P_i^*(x,y) + xd P_{i+1}^*(x,y),$$

and by the fact that $P_{i+1}^*(x,y) = P_i^*(x,y) - P_i(x,y)$ we obtain (4). At the end, it is not hard to see that $P(12\cdots ii|x,y) = x\ell P_i^*(x,y)$, which completes the proof. $\qquad\square$

Lemma 2.2. *For all $i \geq 1$, $P_i(x,y)$ is equal to $x^i y^i r^{i-1}$ plus*

$$\left(1 - \frac{r(1+x(d-\ell))}{r-d} + \frac{d(1+x(d-\ell))}{r-d}\left(\frac{1+x(r-\ell)}{1+x(d-\ell)}\right)^i\right)P_i^*(x,y),$$

with $P_0(x,y) = 1$.

Proof. Lemma 2.1 together with (2) gives

$$P_i(x,y)$$
$$= x^i y^i r^{i-1} + x\ell P_i^*(x,y)$$
$$+ \sum_{j=1}^{i-1}\left(\frac{1+x(d-\ell)}{1+x(r-\ell)}\right)^{j-1} \times$$
$$x\left(\frac{x^i y^i r^{i-1}(d-r) + (xdr - xr\ell + d)P_i^*(x,y) + (r-d)P_i(x,y)}{1+x(r-\ell)}\right).$$

After several simple algebraic operations the lemma holds. $\qquad\square$

In order to obtain an explicit formula from the recurrence relation in the statement of Lemma 2.2, we need to study the following types of recurrence relations.

$$f_\ell = a_\ell + b_\ell \sum_{j \geq \ell} f_j, \quad \ell = 0, 1, \dots. \tag{6}$$

The above recurrence relation gives

$$b_{\ell+1} f_\ell - b_\ell f_{\ell+1} = b_{\ell+1} a_\ell - b_\ell a_{\ell+1} + b_\ell b_{\ell+1} f_\ell,$$

which implies that (6) is equivalent to

$$f_\ell = \frac{a_\ell b_{\ell+1} - a_{\ell+1} b_\ell}{b_{\ell+1}(1 - b_\ell)} + \frac{b_\ell}{b_{\ell+1}(1 - b_\ell)} f_{\ell+1}, \quad \ell = 0, 1, \dots. \tag{7}$$

Lemma 2.3. *Let g_ℓ be any sequence satisfying the recurrence relation $g_\ell = \alpha_\ell + \beta_\ell g_{\ell+1}$, for $\ell \geq 0$. Then*

$$g_\ell = \sum_{i \geq \ell} \left(\alpha_i \prod_{j=\ell}^{i-1} \beta_j \right).$$

Proof. An infinite number of applications of this recurrence relation gives

$$
\begin{aligned}
g_\ell &= \alpha_\ell + \beta_\ell g_{\ell+1} \\
&= \alpha_\ell + \alpha_{\ell+1} \beta_\ell + \beta_\ell \beta_{\ell+1} g_{\ell+2} \\
&= \alpha_\ell + \alpha_{\ell+1} \beta_\ell + \alpha_{\ell+2} \beta_\ell \beta_{\ell+1} + \beta_\ell \beta_{\ell+1} \beta_{\ell+2} g_{\ell+3} \\
&= \cdots \\
&= \sum_{i \geq \ell} \left(\alpha_i \prod_{j=\ell}^{i-1} \beta_j \right),
\end{aligned}
$$

as required. $\qquad \square$

Define $f = \sum_{\ell \geq 0} f_\ell$. Then, applying the above lemma on (7) we obtain an explicit formula for the sequence f, where f_ℓ satisfies (6).

Proposition 2.4. *Let f_ℓ be any sequence satisfying (6) and $f = \sum_{\ell \geq 0} f_\ell$. Then $f = \sum_{i \geq 0} \dfrac{a_i}{\prod_{j=0}^{i}(1 - b_j)}$.*

Proof. Let f_ℓ be any sequence satisfying (6). Lemma 2.1 together with (7) implies that for all $\ell \geq 0$,

$$f_\ell = \sum_{i \geq \ell} \left(\frac{a_i b_{i+1} - a_{i+1} b_i}{b_{i+1}(1 - b_i)} \prod_{j=\ell}^{i-1} \frac{b_j}{b_{j+1}(1 - b_j)} \right).$$

Summing over all possible values of ℓ we get

$$
\begin{aligned}
f &= \sum_{i \geq 0} \frac{a_i b_{i+1} - a_{i+1} b_i}{b_{i+1}(1 - b_i)} \left[1 + \sum_{j=0}^{i-1} \left(\frac{b_j}{b_i \prod_{k=j}^{i-1}(1 - b_k)} \right) \right] \\
&= \frac{a_0 b_1 - a_1 b_0}{b_1(1 - b_0)} + \sum_{i \geq 1} \frac{a_i b_{i+1} - a_{i+1} b_i}{b_{i+1}} \left[\frac{1}{1 - b_i} + \sum_{j=0}^{i-1} \left(\frac{b_j}{b_i \prod_{k=j}^{i}(1 - b_k)} \right) \right] \\
&= \frac{a_0 b_1 - a_1 b_0}{b_1(1 - b_0)} + \sum_{i \geq 1} \left(\frac{a_i b_{i+1} - a_{i+1} b_i}{b_{i+1}} \sum_{j=0}^{i} \frac{b_j}{b_i \prod_{k=j}^{i}(1 - b_k)} \right) \\
&= \frac{a_0 b_1 - a_1 b_0}{b_1(1 - b_0)} + \sum_{i \geq 1} \left(\frac{a_i b_{i+1} - a_{i+1} b_i}{b_{i+1} b_i \prod_{j=0}^{i}(1 - b_j)} \sum_{j=0}^{i} b_j \prod_{k=0}^{j-1}(1 - b_k) \right).
\end{aligned}
$$

Using the fact that $1 - \sum_{j=0}^{i} b_j \prod_{k=0}^{j-1}(1 - b_k) = \prod_{j=0}^{i}(1 - b_j)$, we obtain that

$$
f = \frac{a_0 b_1 - a_1 b_0}{b_1(1 - b_0)} + \sum_{i \geq 1} \left(\frac{a_i b_{i+1} - a_{i+1} b_i}{b_{i+1} b_i \prod_{j=0}^{i}(1 - b_j)} \left(1 - \prod_{k=0}^{i}(1 - b_k) \right) \right),
$$

which is equivalent to

$$
\begin{aligned}
f &= \frac{a_0 b_1 - a_1 b_0}{b_1(1 - b_0)} + \sum_{i \geq 1} \frac{a_i b_{i+1} - a_{i+1} b_i}{b_{i+1} b_i \prod_{j=0}^{i}(1 - b_j)} - \sum_{i \geq 1} \left(\frac{a_i}{b_i} - \frac{a_{i+1}}{b_{i+1}} \right) \\
&= \frac{a_0 b_1 - a_1 b_0}{b_1(1 - b_0)} - \frac{a_1}{b_1} + \sum_{i \geq 1} \frac{a_i b_{i+1} - a_{i+1} b_i}{b_{i+1} b_i \prod_{j=0}^{i}(1 - b_j)} \\
&= \frac{a_0 - \frac{a_1}{b_1}}{1 - b_0} + \sum_{i \geq 1} \frac{\frac{a_i}{b_i} - \frac{a_{i+1}}{b_{i+1}}}{\prod_{j=0}^{i}(1 - b_j)} \\
&= \frac{a_0 - \frac{a_1}{b_1}}{1 - b_0} + \sum_{i \geq 1} \frac{a_i}{b_i \prod_{j=0}^{i}(1 - b_j)} - \sum_{i \geq 2} \frac{a_i}{b_i \prod_{j=0}^{i-1}(1 - b_j)} \\
&= \frac{a_0}{1 - b_0} + \sum_{i \geq 1} \frac{a_i}{b_i \prod_{j=0}^{i}(1 - b_j)} - \sum_{i \geq 1} \frac{a_i}{b_i \prod_{j=0}^{i-1}(1 - b_j)} \\
&= \frac{a_0}{1 - b_0} + \sum_{i \geq 1} \frac{a_i}{\prod_{j=0}^{i}(1 - b_j)} = \sum_{i \geq 0} \frac{a_i}{\prod_{j=0}^{i}(1 - b_j)},
\end{aligned}
$$

as required. $\square$

The proof of Theorem 1.1 can be obtained directly from Lemma 2.2 together with Proposition 2.4.

3 Applications

In this section we give a brief indication of the practical combinatorial enumerations embodied by Theorem 1.1.

Assume $d, r \to 1$, then Theorem 1.1 gives that the generating function for the number of partitions of $[n]$ with $k \geq 1$ blocks according to n and number of levels is given by

$$\frac{x^k}{\prod_{j=1}^{k}(1 - x(\ell + j - 1))} = \frac{x}{(1 - x\ell)} \frac{\frac{x^{k-1}}{(1-\ell x)^{k-1}}}{\prod_{j=1}^{k-1}(1 - jx/(1 - x\ell))},$$

which implies that (see [9, Page 57])

$$\frac{x^k}{\prod_{j=1}^{k}(1 - x(\ell + j - 1))} = \sum_{n \geq 1} S(n, k - 1) \frac{x^{n+1}}{(1 - x\ell)^{n+1}},$$

where $S(n, k)$ is the Stirling number of the second kind. Hence, we have the following result.

Corollary 3.1. *The number of partitions of* $[n]$ *with* k *blocks and* m *levels is* $\binom{n-1}{m} S(n - 1 - m, k - 1)$.

Proof. We give a direct proof. Let $P_n(k, m)$ denote the number of partitions of $[n]$ with k blocks and m levels. We obtain a recursion for $w(n, k, m) = |P_n(k, m)|$ by considering when the last letter π_n in a partition $(\pi_1, \ldots, \pi_n)$ forms a level.

An element of $P_n(k, m)$ in which the last letter does not form a level is obtained by inserting $\pi_n = k$ at the end of an element of $P_{n-1}(k - 1, m)$, or by inserting $\pi_n = c$ at the end of a $(\pi_1, \ldots, \pi_{n-1}) \in P_{n-1}(k, m)$ where c is a distinct letter in $(\pi_1, \ldots, \pi_{n-2})$. These partitions are enumerated by $w(n - 1, k - 1, m) + (k - 1)w(n - 1, k, m)$.

But an element of $P_n(k, m)$ in which the last letter forms a level is obtained by inserting another copy of π_{n-1} at the end of a $(\pi_1, \ldots, \pi_{n-1}) \in P_{n-1}(k, m - 1)$. Such partitons are enumerated by $w(n - 1, k, m - 1)$. Thus the recursion is

$$w(n, k, m) = w(n-1, k-1, m) + (k-1)w(n-1, k, m) + w(n-1, k, m-1)$$

for $n \geq k > m > 0$, and with the starting value $w(n, k, 0) = S(n-1, k-1)$ (see Corollary 3.2), it is routine to verify directly that the solution is $\binom{n-1}{m} S(n - 1 - m, k - 1)$. $\qquad\square$

Corollary 3.1 implies the following result.

Corollary 3.2. *There is a bijection between the set of partitions of $[n]$ with k blocks and the set of partitions of $[n+1]$ with $k+1$ blocks and no levels.*

Proof. Let $P_n(k)$ denote the set of partitions of $[n]$ with k blocks, and $T(n,k)$ the set of partitions of $[n]$ with k blocks and no levels. Note that an ℓ-level means a subword of ℓ identical letters, $\ell \geq 2$. We describe a bijection between $P_n(k)$ and $T(n+1, k+1)$.

To obtain $q \in T(n+1, k+1)$ from $p = (\pi_1, \dots, \pi_n) \in P_n(n, k)$ transform p as follows

(1) if $p \in T(n, k)$, insert the letter $k+1$ at the end of p to obtain q;

(2) otherwise, proceed as follows:

    (i) replace each member π_j of an ℓ-level with c if j is even (respectively, odd) and ℓ is odd (respectively, even), where $c = max(\pi_1, \dots, \pi_i) + 1$ and $\pi_1, \dots, \pi_i$, is the subword immediately preceding the first π_j to be replaced, $i = j - 1$. ($c = 1$ iff the first ℓ-level begins with $\pi_1 = 1$ and ℓ is even). It may be necessary to tag each designated c for step (ii);

    (ii) add 1 to all other letters $\geq c$ on the right of the first c; and

    (iii) insert c at the end of the resulting word to obtain q;

The mapping is reversible since the last letter c of the image indicates the source in an obvious way.

For example, $p = 12132312 \in P_8(3)$ maps to $121323124 \in T(9,4)$.

For $11112133321 \in P_{11}(3)$ we have the following transformations

$$11112133321 \to \bar{1}1\bar{1}12131\bar{3}21 \to \bar{1}2\bar{1}23241\bar{4}32 \to 121232414321$$

which lies in $T(12,4)$. Similarly, $12111112233 \in P_{11}(3)$ gives

$$12111112233 \to 121\bar{3}1\bar{3}13\bar{2}\bar{3}3 \to 121\bar{3}1\bar{3}13\bar{2}\bar{3}4 \to 121313132343$$

which also lies in $T(12,4)$. $\qquad\square$

Assume $\ell = 1$, $d \to 1/v$ and $r \to v$, then Theorem 1.1 gives that the generating function for the number of partitions of $[n]$ with $k \geq 1$ blocks according to n and number of rises-descents is given by

$$f(x, v) = \sum_{n \geq 0} \sum_{\pi \in P_n} x^n v^{\text{rises}(\pi) - \text{descents}(\pi)}$$

$$= \frac{x^k v^{k-1}}{(1 + x(1/v - 1))^k \prod_{j=1}^{k} \frac{v - \frac{1}{v}\left(\frac{1 + x(v-1)}{1 + x(1/v - 1)}\right)^j}{v - 1/v}}.$$

Hence, after simple algebraic operations we have that

$$\frac{\partial}{\partial v} f(x,v) \mid_{v=1} = \frac{(k-1)(kx-2)x^k}{2\prod_{j=1}^{k}(1-jx)} = (k-1)(kx-2)\sum_{j\geq 0} S(j,k)x^j,$$

which implies that (see [9, Page 57])

$$\frac{\partial}{\partial v} f(x,v) \mid_{v=1} = \frac{1}{2}(k-1)(kx-2)\sum_{j\geq 1} S(j,k)x^j. \tag{8}$$

Hence, we can state the following result.

Corollary 3.3. *We have*

$$\sum_{\pi} rises(\pi) - descents(\pi) = (k-1)S(n,k) - \binom{k}{2}S(n-1,k),$$

where the sum is over all partitions of $[n]$ *with exactly* k *blocks. Moreover,*

$$\sum_{\pi} rises(\pi) - descents(\pi) = \frac{1}{2}(B_{n+1} - B_n - B_{n-1}),$$

where the sum is over all the partitions of $[n]$, $S(n,k)$ *is the Stirling number of the second kind, and* B_n *is the n-th Bell number (see [2]).*

Proof. The first sum is holds immediately from (8). Using the fact that $S(n,k) = S(n-1,k-1) + kS(n-1,k)$ we get that

$$(k-1)S(n,k) - \binom{k}{2}S(n-1,k)$$

$$= \frac{1}{2}(S(n+1,k) - S(n,k-1) - S(n-1,k-2)).$$

Thus,

$$\sum_{\pi} rises(\pi) - descents(\pi)$$

$$= \sum_{k\geq 1} \frac{1}{2}(S(n+1,k) - S(n,k-1) - S(n-1,k-2))$$

$$= \frac{1}{2}(B_{n+1} - B_n - B_{n-1}),$$

where the sum is over all the partitions of $[n]$. $\square$

Assume $d=1$, $\ell \to 1/v$ and $r \to v$, then Theorem 1.1 gives that the generating function for the number of partitions of $[n]$ with $k \geq 1$ blocks according to n and number of rises-levels is given by

$$g(x, v) = \sum_{n \geq 0} \sum_{\pi \in P_n} x^n v^{\text{rises}(\pi) - \text{levels}(\pi)}$$

$$= \frac{x^k v^{k-1}}{(1 + x(1 - 1/v))^k \prod_{j=1}^{k} \frac{v - (\frac{1 + x(v - 1/v)}{1 + x(1 - 1/v)})^j}{v - 1/v}} .$$

Hence, after simple algebraic operations we have that

$$\frac{\partial}{\partial v} g(x, v) \big|_{v=1} = \frac{(k - 1)x^k - kx^{k+1} + x^{k+2} \sum_{j=1}^{k} \frac{j(j-3)}{2(1-jx)}}{\prod_{j=1}^{k}(1 - jx)} .$$

Expanding at $x = 0$ with using the fact that

$$\frac{x^k}{\prod_{j=1}^{k}(1 - jx)} = \sum_{n \geq 1} S(n, k) x^n$$

(see [9, Page 57]) we get the following result.

Corollary 3.4. *Let* $1 \leq k \leq n$. *Then*

$$\sum_{\pi} rises(\pi) - levels(\pi)$$

$$= (k - 1)S(n, k) - kS(n - 1, k)$$

$$+ \sum_{i=0}^{n-3} \left(\sum_{j=1}^{k} \frac{j^{i+1}(j - 3)}{2} \right) S(n - 2 - i, k),$$

where the sum is over all partitions π *of* $[n]$ *with exactly* k *blocks.*

4 Conclusions

We have presented the enumeration of set partitions according to number of rises, number of levels and number of descents.

We are unable to provide combinatorial proofs of Corollary 3.3 and Corollary 3.4.

One approach towards extending our work is to consider the enumeration of the statistics by taking t letters at a time, $t \geq 2$. This paper just addressed the $t = 2$ case. Thus one could seek the enumeration of set partitions relative to $3-$rises, $3-$levels and $3-$descents. For instance, $\pi = 121112341$ has two $3-$rises, one $3-$level and no $3-$descent. As an illustration let $Q_{n,k}$ denote the subset of partitions of $[n]$ into k blocks in which each element contains only 3−levels, so that (example) $\pi \in Q_{9,4}$ but neither 121222334 nor 111233334 belong to $Q_{9,4}$. Then it can be shown that the number of elements with exactly m copies of 3−levels is given by

$$\binom{n-2m}{m} S(n-2m-1, k-1)$$

The complete enumeration could be undertaken but a result as nice as Theorem 1.1 seems unlikely.

References

[1] H. Davenport and A. Schinzel. A combinatorial problem connected with differential equations. *Amer. J. Math.*, 87:684–694, 1965.

[2] R. L. Graham, D. E. Knuth, and O. Patashnik. *Concrete mathematics*. Addison-Wesley Publishing Company, Reading, MA, second edition, 1994.

[3] V. Jelínek and T. Mansour. On pattern-avoiding partitions. *Electron. J. Combin.*, 15(1):Research paper 39, 52, 2008.

[4] M. Klazar. On *abab*-free and *abba*-free set partitions. *European J. Combin.*, 17(1):53–68, 1996.

[5] M. Klazar. On trees and noncrossing partitions. *Discrete Appl. Math.*, 82(1-3):263–269, 1998.

[6] P. A. MacMahon. *Combinatory Analysis*. Cambridge University Press, London, 1915/16.

[7] R. C. Mullin and R. G. Stanton. A map-theoretic approach to Davenport-Schinzel sequences. *Pacific J. Math.*, 40:167–172, 1972.

[8] B. E. Sagan. Pattern avoidance in set partitions. arXiv:math/0604292v1 [math.CO].

[9] R. P. Stanley. *Enumerative combinatorics. Vol. 1*, volume 49 of *Cambridge Studies in Advanced Mathematics*. Cambridge University Press, Cambridge, 1997.

[10] M. Wachs and D. White. p, q-Stirling numbers and set partition statistics. *J. Combin. Theory Ser. A*, 56(1):27–46, 1991.

Restricted patience sorting and barred pattern avoidance

Alexander Burstein†

Department of Mathematics, Howard University, Washington, DC 20059, USA

Isaiah Lankham‡

Department of Mathematics, Simpson University, Redding, CA 96003, USA

Abstract

Patience Sorting is a combinatorial algorithm that can be viewed as an iterated, non-recursive form of the Schensted Insertion Algorithm. In recent work the authors have shown that Patience Sorting provides an algorithmic description for permutations avoiding the barred (generalized) permutation pattern 3-$\bar{1}$-42. Motivated by this and a recently formulated geometric form for Patience Sorting in terms of certain intersecting lattice paths, we study the related themes of restricted input and avoidance of similar barred permutation patterns. One such result is to characterize those permutations for which Patience Sorting is an invertible algorithm as the set of permutations simultaneously avoiding the barred patterns 3-$\bar{1}$-42 and 3-$\bar{1}$-24. We then enumerate this avoidance set, which involves convolved Fibonacci numbers.

1 Introduction

The term *Patience Sorting* was introduced in 1962 by C. L. Mallows [12] while studying a card sorting algorithm invented by A. S. C. Ross. Given a shuffled deck of cards $\sigma = c_1 c_2 \cdots c_n$ (which we take to be a permutation $\sigma \in \mathfrak{S}_n$), Ross proposed the following two-part algorithm.

† The work of the first author was supported in part by the U.S. National Security Agency Young Investigator Grant H98230-06-1-0037.
‡ The work of the second author was supported in part by the U.S. National Science Foundation under Grants DMS-0135345, DMS-0304414, and DMS-0553379.

Algorithm 1 Two-part patience sorting algorithm

Step 1: Use what Mallows called a "patience sorting procedure" to form the subsequences $r_1, r_2, \ldots, r_m$ of σ (called *piles*) as follows:

- Place the first card c_1 from the deck into a pile r_1 by itself.
- For each remaining card c_i $(i = 2, \ldots, n)$, consider the cards $d_1, d_2, \ldots, d_k$ atop the piles $r_1, r_2, \ldots, r_k$ that have already been formed.

  - If $c_i > \max\{d_1, d_2, \ldots, d_k\}$, then put c_i into a new right-most pile r_{k+1} by itself.

  - Otherwise, find the left-most card d_j that is larger than c_i, and put the card c_i atop pile r_j.

Step 2: Gather the cards up one at a time from these piles in ascending order.

We call **Step 1** of the above algorithm *Patience Sorting* and denote by $R(\sigma) = \{r_1, r_2, \ldots, r_m\}$ the *pile configuration* associated to the permutation $\sigma \in \mathfrak{S}_n$. Moreover, given any pile configuration R, one forms its *reverse patience word* $RPW(R)$ by listing the piles in R "from bottom to top, left to right" (i.e., by reversing the so-called "far-eastern reading"; see Example 1.1 below). In [5] these words are characterized as being exactly the elements of the avoidance set $S_n(3\text{-}\bar{1}\text{-}42)$. That is, reverse patience words are permutations avoiding the generalized pattern 2-31 unless every occurrence of 2-31 is also part of an occurrence of the generalized pattern 3-1-42. (A review of generalized permutation patterns can be found in Section 1.2 below).

We illustrate the formation of both $R(\sigma)$ and $RPW(R)$ in the following example.

Example 1.1. Let $\sigma = 64518723 \in \mathfrak{S}_8$. Then, in forming the pile configuration $R(\sigma)$ under Patience Sorting, we write the constituent piles vertically and bottom-justified (with respect to the largest value in each pile) as follows:

First, use **6** to form a new pile:

6

Then place **4** atop this new pile:

4
6

Use **5** to form a new pile:

4
6 **5**

Then place **1** atop the left-most pile:

1
4
6 5

Use **8** to form a new pile:

1
4
6 5 **8**

Then place **7** atop this new pile:

1
4   7
6 5 8

Place **2** atop the middle pile:

1
4 **2** 7
6 5 8

Finally, place **3** atop the right-most pile:

1   **3**
4 2 7
6 5 8

It follows that $R(\sigma) = \begin{matrix} 1 & & 3 \\ 4 & 2 & 7 \\ 6 & 5 & 8 \end{matrix}$, with piles $r_1 = 641$, $r_2 = 52$, and $r_3 = 873$. Moreover, by reading up the columns of $R(\sigma)$ from left to right, $RPW(R(\sigma)) = 64152873 \in S_8(3\text{-}\bar{1}\text{-}42)$.

Given $\sigma \in \mathfrak{S}_n$, the formation of $R(\sigma)$ can be viewed as an iterated, non-recursive form of the Schensted Insertion Algorithm for interposing values into the rows of a standard Young tableau (see [2]). In [5] the authors augment the formation of $R(\sigma)$ so that the resulting extension of Patience Sorting becomes a full, non-recursive analogue of the celebrated Robinson-Schensted-Knuth (or RSK) Correspondence. As with RSK, this Extended Patience Sorting Algorithm (given as Algorithm 2 in Section 1.1 below) takes a simple idea — that of placing cards into piles — and uses it to build a bijection between elements of the symmetric group $\mathfrak{S}_n$ and certain pairs of combinatorial objects. In the case of RSK, one uses the Schensted Insertion Algorithm to build a bijection with (unrestricted) pairs of standard Young tableau having the same shape (see [15]). However, in the case of Patience Sorting, one achieves a bijection between permutations and (somewhat more restricted) pairs of pile configurations having the same shape. We denote this latter bijection by $\sigma \overset{PS}{\longleftrightarrow} (R(\sigma), S(\sigma))$ and call $R(\sigma)$ (resp. $S(\sigma)$) the *insertion piles* (resp. *recording piles*) corresponding to σ. Collectively, we also call $(R(\sigma), S(\sigma))$ the *stable pair* of pile configurations corresponding to σ and characterize such pairs in [5] using a somewhat involved pattern avoidance condition on their reverse patience words. (The rationale behind the term "stable pair" is also explained in [5].)

Barred (generalized) permutation patterns like 3-$\bar{1}$-42 arise quite naturally when studying Patience Sorting. We discuss and enumerate the avoidance classes for several related patterns in Section 2. Then, in Section 3, we examine properties of Patience Sorting under restricted input that can be characterized using such patterns. One such characterization, discussed in Section 3.1, is for the crossings in the initial iteration of the Geometric Patience Sorting Algorithm given by the authors in [6]. This geometric form for the Extended Patience Sorting Algorithm is naturally dual to Geometric RSK (originally defined in [19]) and gives, among other things, a geometric interpretation for the stable pairs of 3-$\bar{1}$-42-avoiding permutations corresponding to a permutation under Extended Patience Sorting. However, unlike the geometric form for RSK, the shadow lines in Geometric Patience Sorting are allowed to cross. While a complete characterization for these crossings is given in [6] in terms of the pile configurations formed, this new result is the first step in providing a characterization for the permutations involved in terms of barred pattern avoidance.

We close this introduction by describing both the Extended and Geometric Patience Sorting Algorithms. We also briefly review the notation of generalized permutation patterns.

1.1 Extended and Geometric Patience Sorting

C. L. Mallows' original "patience sorting procedure" can be extended to a full bijection between the symmetric group $\mathfrak{S}_n$ and certain restricted pairs of pile configurations using the following algorithm (which was first introduced in [5]).

Algorithm 2 Extended patience sorting algorithm.

Given $\sigma = c_1 c_2 \cdots c_n \in \mathfrak{S}_n$, inductively build *insertion piles* $R(\sigma) = \{r_1, r_2, \ldots, r_m\}$ and *recording piles* $S(\sigma) = \{s_1, s_2, \ldots, s_m\}$ as follows:

(i) First form a new pile r_1 using the card c_1, and set $s_1 = 1$.

(ii) Then, for each $i = 2, \ldots, n$, consider the cards $d_1, d_2, \ldots, d_k$ atop the piles $r_1, r_2, \ldots, r_k$ that have already been formed.

    (a) If $c_i > \max_{1 \leq j \leq k} \{d_j\}$, then form a new pile r_{k+1} using c_i, and set $s_{k+1} = i$.

    (b) Otherwise, find the left-most card d_j that is larger than c_i, and place the card c_i atop pile r_j while simultaneously placing i at the bottom of pile s_j. In other words, set

$$d_j = c_i, \quad \text{where} \quad j = \min_{1 \leq m \leq k} \{m \mid c_i < d_m\},$$

    and insert i at the bottom of pile s_j.

Note that the pile configurations that comprise a resulting stable pair must have the same "shape", which we define as follows:

Definition 1.2. Given a pile configuration $R = \{r_1, r_2, \ldots, r_m\}$ on n cards, the *shape* of R is defined to be the composition $\mathrm{sh}(R) = (|r_1|, |r_2|, \ldots, |r_m|)$ of n. We denote this relationship by $\mathrm{sh}(R) \models n$.

The idea behind Algorithm 2 is that we are using the auxiliary pile configuration $S(\sigma)$ to implicitly label the order in which the elements of the permutation $\sigma \in \mathfrak{S}_n$ are added to the usual Patience Sorting pile configuration $R(\sigma)$. (Consequently, we call $R(\sigma)$ the "insertion piles" of

σ and $S(\sigma)$ the "recording piles" of σ by analogy to RSK.) This information then allows us to uniquely reconstruct σ by reversing the order in which the cards were played. As with normal Patience Sorting, we visualize the pile configurations $R(\sigma)$ and $S(\sigma)$ by listing their constituent piles vertically as illustrated in the following example.

Example 1.3. Given $\sigma = 64518723 \in \mathfrak{S}_8$ from Example 1.1 above, we simultaneously form the following pile configurations with shape $\text{sh}(R(\sigma)) = \text{sh}(S(\sigma)) = (3,2,3)$ under Extended Patience Sorting (Algorithm 2):

$$
R(\sigma) = \begin{matrix} 1 & & 3 \\ 4 & 2 & 7 \\ 6 & 5 & 8 \end{matrix} \quad \text{and} \quad S(\sigma) = \begin{matrix} 1 & & 5 \\ 2 & 3 & 6 \\ 4 & 7 & 8 \end{matrix}.
$$

Note that the insertion piles $R(\sigma)$ are the same as those formed in Example 1.1 and that the reverse patience word of $S(\sigma)$ satisfies

$$
RPW(S(64518723)) = 42173865 \in S_8(3\text{-}\bar{1}\text{-}42).
$$

Given a permutation $\sigma \in \mathfrak{S}_n$, the Extended Patience Sorting Algorithm is not the only method of forming the pile configurations $R(\sigma)$ and $S(\sigma)$. In particular, one can realize the elements in each pile of $R(\sigma)$ and $S(\sigma)$ as points in the diagram of σ that are identified via specially formed lattice paths. This provides a Geometric Patience Sorting Algorithm (first introduced in [6]) that resembles the geometric form for RSK given by X. G. Viennot in [19].

In order to describe this geometric form for the Extended Patience Sorting Algorithm, we begin with the following series of definitions.

Definition 1.4. Given a lattice point $(m,n) \in \mathbb{Z}^2$, we define the *(southwest) shadow* of (m,n) to be the quarter space

$$
U(m,n) = \left\{ (x,y) \in \mathbb{R}^2 \mid x \le m, \ y \le n \right\}.
$$

As with the northeasterly-oriented shadows that Viennot used when building his geometric form for RSK, the most important use of these southwesterly-oriented shadows is in building lattice paths called *shadowlines* (as illustrated in Figure 1(a)) according to the following definition.

Definition 1.5. The *(southwest) shadowline* of (m_1, n_1), (m_2, n_2), $\cdots$,

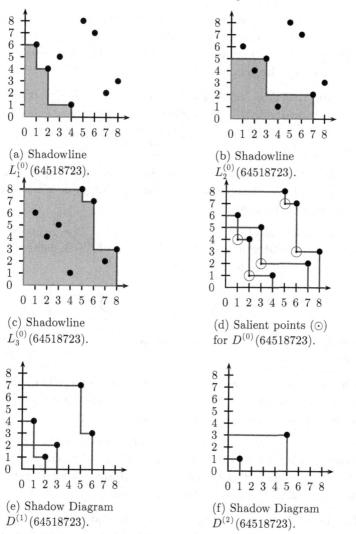

(a) Shadowline
$L_1^{(0)}(64518723)$.

(b) Shadowline
$L_2^{(0)}(64518723)$.

(c) Shadowline
$L_3^{(0)}(64518723)$.

(d) Salient points ($\odot$)
for $D^{(0)}(64518723)$.

(e) Shadow Diagram
$D^{(1)}(64518723)$.

(f) Shadow Diagram
$D^{(2)}(64518723)$.

Fig. 1. Examples of Shadowline and Shadow Diagram Construction.

$(m_k, n_k) \in \mathbb{Z}^2$ is defined to be the boundary of the union of the shadows $U(m_1, n_1), U(m_2, n_2), \ldots, U(m_k, n_k)$.

In particular, we wish to associate to each permutation a certain collection of (southwest) shadowlines called its *(southwest) shadow diagram*. However, unlike the northeasterly-oriented shadowlines used to define the northeast shadow diagrams of Geometric RSK, these southwest shadowlines are allowed to intersect as illustrated in Figure 1(d)–(e). (We characterize those permutations having intersecting shadowlines under Definition 1.6 in Theorem 3.6 below.)

Definition 1.6. The *(southwest) shadow diagram* $D^{(0)}(\sigma)$ of the permutation $\sigma = \sigma_1\sigma_2\cdots\sigma_n \in \mathfrak{S}_n$ consists of the (southwest) shadowlines $D^{(0)}(\sigma) = \left\{L_1^{(0)}(\sigma), L_2^{(0)}(\sigma), \ldots, L_k^{(0)}(\sigma)\right\}$ formed as follows:

- $L_1^{(0)}(\sigma)$ is the shadowline for those lattice points $(x,y) \in \{(1,\sigma_1), (2,\sigma_2), \ldots, (n,\sigma_n)\}$ such that the shadow $U(x,y)$ does not contain any other lattice points.

- While at least one of the points $(1,\sigma_1), (2,\sigma_2), \ldots, (n,\sigma_n)$ is not contained in the shadowlines $L_1^{(0)}(\sigma), L_2^{(0)}(\sigma), \ldots, L_j^{(0)}(\sigma)$, define $L_{j+1}^{(0)}(\sigma)$ to be the shadowline for the points

$$(x,y) \in A := \left\{(i,\sigma_i) \mid (i,\sigma_i) \notin \bigcup_{k=1}^{j} L_k^{(0)}(\sigma)\right\}$$

such that the shadow $U(x,y)$ does not contain any other lattice points from the set A.

In other words, we define a shadow diagram by recursively eliminating points in the permutation diagram until every point has been used to define a shadowline (as illustrated in Figure 1(a)–(c)).

One can prove (see [5]) that the ordinates of the points used to define each shadowline in the shadow diagram $D^{(0)}(\sigma)$ are exactly the *left-to-right minima subsequences* (a.k.a. *basic subsequences*) in the permutation $\sigma \in \mathfrak{S}_n$. These are defined as follows:

Definition 1.7. Let $\pi = \pi_1\pi_2\cdots\pi_l$ be a partial permutation on the set $[n] = \{1, 2, \ldots, n\}$. Then the *left-to-right minima* (resp. *maxima*) *subsequence* of π consists of those components π_j of π such that

$$\pi_j = \min\{\pi_i \mid 1 \le i \le j\} \quad (\text{resp. } \pi_j = \max\{\pi_i \mid 1 \le i \le j\}).$$

We then inductively define the left-to-right minima (resp. maxima) subsequences $s_1, s_2, \ldots, s_k$ of the permutation σ by taking s_1 to be the left-to-right minima (resp. maxima) subsequence for σ itself and then each remaining subsequence s_i to be the left-to-right minima (resp. maxima) subsequence for the partial permutation obtained by removing the elements of $s_1, s_2, \ldots, s_{i-1}$ from σ.

Finally, a sequence $D(\sigma) = \left(D^{(0)}(\sigma), D^{(1)}(\sigma), D^{(2)}(\sigma), \ldots \right)$ of shadow diagrams can be produced for a given permutation $\sigma \in \mathfrak{S}_n$ by recursively applying Definition 1.6 to the southwest corners (called *salient points*) of a given set of shadowlines (as illustrated in Figure 1(d)–(f)). The only difference is that, with each iteration, newly formed shadowlines can only connect salient points along the same pre-existing shadowline. This gives rise to the terminology in the following definition.

Definition 1.8. Given a permutation $\sigma \in \mathfrak{S}_n$, we call $D^{(k)}(\sigma)$ the kth *iterate* of the *exhaustive shadow diagram* $D(\sigma)$ for σ.

Given a permutation $\sigma \in \mathfrak{S}_n$, one can uniquely reconstruct the pile configurations $R(\sigma)$ and $S(\sigma)$ from the exhaustive shadow diagram $D(\sigma)$ by intersecting each iterate shadow diagram with the x- and y-axes in a certain canonical order (as detailed in [6]). Consequently, $D(\sigma)$ can be considered the result of applying a geometric from of Extended Patience Sorting to σ, and so we call the formation of $D(\sigma)$ the *Geometric Patience Sorting Algorithm*.

1.2 Generalized Pattern Avoidance

We first recall the following definition.

Definition 1.9. Let $\sigma = \sigma_1 \sigma_2 \cdots \sigma_n \in \mathfrak{S}_n$ and $\pi \in \mathfrak{S}_m$ with $m \le n$. Then we say that σ *contains* the *(classical) permutation pattern* π if there exists a subsequence $(\sigma_{i_1}, \sigma_{i_2}, \ldots, \sigma_{i_m})$ of σ (meaning $i_1 < i_2 < \cdots < i_m$) such that the word $\sigma_{i_1} \sigma_{i_2} \ldots \sigma_{i_m}$ is order-isomorphic to π. I.e., each $\sigma_{i_j} < \sigma_{i_{j+1}}$ if and only if $\pi_j < \pi_{j+1}$.

Note, though, that the elements in the subsequence $(\sigma_{i_1}, \sigma_{i_2}, \ldots, \sigma_{i_m})$ are not required to be contiguous in σ. This motivates the following refinement of Definition 1.9.

Definition 1.10. A *generalized permutation pattern* is a classical permutation pattern π in which one assumes that every element in the subsequence $(\sigma_{i_1}, \sigma_{i_2}, \ldots, \sigma_{i_m})$ of σ must be taken contiguously unless a dash is inserted between the corresponding order-isomorphic elements of the pattern π.

Finally, if σ does not contain a subsequence that is order-isomorphic to π, then we say that σ *avoids* the pattern π. This motivated the following definition.

Definition 1.11. Given any collection $\pi^{(1)}, \pi^{(2)}, \ldots, \pi^{(k)}$ of permutation patterns (classical or generalized), we denote by

$$S_n(\pi^{(1)}, \pi^{(2)}, \ldots, \pi^{(k)}) = \bigcap_{i=1}^{k} S_n(\pi^{(i)}) = \bigcap_{i=1}^{k} \left\{ \sigma \in \mathfrak{S}_n \mid \sigma \text{ avoids } \pi^{(i)} \right\}$$

the *avoidance set* for the patterns $\pi^{(1)}, \pi^{(2)}, \ldots, \pi^{(k)}$, which consists of all permutations $\sigma \in \mathfrak{S}_n$ such that σ simultaneously avoids each of $\pi^{(1)}, \pi^{(2)}, \ldots, \pi^{(k)}$. Furthermore, the set

$$\bigcup_{n \geq 1} S_n(\pi^{(1)}, \pi^{(2)}, \ldots, \pi^{(k)})$$

is called the *(pattern) avoidance class* with *basis* $\{\pi^{(1)}, \pi^{(2)}, \ldots, \pi^{(k)}\}$.

More information about permutation patterns in general can be found in [4].

2 Barred and Unbarred Generalized Pattern Avoidance

An important further generalization of the notion of generalized permutation pattern requires that the context in which the generalized pattern occurs be taken into account. The resulting concept of *barred permutation patterns*, along with its accompanying bar notation, first arose within the study of stack-sortability of permutations by J. West [20]. (West's barred patterns, though, were based upon the definition of a classical pattern and not upon the definition of a generalized pattern as below.) Given how naturally these barred patterns arise in the study of Patience Sorting (as illustrated in both [5] and Section 3 below), we initiate their systematic study in this section.

Definition 2.1. A *barred (generalized) permutation pattern* β is a generalized permutation pattern in which overbars are used to indicate that barred values cannot occur at the barred positions. As before, we denote by $S_n(\beta^{(1)}, \ldots, \beta^{(k)})$ the set of all permutations $\sigma \in \mathfrak{S}_n$ that simultaneously avoid $\beta^{(1)}, \ldots, \beta^{(k)}$ (i.e., permutations that contain no subsequence that is order-isomorphic to any of the $\beta^{(1)}, \ldots, \beta^{(k)}$).

Example 2.2. A permutation $\sigma = \sigma_1 \sigma_2 \cdots \sigma_n \in \mathfrak{S}_n \notin S_n(3\text{-}\bar{5}\text{-}2\text{-}4\text{-}1)$ contains an occurrence of the barred permutation pattern $3\text{-}\bar{5}\text{-}2\text{-}4\text{-}1$ if it contains an occurrence of the generalized pattern $3\text{-}2\text{-}4\text{-}1$ (i.e., contains a subsequence $(\sigma_{i_1}, \sigma_{i_2}, \sigma_{i_3}, \sigma_{i_4})$ that is order-isomorphic to the classical pattern 3241) in which no value larger than the element playing the role of "4" is allowed to occur between the elements playing the roles of "3" and "2". (This is one of the two basis elements for the pattern avoidance class used to characterize the set of 2-stack-sortable permutations [10, 11, 20]. The other pattern is 2-3-4-1, i.e., the classical pattern 2341.)

Despite the added complexity involved in avoiding barred permutation patterns, it is still sometimes possible to characterize the avoidance class for a barred permutation pattern in terms of an unbarred generalized permutation pattern. The following theorem (which summarizes results that first appeared in [5]) gives such a characterization for the pattern 3-$\bar{1}$-42. (Note, though, that there is no equivalent characterization for such barred permutation patterns as 1$\bar{3}$-42 and 3-$\bar{5}$-2-4-1.)

Theorem 2.3. *Let* $B_n = \frac{1}{e} \sum_{k \geq 0} \frac{k^n}{k!}$ *denote the* n^{th} *Bell number. Then*

(i) $S_n(3\text{-}\bar{1}\text{-}42) = S_n(3\text{-}\bar{1}\text{-}4\text{-}2) = S_n(23\text{-}1)$.
(ii) $|S_n(3\text{-}\bar{1}\text{-}42)| = B_n$.

Remark 2.4. We emphasize the following important consequences of Theorem 2.3.

(i) Even though $S_n(3\text{-}\bar{1}\text{-}42) = S_n(23\text{-}1)$ by Theorem 2.3(1), it is more natural to use avoidance of the barred pattern 3-$\bar{1}$-42 in studying Patience Sorting. As shown in [5] and elaborated upon in Section 3 below, $S_n(3\text{-}\bar{1}\text{-}42)$ is the set of equivalence classes of $\mathfrak{S}_n$ modulo equivalence relation generated by the pattern equivalence 3-$\bar{1}$-42 $\sim$ 3-$\bar{1}$-24. (I.e., two permutations $\sigma, \tau \in \mathfrak{S}_n$ satisfy $\sigma \sim \tau$ if the elements creating an occurrence of one of these patterns in σ form an occurrence of the other pattern in τ.) Moreover, each permutation $\sigma \in \mathfrak{S}_n$ in a given equivalence class has the same pile configuration $R(\sigma)$ under Patience Sorting, a description of which is significantly more difficult to describe for occurrences of the unbarred generalized permutation pattern 23-1.

(ii) A. Marcus and G. Tardos proved in [13] that the avoidance set $S_n(\pi)$ for any classical pattern π grows at most exponentially fast as $n \to \infty$. (This was previously known as the Stanley-Wilf Conjecture.) The Bell numbers, though, satisfy $\log B_n = n(\log n - \log \log n + O(1))$ and so exhibit superexponential growth. (See [18] for more information about Bell numbers.) While it was previously known that the Stanley-Wilf Conjecture does not extend to generalized permutation patterns (see, e.g., [7]), it took

Theorem 2.3(2) (originally proven in [5] using Patience Sorting) to provide the first verification that one also cannot extend the Stanley-Wilf Conjecture to barred generalized permutation patterns.

A further abstraction of barred permutation pattern avoidance (called *interval pattern avoidance*) was recently given by A. Woo and A. Yong in [21, 22]. The result in Theorem 2.3(2) has led A. Woo to conjecture to the second author that the Stanley-Wilf ex-Conjecture also does not extend to this new notion of pattern avoidance.

We conclude this section with a simple corollary to Theorem 2.3 that gives similar equivalences and enumerations for some barred permutation patterns that also arise naturally in the study of Patience Sorting (see Proposition 3.2 and Theorem 3.6 in Section 3 below).

Corollary 2.5. *Following the notation in Theorem 2.3,*

(i) $S_n(31\text{-}\bar{4}\text{-}2) = S_n(3\text{-}1\text{-}\bar{4}\text{-}2) = S_n(3\text{-}12)$

(ii) $S_n(\bar{2}\text{-}41\text{-}3) = S_n(\bar{2}\text{-}4\text{-}1\text{-}3) = S_n(2\text{-}4\text{-}1\text{-}\bar{3}) = S_n(2\text{-}41\text{-}\bar{3})$

(iii) $|S_n(\bar{2}\text{-}41\text{-}3)| = |S_n(31\text{-}\bar{4}\text{-}2)| = |S_n(3\text{-}\bar{1}\text{-}42)| = B_n$.

Proof. (Sketches)

(i) Take reverse complements in $S_n(3\text{-}\bar{1}\text{-}42)$ and apply Theorem 2.3.

(ii) Similar to (1). (Note that (2) is also proven in [1].)

(iii) This follows from the fact that the patterns $3\text{-}1\text{-}\bar{4}\text{-}2$ and $\bar{2}\text{-}4\text{-}1\text{-}3$ are inverses of each other.   □

3 Patience Sorting under Restricted Input

3.1 Patience Sorting on Restricted Permutations

The similarities between the Extended Patience Sorting Algorithm (Algorithm 2) and RSK applied to permutations are perhaps most observable in the following simple proposition.

Proposition 3.1. *Let $1_k = 1\text{-}2\text{-}\cdots\text{-}k$ and $J_k = k\text{-}\cdots\text{-}2\text{-}1$ be the classical monotone permutation patterns. Then there is a bijection between*

(i) $S_n(1_{k+1})$ *and stable pairs of pile configurations with the same composition shape* $(\gamma_1, \gamma_2, \ldots, \gamma_m) \models n$ *but also with at most k piles (i.e., $m \leq k$).*

(ii) $S_n(J_{k+1})$ *and stable pairs of pile configurations with the same composition shape* $(\gamma_1, \gamma_2, \ldots, \gamma_m) \models n$ *but also with no pile having more than k cards in it (i.e., $\gamma_i \leq k$ for each $i = 1, 2, \ldots, m$).*

Proof.

(i) Given $\sigma \in \mathfrak{S}_n$, a bijection is formed in [2] between the set of piles $R(\sigma) = \{r_1, r_2, \ldots, r_k\}$ formed under Patience Sorting and the components of a particular longest increasing subsequence in σ. Since avoiding the monotone pattern 1_{k+1} is equivalent to restricting the length of the longest increasing subsequence in a permutation, the result then follows.

(ii) Follows from (1) by reversing each of the permutations in $S_n(1_{k+1})$ in order to form $S_n(J_{k+1})$. □

According to Proposition 3.1, Extended Patience Sorting can be used to efficiently compute the length of both the longest increasing and longest decreasing subsequences in a given permutation. In particular, one can compute these lengths without examining every subsequence of a permutation, just as with RSK. However, while both RSK and Patience Sorting can be used to implement this computation in $O(n \log(n))$ time, an extension of this technique is given in [3] that also simultaneously tabulates all of the longest increasing or decreasing subsequences without incurring any additional asymptotic computational cost.

As mentioned in Section 2 above, Patience Sorting also has immediate connections to certain barred permutation patterns. We illustrate this in the following proposition.

Proposition 3.2.

(i) $S_n(3\text{-}\bar{1}\text{-}42) = \{RPW(R(\sigma)) \mid \sigma \in \mathfrak{S}_n\}$. *In particular, given $\sigma \in S_n(3\text{-}\bar{1}\text{-}42)$, the entries in each column of the insertion piles $R(\sigma)$ (when read from bottom to top) occupy successive positions in the permutation σ.*

(ii) $S_n(\bar{2}\text{-}41\text{-}3) = \{RPW(R(\sigma))^{-1} \mid \sigma \in \mathfrak{S}_n\}$. *In particular, given $\sigma \in S_n(\bar{2}\text{-}41\text{-}3)$, the columns of the insertion piles $R(\sigma)$ (when read from top to bottom) contain successive values.*

Proof. Part (1) is proven in [5], and part (2) follows immediately by taking inverses in (1).   □

As an immediate corollary, we can characterize an important category of classical permutation patterns in terms of barred permutation patterns.

Definition 3.3. Given a composition $\gamma = (\gamma_1, \gamma_2, \ldots, \gamma_m) \models n$, the *(classical) layered permutation pattern* $\pi_\gamma \in \mathfrak{S}_n$ is the permutation

$$\gamma_1 \cdots 1(\gamma_1 + \gamma_2) \cdots (\gamma_1 + 1) \cdots n \cdots (\gamma_1 + \gamma_2 + \cdots + \gamma_{m-1} + 1).$$

Example 3.4. Given $\gamma = (3, 2, 3) \models 8$, the corresponding layered pattern is $\pi_{(3,2,3)} = \widetilde{32154876} \in \mathfrak{S}_8$ (following the notation in [14]). Moreover, applying Extended Patience Sorting (Algorithm 2) to $\pi_{(3,2,3)}$:

$$R(\pi_{(3,2,3)}) = \begin{matrix} 1 & & 6 \\ 2 & 4 & 7 \\ 3 & 5 & 8 \end{matrix} \quad \text{and} \quad S(\pi_{(3,2,3)}) = \begin{matrix} 1 & & 6 \\ 2 & 4 & 7 \\ 3 & 5 & 8 \end{matrix}.$$

Note, in particular, that $\pi_{(3,2,3)}$ satisfies both of the conditions in Proposition 3.2, which illustrates the following characterization of layered patterns.

Corollary 3.5. $S_n(3\text{-}\bar{1}\text{-}42, \bar{2}\text{-}41\text{-}3)$ *is the set of layered patterns in* $\mathfrak{S}_n$.

Proof. Apply Proposition 3.2 noting that $S_n(3\text{-}\bar{1}\text{-}42, \bar{2}\text{-}41\text{-}3) = S_n(23\text{-}1, 31\text{-}2)$, as considered in [8].   □

As a consequence of this interaction between Patience Sorting and barred permutation patterns, we can now explicitly characterize those permutations for which the initial iteration of Geometric Patience Sorting (as defined in Section 1.1 above) yields non-crossing lattice paths.

Theorem 3.6. *The set* $S_n(3\text{-}\bar{1}\text{-}42, 31\text{-}\bar{4}\text{-}2)$ *consists of all reverse patience words having non-intersecting shadow diagrams (i.e., no shadowlines cross in the* 0th *iterate shadow diagram). Moreover, given a permutation* $\sigma \in S_n(3\text{-}\bar{1}\text{-}42, 31\text{-}\bar{4}\text{-}2)$, *the values in the bottom rows of* $R(\sigma)$ *and* $S(\sigma)$ *increase from left to right.*

Proof. From Theorem 2.3 and Corollary 2.5, $R(S_n(3\text{-}\bar{1}\text{-}42, 31\text{-}\bar{4}\text{-}2)) = R(S_n(23\text{-}1, 3\text{-}12))$ consists exactly of set partitions of $[n] = \{1, 2, \ldots, n\}$ whose components can be ordered so that both the minimal and maximal elements of the components simultaneously increase. (These are called *strongly monotone partitions* in [9]).

Let $\sigma \in S_n(3\text{-}\bar{1}\text{-}42, 31\text{-}\bar{4}\text{-}2)$. Since σ avoids $3\text{-}\bar{1}\text{-}42$, we have that $\sigma = RPW(R(\sigma))$ by Theorem 2.3. Thus, the ith shadowline $L_i^{(0)}(\sigma)$ is the boundary of the union of shadows generated by the ith left-to-right minima subsequence s_i of σ. Now, write $s_i = \varsigma_i a_i$, where $a_k > \cdots > a_2 > a_1$ form the right-to-left minima subsequence of σ (reversing the order of the elements in Definition 1.7) and where ς_i is a (possibly empty) decreasing sequence, and let b_i be the ith left-to-right maximum of σ. If ς_i is empty, then $b_i = a_i$. Otherwise, b_i is the leftmost (i.e., maximal) entry of $\varsigma_i a_i$, and so $\varsigma_i a_i = b_i \varsigma_i' a_i$ for some (possibly empty) decreasing subsequence ς_i'. Moreover, since b_i is the ith left-to-right maximum of σ, it must be at the bottom of the ith column of $R(\sigma)$. (Similarly, a_i is at the top of the ith column.) So the bottom rows of both $R(\sigma)$ and $S(\sigma)$ must be in increasing order.

Now consider the ith and j^{th} shadowlines $L_i^{(0)}(\sigma)$ and $L_j^{(0)}(\sigma)$ of σ, respectively, where $i < j$. We have that $b_i < b_j$, from which the initial horizontal segment of the ith shadowline is lower than that of the jth shadowline. Moreover, a_i is to the left of b_j, so the remaining segment of the ith shadowline is completely to the left of the remaining segment of the jth shadowline. Thus, $L_i^{(0)}(\sigma)$ and $L_j^{(0)}(\sigma)$ do not intersect.   □

In [6] the authors actually give the following stronger result:

Theorem 3.7. *Each iterate* $D^{(m)}(\sigma)$ *($m \geq 0$) of the exhaustive shadow diagram for $\sigma \in \mathfrak{S}_n$ is free from crossings if and only if every row in both $R(\sigma)$ and $S(\sigma)$ is monotone increasing from left to right.*

However, this only characterizes the output of the Extended Patience Sorting Algorithm involved. As such, Theorem 3.6 provides the first step toward characterizing those permutations that result in non-crossing lattice paths under Geometric Patience Sorting.

We conclude this section by noting that, while the strongly monotone condition implied by simultaneously avoiding $3\text{-}\bar{1}\text{-}42$ and $31\text{-}\bar{4}\text{-}2$ is necessary to alleviate such crossings, it is clearly not sufficient. (The problem lies with what we call "polygonal crossings" in the shadow diagrams in [6], which occur in permutations like $\sigma = 45312$.) Thus, to avoid crossings at all iterations of Geometric Patience Sorting, we need to impose further "ordinally increasing" conditions on the set partition associated to a given permutation under Patience Sorting. In particular, in addition to requiring just the minima and maxima elements in the set partition to increase as in the strongly monotone partitions encountered in the proof of Theorem 3.6, it is necessary to require that every record value simultaneously increase under an appropriate ordering of the blocks. That is, under a single ordering of these blocks, we must simultaneously have that the largest elements in each block increase, then the next largest elements, then the next-next largest elements, and so on. E.g., the partition $\{\{5, 3, 1\}, \{6, 4, 2\}\}$ of the set $[6] = \{1, 2, \ldots, 6\}$ satisfies this condition.

3.2 Invertibility of Patience Sorting

It is clear that many permutations can correspond to the same pile configuration under the Patience Sorting Algorithm. E.g., $R(3142) = R(3412)$. (As proven in [5], two permutations give rise to the same pile configuration under Patience Sorting if and only if they have the same left-to-right minima subsequences; e.g., 3142 and 3412 both have the left-to-right minima subsequences 31 and 42). In this section we use barred permutation patterns to characterize permutations for which this does not hold. We also establish a non-trivial enumeration for the resulting avoidance sets.

Theorem 3.8. *A pile configuration pile R has a unique preimage $\sigma \in \mathfrak{S}_n$ under Patience Sorting if and only if $\sigma \in S_n(3\text{-}\bar{1}\text{-}42, 3\text{-}\bar{1}\text{-}24)$.*

Proof. It is clear that every pile configuration R has at least its reverse patience word $RPW(R)$ as a preimage under Patience Sorting. By Proposition 3.2, reverse patience words are exactly those permutations that avoid the barred pattern $3\text{-}\bar{1}\text{-}42$. Furthermore, as shown in [5], two permutations have the same insertion piles under Extended Patience Sorting (Algorithm 2) if and only if one can be obtained from the other by a sequence of order-isomorphic exchanges of the form

$$3\text{-}\bar{1}\text{-}24 \rightsquigarrow 3\text{-}\bar{1}\text{-}42 \quad \text{or} \quad 3\text{-}\bar{1}\text{-}42 \rightsquigarrow 3\text{-}\bar{1}\text{-}24.$$

(I.e., the occurrence of one pattern is reordered to form an occurrence of the other pattern.) Thus, it is easy to see that R has the unique preimage $RPW(R)$ if and only if $RPW(R)$ avoids both $3\text{-}\bar{1}\text{-}42$ and $3\text{-}\bar{1}\text{-}24$. □

Given this pattern avoidance characterization of invertibility, we have the following recurrence relation for the size of the corresponding avoidance sets.

Lemma 3.9. *Set $f(n) = |S_n(3\text{-}\bar{1}\text{-}42, 3\text{-}\bar{1}\text{-}24)|$ and, for $k \leq n$, denote by $f(n,k)$ the cardinality*

$$f(n,k) = \#\left\{\sigma \in S_n(3\text{-}\bar{1}\text{-}42, 3\text{-}\bar{1}\text{-}24) \mid \sigma(1) = k\right\}.$$

Then $f(n) = \sum_{k=1}^{n} f(n,k)$, and $f(n,k)$ satisfies the four-part recurrence relation

$$f(n,0) = 0 \quad for\ n \geq 1 \tag{1}$$

$$f(n,1) = f(n,n) = f(n-1) \quad for\ n \geq 1 \tag{2}$$

$$f(n,2) = 0 \quad for\ n \geq 3, \tag{3}$$

and for $n \geq 3$,

$$f(n,k) = f(n,k-1) + f(n-1,k-1) + f(n-2,k-2) \tag{4}$$

subject to the initial conditions $f(0,0) = f(0) = 1$.

Proof. Note first that Equation (1) is the obvious boundary condition for $k = 0$.

Now, suppose that the first component of $\sigma \in S_n(3\text{-}\bar{1}\text{-}42, 3\text{-}\bar{1}\text{-}24)$ is either $\sigma(1) = 1$ or $\sigma(1) = n$. Then $\sigma(1)$ cannot be part of any occurrence of 3-$\bar{1}$-42 or 3-$\bar{1}$-24 in σ. Thus, upon removing $\sigma(1)$ from σ, and subtracting one from each component if $\sigma(1) = 1$, a bijection is formed with $S_{n-1}(3\text{-}\bar{1}\text{-}42, 3\text{-}\bar{1}\text{-}24)$. Therefore, Equation (2) follows.

Next, suppose that the first component of $\sigma \in S_n(3\text{-}\bar{1}\text{-}42, 3\text{-}\bar{1}\text{-}24)$ is $\sigma(1) = 2$. Then the first column of $R(\sigma)$ must be $r_1 = 21$ regardless of where 1 occurs in σ. Therefore, $R(\sigma)$ has the unique preimage σ if and only if $\sigma = 21 \in \mathfrak{S}_2$, and from this Equation (3) follows.

Finally, suppose that $\sigma \in S_n(3\text{-}\bar{1}\text{-}42, 3\text{-}\bar{1}\text{-}24)$ with $3 \leq k \leq n$. Since σ avoids 3-$\bar{1}$-42, σ is a RPW by Proposition 3.2, and hence the left prefix of σ from k to 1 is a decreasing subsequence. Let σ' be the permutation obtained by interchanging the values k and $k-1$ in σ. Then the only instances of the patterns 3-$\bar{1}$-42 and 3-$\bar{1}$-24 in σ' must involve both k and $k-1$. Note that the number of σ for which no instances of these patterns are created by interchanging k and $k-1$ is $f(n,k-1)$.

There are now two cases in which an instance of the barred pattern 3-$\bar{1}$-42 or 3-$\bar{1}$-24 will be created in σ' by this interchange:

Case 1. If $k-1$ occurs between $\sigma(1) = k$ and 1 in σ, then $\sigma(2) = k-1$, so interchanging k and $k-1$ will create an instance of the pattern 23-1 via the subsequence $(k-1,k,1)$ in σ'. Thus, by Theorem 2.3, σ' contains 3-$\bar{1}$-42 from which $\sigma' \in S_n(3\text{-}\bar{1}\text{-}42)$ if and only if $k-1$ occurs after 1 in σ. Note also that if $\sigma(2) = k - 1$, then removing k from σ yields a bijection with permutations in $S_{n-1}(3\text{-}\bar{1}\text{-}42, 3\text{-}\bar{1}\text{-}24)$ that start with $k - 1$. Therefore, the number of permutations counted in Case 1 is $f(n - 1, k - 1)$.

Case 2. If $k - 1$ occurs to the right of 1 in σ, then σ' both contains the subsequence $(k - 1, 1, k)$ and avoids the pattern $3\text{-}\bar{1}\text{-}42$, so it must also contain the pattern $3\text{-}\bar{1}\text{-}24$. If an instance of $3\text{-}\bar{1}\text{-}24$ in σ' involves both $k - 1$ and k, then $k - 1$ and k must play the roles of "3" and "4", respectively. Moreover, if the value ℓ preceding k is not 1, then the subsequence $(k - 1, 1, \ell, k)$ is an instance of $3\text{-}1\text{-}24$, so $(k - 1, \ell, k)$ is not an instance of $3\text{-}\bar{1}\text{-}24$. Therefore, for σ' to contain $3\text{-}\bar{1}\text{-}24$, k must follow 1 in σ', and so $k - 1$ follows 1 in σ. Similarly, if the letter preceding 1 is some $m < k$, then the subsequence $(m, 1, k - 1)$ is an instance of $3\text{-}\bar{1}\text{-}24$ in σ, which is impossible. Therefore, k must precede 1 in σ, from which σ must start with the initial segment $(k, 1, k - 1)$. It follows that removing the values k and 1 from σ and then subtracting 1 from each component yields a bijection with permutations in $S_{n-2}(3\text{-}\bar{1}\text{-}42, 3\text{-}\bar{1}\text{-}24)$ that start with $k - 2$. Thus, the number of permutations counted in Case 2 is then exactly $f(n - 2, k - 2)$, which yields Equation (4). $\square$

If we denote by

$$\Phi(x, y) = \sum_{n=0}^{\infty} \sum_{k=0}^{n} f(n, k) x^n y^k$$

the bivariate generating function for the sequence $\{f(n, k)\}_{n \geq k \geq 0}$, then Equation (4) implies that

$$(1 - y - xy - x^2 y^2) \Phi(x, y)$$
$$= 1 - y - xy + xy^2 - xy^2 \Phi(xy, 1) + xy(1 - y - xy) \Phi(x, 1). \quad (5)$$

Moreover, we can use the kernel method by choosing y such that $1 - y - xy - x^2 y^2 = 0$ and solving for $\Phi(x, 1)$. Then Equation (5) implies

$$x + 1 + \frac{\sqrt{1 + 2x + 5x^2} - x - 1}{2} F(x)$$
$$- F\left(\frac{\sqrt{1 + 2x + 5x^2} - x - 1}{2x}\right) = 0, \quad (6)$$

where $F(x) = \sum_{n \geq 0} f(n) x^n = \Phi(x, 1)$ is the generating function for the sequence $\{f(n)\}_{n \geq 0}$. Now let

$$t = \frac{\sqrt{1 + 2x + 5x^2} - x - 1}{2x};$$

then $xt^2 + (x+1)t - x = 0$, or, equivalently,

$$x = \frac{t}{1 - t - t^2}.$$

Equation (6) thus implies that

$$F(t) = \frac{t^2}{1 - t - t^2} F\left(\frac{t}{1 - t - t^2}\right) + \frac{1 - t^2}{1 - t - t^2} \tag{7}$$

This, in turn, implies the following main enumerative result about invertibility of Patience Sorting.

Theorem 3.10. *Denote by F_n the nth Fibonacci number (with $F_0 = F_1 = 1$) and by*

$$a(n, k) = \sum_{\substack{n_1, \ldots, n_{k+1} \geq 0 \\ n_1 + \cdots + n_{k+1} = n - k - 2}} F_{n_1} F_{n_2} \cdots F_{n_{k+1}}$$

the convolved Fibonacci numbers for $n \geq k+2$, $k \geq 0$ (where $a(n, k) := 0$ otherwise). Then, defining

$$X = \begin{bmatrix} f(0) \\ f(1) \\ f(2) \\ f(3) \\ f(4) \\ \vdots \end{bmatrix}, \quad F = \begin{bmatrix} 1 \\ F_0 \\ F_1 \\ F_2 \\ F_3 \\ \vdots \end{bmatrix} = \begin{bmatrix} F_0 \\ F_1 \\ F_2 \\ F_3 \\ F_4 \\ \vdots \end{bmatrix} - \begin{bmatrix} 0 \\ 0 \\ F_0 \\ F_1 \\ F_2 \\ \vdots \end{bmatrix},$$

and

$$\mathbf{A} = (a(n, k))_{n, k \geq 0} = \begin{bmatrix} 0 \\ 0 & 0 \\ a(2,0) & 0 & 0 \\ a(3,0) & a(3,1) & 0 & 0 \\ a(4,0) & a(4,1) & a(4,2) & 0 & 0 \\ \vdots & \vdots & \vdots & \vdots & \vdots & \ddots \end{bmatrix},$$

we have that $X = (\mathbf{I} - \mathbf{A})^{-1} F$, where $\mathbf{I}$ is the infinite identity matrix and $\mathbf{A}$ is lower triangular.

Proof. Letting

$$u(t) = \frac{t}{1 - t - t^2},$$

$$v(t) = tu(t) = \frac{t^2}{1 - t - t^2},$$

and

$$w(t) = 1 + u(t) = \frac{1 - t^2}{1 - t - t^2},$$

we can rewrite Equation (7) as

$$F(t) = v(t)F(u(t)) + w(t).$$

This, by a standard Riordan array argument (see [16]), is equivalent to the matrix equation

$$X = \mathbf{A}X + F,$$

where X is the vector of coefficients of the power series of $F(t)$, F is the vector of coefficients of the power series of $w(t)$, and $\mathbf{A} = (a(n, k))_{n,k \geq 0}$ is the matrix such that the generating function of the entries in each column $k \geq 0$ is $v(t)u(t)^k$. In other words,

$$\sum_{n \geq 0} a(n, k)t^n = \frac{t^{k+2}}{(1 - t - t^2)^{k+1}}.$$

It follows that the coefficients $a(n, k)$ in the above equation are convolved Fibonacci numbers (sequence A037027 of [17]) and form the so-called skew Fibonacci-Pascal triangle in the matrix $\mathbf{A} = (a(n, k))_{n,k \geq 0}$. In particular, the sequence of entries in column $k \geq 0$ of $\mathbf{A}$ is $k + 2$ zeros followed by the kth convolution of the sequence $\{F_n\}_{n \geq 0}$.

Therefore, since $\mathbf{I} - \mathbf{A}$ is clearly invertible, the result follows.   □

Note that a direct bijective proof of Theorem 3.10 is also given after the following remark.

Remark 3.11. Since $\mathbf{A}$ is strictly lower triangular with zero main diagonal and zero sub-diagonal, it follows that multiplication of a matrix $\mathbf{B}$ by $\mathbf{A}$ shifts the position of the highest nonzero diagonal in $\mathbf{B}$ down by two rows. Thus, $(\mathbf{I} - \mathbf{A})^{-1} = \sum_{n \geq 0} \mathbf{A}^n$ as a Neumann series, and so all nonzero entries of $(\mathbf{I} - \mathbf{A})^{-1}$ are positive integers.

In particular, one can explicitly compute

$$
\mathbf{A} = \begin{bmatrix}
0 \\
0 & 0 \\
1 & 0 & 0 \\
1 & 1 & 0 & 0 \\
2 & 2 & 1 & 0 & 0 \\
3 & 5 & 3 & 1 & 0 & 0 \\
5 & 10 & 9 & 4 & 1 & 0 & 0 \\
8 & 20 & 22 & 14 & 5 & 1 & 0 & 0 \\
\vdots & \vdots & \vdots & \vdots & \vdots & \vdots & \vdots & \vdots & \ddots
\end{bmatrix}
$$

$$
\implies \quad (\mathbf{I} - \mathbf{A})^{-1} = \begin{bmatrix}
1 \\
0 & 1 \\
1 & 0 & 1 \\
1 & 1 & 0 & 1 \\
3 & 2 & 1 & 0 & 1 \\
7 & 6 & 3 & 1 & 0 & 1 \\
21 & 16 & 10 & 4 & 1 & 0 & 1 \\
66 & 50 & 30 & 15 & 5 & 1 & 0 & 1 \\
\vdots & \vdots & \vdots & \vdots & \vdots & \vdots & \vdots & \vdots & \ddots
\end{bmatrix}
$$

from which the first few values of the sequence $\{f(n)\}_{n \geq 0}$ are immediately calculable as

$$1, 1, 2, 4, 9, 23, 66, 209, 718, 2645, 10373, 43090, 188803, 869191, \ldots .$$

We now close this section with the following bijective proof, which provides a more direct explanation for the convolved Fibonacci numbers in Theorem 3.10.

Bijective Proof of Theorem 3.10. Suppose we are given a permutation σ uniquely determined by $R = R(\sigma)$. Then $\sigma = RPW(R)$. Say σ starts with k. If $k > 1$, then the left prefix of σ from k to 1 is a decreasing subsequence. Also, if l precedes 1 in σ and k' follows 1, then $k' < l$ (since σ avoids 3-$\bar{1}$-24). Hence, if the first basic subsequence has more than one term, then the first entry k of the first basic subsequence is greater than the first entry k' of the second basic subsequence.

Therefore, it is easy to see by induction on the number of basic sub-sequences of σ that the first entries in the basic subsequences of σ (i.e., the entries of the bottom row of R) decrease from left-to-right until we reach a basic subsequence of length one (i.e., a column of R of height one). This is the value z between the first two consecutive ascents in σ. Note that the terms of σ to the right of z must all be greater than z since otherwise σ will contain the pattern 23-1 and hence also the pattern 3-$\bar{1}$-42.

It follows that we can represent R as a (possibly empty) sequence $R_1', R_2', \ldots$ of sub-pile configurations, each starting on the left either at the beginning or after a column of height one and ending on the right at the next column of height one, followed by a (possibly empty) sub-pile configuration R'' that has no columns of height one. Note also that the left prefix of σ that corresponds to each sub-pile configuration R_i' must start with its largest letter. In particular, k is the largest entry in R_1'. Moreover, without loss of generality, we need only consider R_1'. (This follows from the fact that $R_2', R_3', \ldots$ can be obtained using a recursive procedure similar the one used to form shadow diagrams in Definition 1.6.) For example,

$$R(7\ 1\ 2\ 9\ 3\ 6\ 4\ 5\ 10\ 8) = \begin{matrix} 1 & & 3 & 4 & & 8 \\ 7 & 2 & 9 & 6 & 5 & 10 \end{matrix},$$

with

$$R_1' = \begin{matrix} 1 \\ 7 & 2 \end{matrix}, \quad R_2' = \begin{matrix} 3 & 4 \\ 9 & 6 & 5 \end{matrix}, \quad \text{and } R'' = \begin{matrix} 8 \\ 10 \end{matrix}.$$

Now, let C be a sub-pile configuration having only one preimage under Patience Sorting such that the bottommost entry in the leftmost column of C is equal to $k+1$. If C has $k+1-m$ entries and exactly one column (namely, its rightmost column) of height one, then we say that C is a sub-pile configuration of *Type I*. Similarly, if C has $k+1$ entries and no columns of height one, then we say that C is a sub-pile configuration of *Type II*.

Let $d(k+1, m)$ to denote the number of sub-pile configurations of Type I, and let $g(k)$ to denote the number of sub-pile configurations of Type II. Then, setting

$$f = [f(0), f(1), f(2), \ldots]^T,$$

$$D = (d(k+1, m))_{k, m \geq 0},$$

and

$$g = [g(0), g(1), g(2), \ldots]^T,$$

it follows from our previous observation about the structure of R that

$$f = g + Dg + D^2 g + \cdots = (I - D)^{-1} g.$$

Consequently, we need only now prove that $D = \mathbf{A}$ and that $g = F$.

Let C_I be a sub-pile configuration of R of Type I, and set $\tau = RPW(C_I)$. From the discussion in the first paragraph of the proof, it follows that the values of the entire second basic subsequence in τ are contained between the smallest and second smallest terms of the first basic subsequence in τ. Proceeding by induction, a similar statement holds for the jth and $(j+1)^{\text{st}}$ basic subsequences in τ, for any $j \geq 1$.

Moreover, the smallest terms in the basic subsequences of τ are exactly the right-to-left minima of τ and are also the topmost entries in the columns of C_I. Consequently, these terms form an increasing subsequence $\{x_i\}$. On the other hand, the remaining terms of τ, together with the terminal entry of τ (which corresponds to the basic subsequence of length one), form a decreasing subsequence $\{y_j\}$, and it is easy to see that

$$k = y_1 > y_2 > \cdots > y_{\text{last}} = z = x_{\text{last}} > \cdots > x_2 > x_1 = 1,$$

where z is the rightmost element of τ. In addition, it should be clear that no basic subsequence in τ, except for the last basic subsequence, can end with two consecutive integers. Thus, for each $i < z$, it follows that i and $i + 1$ must be in different basic subsequence, and so $x_i = i$ for each $i \leq z$.

Now, note that the entry w of R following z must be greater than z since z forms a column. Therefore, no entry u after z can be less than z. (Otherwise, the word zwu would be an instance of the subsequence 23-1 and thus imply that σ contains 3-$\bar{1}$-42 as well, which is impossible.) Suppose also that C_{II} is a sub-pile configuration of R of Type II. Then, using a similar argument, we see that there is exactly one way in which a column of height one can be appended to C_{II} such that C_{II} becomes a sub-pile configuration of Type I. Consequently, the vector g satisfies $g(0) = g(1) = 1$, and, for $n \geq 2$, we have that $g(n) = d(n+1, 0)$.

Finally, we again let C_I be a sub-pile configuration of R of Type I and set $\tau = RPW(C_I)$. Then τ has $k - m$ entries between k and 1. (More generally, τ has $k - m$ entries between its greatest and least element.) Thus, τ is missing m elements from σ. Thus, we can map τ into a unique sequence of elements from the set $\{|, 1, 2\}$ as follows: For each $i \geq 1$, let b_i be the number of elements $v \in (y_{i+1}, y_i)$ to the right of y_{i+1} (so that the sequence $\{b_i\}$ has one less term than $\{y_i\}$). If y_{i+1} immediately follows y_i in τ, then map y_i to $b_i - 1$ bars followed by the element 1, and if there is some x_j between y_i and y_{i+1} (in which case there must be exactly one such x_i), then map y_i to $b_i - 1$ bars followed by the element 2.

For example, if we set $k = 9$, $m = 2$, and $z = 3$, then the left prefix 9817523 maps to $12|1|2$, and the left prefix 9716423 maps to $|12|12$. Moreover, both permutations have $(x_1, x_2, x_3) = (1, 2, 3)$, while $(y_1, y_2, y_3, y_4, y_5) = (9, 8, 7, 5, 3)$ in the first case and $(y_1, y_2, y_3, y_4, y_5) = (9, 7, 6, 4, 3)$ in the second.

As one can see, the map defined above is a bijection since it uniquely determines the sequence $\{y_i\}$ and the positions of the x_i's (whose values were uniquely determined to begin with). Note that the last two terms of τ must be $(z - 1, z)$, so the last entry in the image sequence is always 2. Each 1 or 2 counts the number of terms of τ from y_i before y_{i+1}, so the sum of 1's and 2's is the number of terms of τ other than z, i.e., $k + 1 - m - 1 = k - m$. Hence, the sum of all 1's and 2's except the last 2 is $k - m - 2$.

It is well-known that concatenation of objects corresponds to convolution of their counting sequences and that the number of sequences of 1's and 2's whose sum is n is the nth Fibonacci number F_n. Hence, it is easy to see that $d(n, m)$ is the convolved Fibonacci number $a(n, m)$, for $n \geq 1$, and that

$$g(n) = d(n + 1, 0) = a(n + 1, 0) = F_{n-1},$$

for $n \geq 2$. This ends the proof. $\qquad\qquad\qquad\qquad\square$

References

[1] M. H. Albert, S. Linton, and N. Ruškuc. The insertion encoding of permutations. *Electron. J. Combin.*, 12(1):Research paper 47, 31 pp., 2005.

[2] D. Aldous and P. Diaconis. Longest increasing subsequences: from patience sorting to the Baik-Deift-Johansson theorem. *Bull. Amer. Math. Soc. (N.S.)*, 36(4):413–432, 1999.

[3] S. Bespamyatnikh and M. Segal. Enumerating longest increasing subsequences and patience sorting. *Inform. Process. Lett.*, 76(1-2):7–11, 2000.

[4] M. Bóna. *Combinatorics of permutations.* Discrete Mathematics and its Applications (Boca Raton). Chapman & Hall/CRC, Boca Raton, FL, 2004.

[5] A. Burstein and I. Lankham. Combinatorics of patience sorting piles. *Sém. Lothar. Combin.*, 54A:Art. B54Ab, 19 pp., 2005/07.

[6] A. Burstein and I. Lankham. A geometric form for the extended patience sorting algorithm. *Adv. in Appl. Math.*, 36(2):106–117, 2006.

[7] A. Claesson. Generalized pattern avoidance. *European J. Combin.*, 22(7):961–971, 2001.

[8] A. Claesson and T. Mansour. Counting occurrences of a pattern of type $(1,2)$ or $(2,1)$ in permutations. *Adv. in Appl. Math.*, 29(2):293–310, 2002.

[9] A. Claesson and T. Mansour. Enumerating permutations avoiding a pair of Babson-Steingrímsson patterns. *Ars Combin.*, 77:17–31, 2005.

[10] S. Dulucq, S. Gire, and O. Guibert. A combinatorial proof of J. West's conjecture. *Discrete Math.*, 187(1-3):71–96, 1998.

[11] S. Dulucq, S. Gire, and J. West. Permutations with forbidden subsequences and nonseparable planar maps. *Discrete Math.*, 153(1-3):85–103, 1996.

[12] C. L. Mallows. Problem 62-2, patience sorting. *SIAM Review*, 4:148–149, 1962. Solution in Vol. **5** (1963), 375–376.

[13] A. Marcus and G. Tardos. Excluded permutation matrices and the Stanley-Wilf conjecture. *J. Combin. Theory Ser. A*, 107(1):153–160, 2004.

[14] A. Price. *Packing densities of layered patterns*. PhD thesis, Univ. of Pennsylvania, 1997.

[15] B. E. Sagan. *The Symmetric Group*, volume 203 of *Graduate Texts in Mathematics*. Springer-Verlag, New York, second edition, 2001.

[16] L. W. Shapiro, S. Getu, W. J. Woan, and L. C. Woodson. The Riordan group. *Discrete Appl. Math.*, 34(1-3):229–239, 1991.

[17] N. J. A. Sloane. The On-line Encyclopedia of Integer Sequences. Available online at http://www.research.att.com/~njas/sequences/.

[18] R. P. Stanley. *Enumerative combinatorics. Vol. 2*, volume 62 of *Cambridge Studies in Advanced Mathematics*. Cambridge University Press, Cambridge, 1999.

[19] G. Viennot. Une forme géométrique de la correspondance de Robinson-Schensted. In *Combinatoire et représentation du groupe symétrique (Actes Table Ronde CNRS, Univ. Louis-Pasteur Strasbourg, Strasbourg, 1976)*, pages 29–58. Lecture Notes in Math., Vol. 579. Springer, Berlin, 1977.

[20] J. West. *Permutations with forbidden subsequences and stack-sortable permutations*. PhD thesis, M.I.T., 1990.

[21] A. Woo and A. Yong. When is a Schubert variety Gorenstein? *Adv. Math.*, 207(1):205–220, 2006.

[22] A. Woo and A. Yong. Governing singularities of Schubert varieties. *J. Algebra*, 320(2):495–520, 2008.

Permutations with k-regular descent patterns

Anthony Mendes

Department of Mathematics
Cal Poly State University
San Luis Obispo, CA 93407

Jeffrey B. Remmel†

Department of Mathematics
University of California, San Diego
La Jolla, CA 92093

Amanda Riehl

Department of Mathematics
University of Wisconsin Eau Claire, Eau Claire, WI 54702

Abstract

We find generating functions for the permutations in S_{i+nk+j} with descent set $\{i, i+k, i+2k, \ldots, i+nk\}$ for integers i, j, k and n satisfying $k \geq 2$, $0 \leq i \leq k-1$ and $1 \leq j \leq k$. These permutations are said to have *k-regular descent patterns*.

The generating functions are found by introducing homomorphisms on the ring of symmetric functions. For the most general results, we introduce a new class of symmetric functions $p_{n,\alpha_1,\ldots,\alpha_r}$ that depend on r functions α_i. The generating function identities will follow from applying our homomorphisms to identities involving p_{n,α_1,α_2}.

1 Introduction

Let i, j, k, and n be nonnegative integers satisfying $k \geq 2$, $0 \leq i \leq k-1$, and $1 \leq j \leq k$. A permutation with descent set equal to $\{i+k, i+$

† Partially supported by NSF grant 0654060

$2k, \ldots, i + nk\}$ will be called a permutation with a *k-regular descent pattern*. Let $E^{i,j,k}_{i+kn+j}$ be the number of such permutations in S_{i+nk+j}.

In the special case where $k = 2$, $i = 0$, and $j = 2$, $E^{0,2,2}_{2n+2}$ is the number of permutations in S_{2n+2} with descent set $\{2, 4, \ldots, 2n\}$. These are the classical even alternating permutations. André [1, 2] proved that

$$1 + \sum_{n \geq 0} \frac{E^{0,2,2}_{2n+2}}{(2n+2)!} t^{2n+2} = \sec t.$$

Similarly, $E^{0,1,2}_{2n+1}$ counts the number of odd alternating permutations and

$$\sum_{n \geq 0} \frac{E^{0,1,2}_{2n+1}}{(2n+1)!} t^{2n+1} = \tan t.$$

These numbers are also called the Euler numbers. When $k \geq 0$, $E^{0,j,k}_{kn+j}$ are called generalized Euler numbers [14]. There are well-known generating functions for q-analogues of the generalized Euler numbers; see Stanley's book [24], Section 3.16. Various divisibility properties of the q-Euler numbers have been studied in [3, 4, 10] and of the generalized q-Euler numbers in [11, 22]. Prodinger [19] also studied q-analogues of the number $E^{1,2,2}_{2n+1}$ and $E^{1,1,2}_{2n+2}$.

Our goal is to find and refine generating functions for $E^{i,j,k}_{i+kn+j}$. This will be done by applying ring homomorphisms to symmetric function identities. This technique of understanding permutation enumeration through symmetric function identities further advances an already well-documented line of research [5, 13, 16, 17, 18, 25].

The nth elementary symmetric function e_n in the variables $x_1, x_2, \ldots$ is given by

$$E(t) = \sum_{n \geq 0} e_n t^n = \prod_i (1 + x_i t)$$

and the nth homogeneous symmetric function h_n in the variables $x_1, x_2, \ldots$ is given by

$$H(t) = \sum_{n \geq 0} h_n t^n = \prod_i \frac{1}{1 - x_i t}.$$

Although a trivial consequence of these definitions is

$$H(t) = 1/E(-t), \tag{1}$$

a surprisingly large number of results on generating functions for various permutation statistics have been derived by applying a ring homomorphism on simple identities such as (1).

Let $\sigma = \sigma_1 \cdots \sigma_n$ be a permutation in S_n written in one line notation. We define

$$Des(\sigma) = \{i : \sigma_i > \sigma_{i+1}\} \qquad Rise(\sigma) = \{i : \sigma_i < \sigma_{i+1}\}$$
$$des(\sigma) = |Des(\sigma)| \qquad\qquad rise(\sigma) = |Rise(\sigma)|$$
$$inv(\sigma) = \sum_{i<j} \chi(\sigma_i > \sigma_j) \qquad coinv(\sigma) = \sum_{i<j} \chi(\sigma_i < \sigma_j)$$

where for any statement A, $\chi(A) = 1$ if A is true and $\chi(A) = 0$ if A is false. These definitions make sense for any sequence of numbers, not just permutations. Given $\sigma^1, \ldots, \sigma^k \in S_n$, we write

$$comdes(\sigma^1, \ldots, \sigma^k) = \left| \bigcap_{i=1}^{k} Des(\sigma^i) \right|.$$

Standard notation for q and p, q analogues will be used; that is, we let

$$[n] = [n]_{p,q} = \frac{p^n - q^n}{p - q},$$

$$[n]! = [n]_{p,q}! = [n][n-1] \cdots [1],$$

$$\begin{bmatrix} n \\ k \end{bmatrix} = \begin{bmatrix} n \\ k \end{bmatrix}_{p,q} = \frac{[n]!}{[k]![n-k]!}, \qquad \text{and}$$

$$\begin{bmatrix} n \\ \lambda_1, \ldots, \lambda_\ell \end{bmatrix} = \begin{bmatrix} n \\ \lambda_1, \ldots, \lambda_\ell \end{bmatrix}_{p,q} = \frac{[n]!}{[\lambda_1]! \cdots [\lambda_\ell]!}.$$

All of the following results can proved by applying a suitable homomorphism to the identity (1):

$$\sum_{n=0}^{\infty} \frac{u^n}{n!} \sum_{\sigma \in S_n} x^{des(\sigma)} = \frac{1-x}{-x + e^{u(x-1)}}$$

$$\sum_{n=0}^{\infty} \frac{u^n}{(n!)^2} \sum_{(\sigma,\tau) \in S_n \times S_n} x^{comdes(\sigma,\tau)} = \frac{1-x}{-x + J(u(x-1))}$$

$$\sum_{n=0}^{\infty} \frac{u^n}{[n]_q!} \sum_{\sigma \in S_n} x^{des(\sigma)} q^{inv(\sigma)} = \frac{1-x}{-x + e_q(u(x-1))}$$

$$\sum_{n=0}^{\infty} \frac{u^n}{[n]_q![n]_p!} \sum_{(\sigma,\tau)\in S_n \times S_n} x^{comdes(\sigma,\tau)} q^{inv(\sigma)} p^{inv(\tau)}$$

$$= \frac{1-x}{-x + J_{q,p}(u(x-1))}$$

where $[n]_q = [n]_{1,q}$,

$$e_q(u) = \sum_{n=0}^{\infty} \frac{u^n}{[n]_q!} q^{\binom{n}{2}},$$

$$J_{q,p}(u) = \sum_{n=0}^{\infty} \frac{u^n}{[n]_q![n]_p!} q^{\binom{n}{2}} p^{\binom{n}{2}},$$

and $J(u) = J_{1,1}(u)$; see Carlitz [7], Stanley [23] and Fedou & Rawlings [9].

In [16], Mendes used similar methods to derive the generating functions for p,q-analogues of the generating functions for the even and odd alternating permutations. In this paper, we significantly generalize his methods.

Let $C_{i+kn+j}^{i,j,k}$ denote the set of permutations $\sigma \in S_{i+kn+j}$ with $Des(\sigma) \subseteq \{i, i+k, \ldots, i+nk\}$ and $C_{i+kn+j}^{i,j,k} = |C_{i+kn+j}^{i,j,k}|$. Similarly, let $\mathcal{E}_{i+kn+j}^{i,j,k}$ denote the set of permutations $\sigma \in S_{i+kn+j}$ with $Des(\sigma) = \{i, i+k, \ldots, i+nk\}$ so that $E_{i+kn+j}^{i,j,k} = |\mathcal{E}_{i+kn+j}^{i,j,k}|$. Lastly, for $\sigma \in S_{i+kn+j}$, let $Ris_{i,k}(\sigma) = \{s : 0 \leq s \leq n \text{ and } \sigma_{i+sk} < \sigma_{i+sk+1}\}$ and $ris_{i,k}(\sigma) = |Ris_{i,k}(\sigma)|$. Then $E_{i+kn+j}^{i,j,k}$ is the number of $\sigma \in C_{i+kn+j}^{i,j,k}$ such that $Ris_{i,k}(\sigma) = \varnothing$. To generalize the results of Mendes, we will find the generating function for

$$\sum_{n\geq 0} \frac{t^{i+kn+j}}{(i+kn+j)!} \sum_{\sigma \in C_{i+kn+j}^{i,j,k}} x^{ris_{i,k}(\sigma)}. \qquad (2)$$

Setting $x = 0$ in (2) will give the generating function for $E_{i+kn+j}^{i,j,k}$. We will also find p,q-analogues of such generating functions.

To obtain the generating function for (2), we introduce a new class of symmetric functions p_{n,α_1,α_2} which depend on two weight functions α_1 and α_2. Our results will follow by applying a ring homomorphism to a symmetric function identity involving p_{n,α_1,α_2}'s. In fact, our methods provide vast extensions of (2). Moreover, our extension will contain as special cases all of the generating functions for the q-Euler and generalized q-Euler numbers in the papers mentioned above.

In order to fully extend our results, suppose $\boldsymbol{\Sigma} = (\sigma^{(1)}, \ldots, \sigma^{(L)})$ is a sequence of permutations with $\sigma^{(i)} \in \mathcal{C}^{i,j,k}_{i+kn+j}$. Define

$$Comris_{i,k}(\boldsymbol{\Sigma}) = \{s : 0 \leq s \leq n \text{ and for all } 1 \leq t \leq L, \sigma^{(t)}_{i+sk} < \sigma^{(t)}_{i+sk+1})\}$$

and let $comris_{i,k}(\boldsymbol{\Sigma}) = |Comris_{i,k}(\boldsymbol{\Sigma})|$. Given two sequences of indeterminates, $\mathbf{Q} = (q_1, \ldots, q_L)$ and $\mathbf{P} = (p_1, \ldots, p_L)$, let

$$\mathbf{Q}^m = q_1^m \cdots q_L^m, \qquad\qquad \mathbf{P}^m = p_1^m \cdots p_L^m,$$

$$[n]_{\mathbf{P},\mathbf{Q}} = \prod_{i=1}^{L} [n]_{p_i,q_i}, \qquad\qquad [n]_{\mathbf{P},\mathbf{Q}}! = \prod_{i=1}^{L} [n]_{p_i,q_i}!,$$

$$\mathbf{Q}^{inv(\boldsymbol{\Sigma})} = \prod_{i=1}^{L} q_i^{inv(\sigma^{(i)})}, \qquad \mathbf{P}^{coinv(\boldsymbol{\Sigma})} = \prod_{i=1}^{L} p_i^{coinv(\sigma^{(i)})}, \qquad \text{and}$$

$$\begin{bmatrix} n \\ \lambda_1, \ldots, \lambda_k \end{bmatrix}_{\mathbf{P},\mathbf{Q}} = \prod_{i=1}^{L} \begin{bmatrix} n \\ \lambda_1, \ldots, \lambda_k \end{bmatrix}_{p_i,q_i}.$$

In addition, we set

$$e_{\mathbf{P},\mathbf{Q},k}(t) = \sum_{n \geq 0} \frac{t^{kn} \mathbf{P}^{\binom{kn}{2}}}{[kn]_{\mathbf{P},\mathbf{Q}}!}$$

$$e^{(j)}_{\mathbf{P},\mathbf{Q},k}(t) = \sum_{n \geq 1} \frac{t^{kn} \mathbf{P}^{\binom{k(n-1)+j}{2}}}{[k(n-1)+j]_{\mathbf{P},\mathbf{Q}}!}.$$

Our first generalization of (2) is found when $i = 0$ and $j = k$. In this case, we will show that

$$1 + \sum_{n \geq 1} \frac{t^{kn}}{[kn]_{\mathbf{P},\mathbf{Q}}!} \sum_{\boldsymbol{\Sigma} \in (\mathcal{C}^{0,k,k}_{kn})^L} x^{comris_{0,k}(\boldsymbol{\Sigma})} \mathbf{Q}^{inv(\boldsymbol{\Sigma})} \mathbf{P}^{coinv(\boldsymbol{\Sigma})}$$

$$= \frac{1-x}{-x + e_{\mathbf{P},\mathbf{Q},k}(t(x-1)^{1/k})}. \qquad (3)$$

The next case is when $i = 0$ and $1 \leq j \leq k - 1$. Here, we will show that

$$\sum_{n \geq 1} \frac{t^{kn}}{[k(n-1)+j]_{\mathbf{P},\mathbf{Q}}!} \sum_{\boldsymbol{\Sigma} \in (\mathcal{C}^{0,j,k}_{k(n-1)+j})^L} x^{comris_{0,k}(\boldsymbol{\Sigma})} \mathbf{Q}^{inv(\boldsymbol{\Sigma})} \mathbf{P}^{coinv(\boldsymbol{\Sigma})}$$

$$= \frac{-e^{(j)}_{\mathbf{P},\mathbf{Q},k}(t(x-1)^{1/k})}{-x + e_{\mathbf{P},\mathbf{Q},k}(t(x-1)^{1/k})}. \qquad (4)$$

Lastly, in the case in the case where $1 \leq i, j \leq k - 1$, we will prove

$$\sum_{n \geq 2} \frac{t^{kn}}{[i + k(n - 2) + j]_{\mathbf{P,Q}}!}$$

$$\times \sum_{\Sigma \in (C^{i,j,k}_{i+k(n-2)+j})^L} x^{comris_{i,k}(\Sigma)} \mathbf{Q}^{inv(\Sigma)} \mathbf{P}^{coinv(\Sigma)}$$

$$= \sum_{n \geq 2} \frac{x^{n-1} \mathbf{P}^{\binom{i+k(n-2)+j}{2}} t^{kn}}{[i + k(n - 2) + j]_{\mathbf{P,Q}}!} - \sum_{n \geq 2} \frac{(n - 1) x^{n-2} \mathbf{P}^{\binom{i+k(n-2)+j}{2}} t^{kn}}{[i + k(n - 2) + j]_{\mathbf{P,Q}}!}$$

$$+ \frac{e^{(i)}_{\mathbf{P,Q},k}(t(x - 1)^{1/k}) e^{(j)}_{\mathbf{P,Q},k}(t(x - 1)^{1/k})}{(1 - x)\left(-x + e_{\mathbf{P,Q},k}(t(x - 1)^{1/k})\right)}. \quad (5)$$

In section 2, we provide the necessary background on symmetric functions and introduce our new symmetric functions $p_{n,\alpha_1,\dots,\alpha_r}$. In section 3, we shall deal with the cases where $i = 0$ and prove (3) and (4). Finally, in section 4, we will deal with the cases where $1 \leq i \leq k$ and $1 \leq j \leq k$ and prove (5).

2 Symmetric Functions

In this section we give the necessary background on symmetric functions needed for our proofs of (3), (4), and (5).

A symmetric polynomial p in the variables $x_1, \dots, x_N$ is a polynomial over a field F of characteristic 0 with the property that $p(x_1, \dots, x_N) = p(x_{\sigma_1}, \dots, x_{\sigma_N})$ for all $\sigma = \sigma_1 \cdots \sigma_N \in S_N$. A symmetric function in the variables $x_1, x_2, \dots$ may be thought of as a symmetric polynomial in an infinite number of variables. Let Λ be the ring of all symmetric functions (a more formal definition of Λ may be found in [15]). The previously defined elementary symmetric functions e_n and the homogeneous symmetric functions h_n are both elements of Λ.

Let $\lambda = (\lambda_1, \dots, \lambda_\ell)$ be an integer partition; that is, λ is a finite sequence of weakly increasing nonnegative integers. Let $\ell(\lambda)$ denote the number of nonzero integers in λ. If the sum of these integers is n, we say that λ is a partition of n and write $\lambda \vdash n$. For any partition $\lambda = (\lambda_1, \dots, \lambda_\ell)$, let $e_\lambda = e_{\lambda_1} \cdots e_{\lambda_\ell}$. The well-known fundamental theorem of symmetric functions says that $\{e_\lambda : \lambda \text{ is a partition}\}$ is a basis for Λ. Similarly, if we define $h_\lambda = h_{\lambda_1} \cdots h_{\lambda_\ell}$, then $\{h_\lambda : \lambda \text{ is a partition}\}$ is also a basis for Λ.

Since the elementary symmetric functions e_λ and the homogeneous symmetric functions h_λ are both bases for Λ, it makes sense to talk about the coefficient of the homogeneous symmetric functions when written in terms of the elementary symmetric function basis. This coefficient has been shown to equal the size of a certain set of combinatorial objects. A rectangle of height 1 and length n chopped into "bricks" of lengths found in the partition λ is known as a *brick tabloid of shape (n) and type λ*. One brick tabloid of shape (12) and type $(1, 1, 2, 3, 5)$ is displayed below.

Let $\mathcal{B}_{\lambda,n}$ denote the set of all λ-brick tabloids of shape (n) and let $B_{\lambda,n} = |\mathcal{B}_{\lambda,n}|$. Through simple recursions stemming from (1), Eğecioğlu and Remmel proved in [8] that

$$h_n = \sum_{\lambda \vdash n} (-1)^{n-\ell(\lambda)} B_{\lambda,n} e_\lambda. \tag{6}$$

A symmetric function $p_{n,\nu}$ has a relationship with e_λ which is analogous to the relationship between h_n and e_λ. It was first introduced in [13] and [16]. Let ν be a function which maps the set of nonnegative integers into the field F. Recursively define $p_{n,\nu} \in \Lambda_n$ by setting $p_{0,\nu} = 1$ and letting

$$p_{n,\nu} = (-1)^{n-1}\nu(n)e_n + \sum_{k=1}^{n-1} (-1)^{k-1} e_k p_{n-k,\nu}$$

for all $n \geq 1$. By multiplying series, this means that

$$\left(\sum_{n\geq 0}(-1)^n e_n t^n\right)\left(\sum_{n\geq 1} p_{n,\nu} t^n\right) = \sum_{n\geq 1}\left(\sum_{k=0}^{n-1} p_{n-k,\nu}(-1)^k e_k\right) t^n$$

$$= \sum_{n\geq 1}(-1)^{n-1}\nu(n)e_n t^n,$$

where the last equality follows from the definition of $p_{n,\nu}$. Therefore,

$$\sum_{n\geq 1} p_{n,\nu} t^n = \frac{\sum_{n\geq 1}(-1)^{n-1}\nu(n)e_n t^n}{\sum_{n\geq 0}(-1)^n e_n t^n}$$

or, equivalently,

$$1 + \sum_{n\geq 1} p_{n,\nu} t^n = \frac{1 + \sum_{n\geq 1}(-1)^n (e_n - \nu(n)e_n) t^n}{\sum_{n\geq 0}(-1)^n e_n t^n}. \tag{7}$$

When taking $\nu(n) = 1$ for all $n \geq 1$, (7) becomes

$$1 + \sum_{n \geq 1} p_{n,1} t^n = 1 + \frac{\sum_{n \geq 1} (-1)^{n-1} e_n t^n}{\sum_{n \geq 0} (-1)^n e_n t^n}$$

$$= \frac{1}{\sum_{n \geq 0} (-1)^n e_n t^n} = 1 + \sum_{n \geq 1} h_n t^n$$

which implies $p_{n,1} = h_n$. Other special cases for ν give well-known generating functions. For example, if $\nu(n) = n$ for $n \geq 1$, then $p_{n,\nu}$ is the power symmetric function $\sum_i x_i^n$. By taking $\nu(n) = (-1)^k \chi(n \geq k+1)$ for some $k \geq 1$, $p_{n,(-1)^k \chi(n \geq k+1)}$ is the Schur function corresponding to the partition $(1^k, n)$.

This definition of $p_{n,\nu}$ is desirable because of its expansion in terms of elementary symmetric functions. The coefficient of e_λ in $p_{n,\nu}$ has a nice combinatorial interpretation similar to that of the homogeneous symmetric functions. Suppose T is a brick tabloid of shape (n) and type λ and that the final brick in T has length ℓ. Define the weight of a brick tabloid $w_\nu(T)$ to be $\nu(\ell)$ and let

$$w_\nu(B_{\lambda,n}) = \sum_{\substack{T \text{ is a brick tabloid} \\ \text{of shape } (n) \text{ and type } \lambda}} w_\nu(T).$$

When $\nu(n) = 1$ for $n \geq 1$, $B_{\lambda,n}$ and $w_\nu(B_{\lambda,n})$ are the same. By the recursions found in the definition of $p_{n,\nu}$, it may be shown that

$$p_{n,\nu} = \sum_{\lambda \vdash n} (-1)^{n-\ell(\lambda)} w_\nu(B_{\lambda,n}) e_\lambda$$

in almost the exact same way that (6) was proved in [8].

Suppose that we are given r functions $\alpha_i : \mathbb{P} \to R$ where R is some field. We write $T = (b_1, \ldots, b_k)$ if T is brick tabloid of shape (n) where $n = b_1 + \cdots + b_k$ and the sizes of the bricks are $b_1, \ldots, b_k$ as we read from left to right. If $T = (b_1, \ldots, b_k)$ where $k \geq r$, we define

$$w_{\alpha_1,\ldots,\alpha_r}(T) = \alpha_1(b_1) \cdots \alpha_r(b_r) \prod_{i=1}^{k} (-1)^{b_i - 1}.$$

Let $\mathcal{B}_n = \bigcup_{\lambda \vdash n} \mathcal{B}_{\lambda,n}$ denote the set of all brick tabloids of shape (n). Define $p_{n;\alpha_1,\ldots,\alpha_r}$ by

$$p_{n;\alpha_1,\ldots,\alpha_r} = \sum_{\substack{T = (b_1,\ldots,b_k) \in \mathcal{B}_n \\ k \geq r, b_i \geq 1}} w_{\alpha_1,\ldots,\alpha_r}(T) \prod_{i=1}^{k} e_{b_i}.$$

We can partition the brick tabloids $T = (b_1, \ldots, b_k) \in \mathcal{B}_n$ with $k \geq r$ into two classes. First, there is the class of brick tabloids where $k = r$ and, second, there is the class of brick tabloids where $k > r$. This second class can be further classified by the size of the last brick to give the following recursion:

$$p_{n;\alpha_1,\ldots,\alpha_r} = \sum_{\substack{a_1+\cdots+a_r=n \\ a_i \geq 1}} \prod_{i=1}^{r} \alpha_i(a_i)(-1)^{a_i-1} e_{a_i}$$

$$+ \sum_{k=1}^{n-r} (-1)^{k-1} e_k p_{n-k;\alpha_1,\ldots,\alpha_r}.$$

It follows that

$$\sum_{k=0}^{n-r} (-1)^k e_k p_{n-k;\alpha_1,\ldots,\alpha_r} = \sum_{\substack{a_1+\cdots+a_r=n \\ a_i \geq 1}} \prod_{i=1}^{r} \alpha_i(a_i)(-1)^{a_i-1} e_{a_i}$$

which, in turn, implies that

$$\left(\sum_{k\geq 0} (-1)^k e_k t^k \right) \left(\sum_{n\geq r} p_{n;\alpha_1,\ldots,\alpha_r} t^n \right) = \prod_{i=1}^{r} \left(\sum_{n\geq 1} (-1)^{n-1} \alpha_i(n) e_n t^n \right).$$

Therefore,

$$\sum_{n\geq r} p_{n;\alpha_1,\ldots,\alpha_r} t^n = \frac{\prod_{i=1}^{r} \left(\sum_{n\geq 1} (-1)^{n-1} \alpha_i(n) e_n t^n \right)}{\sum_{k\geq 0} (-1)^k e_k t^k}.$$

This is an analogous equation to equation (7).

3 The case where $i = 0$

In this section, we show why (3) and (4) are true. We will start with the situation where $j = k$. Fix $k \geq 2$. Define a homomorphism ξ_k from the ring of symmetric functions Λ to the polynomial ring

$$\mathbb{Q}(q_1, \ldots, q_L, p_1, \ldots, p_L)[x]$$

by setting $\xi(e_j) = 0$ if $j \not\equiv 0 \mod k$ and

$$\xi_k(e_{kn}) = \frac{(-1)^{kn-1}(x-1)^{n-1} \mathbf{P}^{\binom{kn}{2}}}{[kn]_{\mathbf{P},\mathbf{Q}}!}$$

otherwise.

Theorem 3.1. *If $n \not\equiv 0 \mod k$, then $\xi_k(h_n) = 0$. Otherwise,*

$$[kn]_{\mathbf{P},\mathbf{Q}}!\,\xi_k(h_{kn}) = \sum_{\boldsymbol{\Sigma}=(\sigma_1,\ldots,\sigma_L)\in(\mathcal{C}_{kn}^{0,k,k})^L} x^{comris_{0,k}(\boldsymbol{\Sigma})}\mathbf{Q}^{inv(\boldsymbol{\Sigma})}\mathbf{P}^{coinv(\boldsymbol{\Sigma})}.$$

Proof. If $\lambda = (\lambda_1,\ldots,\lambda_t)$ is a partition of n, we let $k\lambda = (k\lambda_1,\ldots,k\lambda_t)$. Similarly, if $T = (b_1,\ldots,b_t) \in \mathcal{B}_{\mu,n}$, then we let $kT = (kb_1,\ldots,kb_t) \in \mathcal{B}_{k\mu,kn}$.

The relationship between the symmetric functions h_n and e_λ gives

$$\xi_k(h_n) = \sum_{\mu \vdash n} (-1)^{n-\ell(\mu)} B_{\mu,n}\xi_k(e_\mu). \tag{8}$$

If n is not equivalent to $0 \mod k$, then every $\mu = (\mu_1,\ldots,\mu_{\ell(\mu)})$ on the right-hand side of (8) must contain a part μ_i which is not equivalent to $0 \mod k$ and hence $\xi_k(e_\mu) = 0$. Thus $\xi_k(h_n) = 0$ in this case. Similarly, if μ is a partition of kn where $n \geq 1$, then $\xi(e_\mu) = 0$ unless μ consists entirely of parts which are equal to $0 \mod k$. Thus in the expansion of $\xi_k(h_{kn})$, we can restrict ourselves to partitions μ of the form $k\lambda$ where λ is a partition of n.

We have that

$$
\begin{aligned}
[kn]_{\mathbf{P},\mathbf{Q}}!\,\xi_k(h_{kn}) &= [kn]_{\mathbf{P},\mathbf{Q}}! \sum_{\mu \vdash n} (-1)^{kn-\ell(\mu)} B_{k\mu,kn}\xi(e_{k\mu}) \\
&= [kn]_{\mathbf{P},\mathbf{Q}}! \sum_{\mu \vdash n} (-1)^{kn-\ell(\mu)} \\
&\quad \times \sum_{\substack{T=(b_1,\ldots,b_{\ell(\mu)}) \\ \in \mathcal{B}_{\mu,n}}} \prod_{i=1}^{\ell(\mu)} \frac{(-1)^{kb_i-1}(x-1)^{b_i-1}\mathbf{P}^{\binom{kb_i}{2}}}{[kb_i]_{\mathbf{P},\mathbf{Q}}!} \\
&= \sum_{\mu \vdash n} \sum_{T=(b_1,\ldots,b_{\ell(\mu)})\in\mathcal{B}_{\mu,n}} \left[\begin{matrix} kn \\ kb_1,\ldots,kb_{\ell(\mu)} \end{matrix} \right]_{\mathbf{P},\mathbf{Q}} \\
&\quad \times \mathbf{P}^{\sum_{i=1}^{\ell(\mu)}\binom{kb_i}{2}}(x-1)^{n-\ell(\mu)}.
\end{aligned}
$$

Fix a brick tabloid $T = (b_1,\ldots,b_{\ell(\mu)}) \in \mathcal{B}_{\mu,n}$. We want to give a combinatorial interpretation to $\mathbf{P}^{\sum_{i=1}^{\ell(\mu)}\binom{b_i}{2}}\left[\begin{matrix} n \\ b_1,\ldots,b_{\ell(\mu)} \end{matrix} \right]_{\mathbf{P},\mathbf{Q}}$. Let $IF(T)$ denote the set of all fillings of the cells of $T = (b_1,\ldots,b_{\ell(\mu)})$ with the numbers $1,\ldots,n$ so that the numbers increase within each brick reading from left to right. We then think of each such filling as a permutation of S_n by reading the numbers from left to right in each row. For example,

| 4 : 6 : 12 | 1 : 5 : 7 : 8 : 10 : 11 | 2 : 3 : 9 |

is an element of $IF(3,6,3)$ whose corresponding permutation is 4 6 12 1 5 7 8 10 11 2 3 9.

Lemma 3.2. *If* $T = (b_1, \ldots, b_{\ell(\mu)})$ *is a brick tabloid in* $\mathcal{B}_{\mu,n}$, *then*

$$p^{\sum_i \binom{b_i}{2}} \begin{bmatrix} n \\ b_1, \ldots, b_{\ell(\mu)} \end{bmatrix}_{p,q} = \sum_{\sigma \in IF(T)} q^{inv(\sigma)} p^{coinv(\sigma)}.$$

Proof. It follows from a result of Carlitz [7] that for positive integers $b_1, \ldots, b_\ell$ which sum to n,

$$\begin{bmatrix} n \\ b_1, \ldots, b_\ell \end{bmatrix}_{p,q} = \sum_{r \in \mathcal{R}(1^{b_1}, \ldots, \ell^{b_\ell})} q^{inv[r]} p^{coinv[r]}$$

where $\mathcal{R}(1^{b_1}, \ldots, \ell^{b_\ell})$ is the set of rearrangements of b_1 1's, b_2 2's, etc.

Consider a rearrangement r of $1^{b_1}, \ldots, \ell^{b_\ell}$ and construct a permutation σ_r by labeling the 1's from left to right with $1, 2, \ldots, b_1$, the 2's from left to right with $b_1 + 1, \ldots, b_1 + b_2$, and in general the i's from left to right with $1 + \sum_{j=1}^{i-1} b_j, \ldots, b_i + \sum_{j=1}^{i-1} b_j$. In this way, σ_r^{-1} starts with the positions of the 1's in r in increasing order, followed by the positions of the 2's in r in increasing order, etc. For example, if $T = (2, 1, 3, 1, 4, 1) \in \mathcal{B}_{(1,1,1,2,3,4)}$ is below

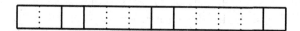

then one possible rearrangement to consider is $r = 5\ 5\ 1\ 5\ 3\ 1\ 2\ 3\ 6\ 3\ 5\ 4$. Below we display σ_r and σ_r^{-1}.

		1	2	3	4	5	6	7	8	9	10	11	12
r	$=$	5	5	1	5	3	1	2	3	6	3	5	4
σ_r	$=$	8	9	1	10	4	2	3	5	12	6	11	7
σ_r^{-1}	$=$	3	6	7	5	8	10	12	1	2	4	11	9

We can think of σ_r^{-1} as a filling of the cells of the brick tabloid $T = (2, 1, 3, 1, 4, 1)$ with the numbers $1, \ldots, 12$ such that the numbers within each brick are increasing, reading from left to right, pictured below.

| 3 ⋮ 6 | 7 | 5 ⋮ 8 ⋮ 10 | 12 | 1 ⋮ 2 ⋮ 4 ⋮ 11 | 9 |

It is then easy to see that

$$\binom{2}{2} + \binom{1}{2} + \binom{3}{2} + \binom{1}{2} + \binom{4}{2} + \binom{1}{2} + coinv(r)$$
$$= coinv(\sigma_r) = coinv(\sigma_r^{-1})$$

and $inv(r) = inv(\sigma_r) = inv(\sigma_r^{-1})$.

In general, for any $T = (b_1, \ldots, b_{\ell(\mu)}) \in \mathcal{B}_{\mu, \alpha}$, the correspondence which takes $r \in \mathcal{R}(1^{b_1}, \ldots, \ell^{b_\ell})$ to σ_r^{-1} shows that

$$p^{\sum_{i=1}^{\ell(\mu)} \binom{b_i}{2}} \begin{bmatrix} n \\ b_1, \ldots, b_{\ell(\mu)} \end{bmatrix}_{p,q} = \sum_{\sigma \in IF(T)} q^{inv(\sigma)} p^{coinv(\sigma)},$$

thereby completing the proof of the lemma. $\qquad\qquad\qquad\square$

It follows that for any $T = (b_1, \ldots, b_{\ell(\mu)}) \in \mathcal{B}_{\mu,n}$,

$$\mathbf{P}^{\sum_{i=1}^{\ell(\mu)} \binom{kb_i}{2}} \begin{bmatrix} kn \\ kb_1, \ldots, kb_{\ell(\mu)} \end{bmatrix}_{\mathbf{P},\mathbf{Q}} = \prod_{i=1}^{L} p_i^{\sum \binom{kb_i}{2}} \begin{bmatrix} kn \\ kb_1, \ldots, kb_{\ell(\mu)} \end{bmatrix}_{p_i, q_i}$$
$$= \prod_{i=1}^{L} \sum_{\sigma^{(i)} \in IF(kT)} q_i^{inv(\sigma^{(i)})} p_i^{coinv(\sigma^{(i)})}.$$

Thus we can interpret $\mathbf{P}^{\sum_{i=1}^{\ell(\mu)} \binom{kb_i}{2}} \begin{bmatrix} kn \\ kb_1, \ldots, kb_{\ell(\mu)} \end{bmatrix}_{\mathbf{P},\mathbf{Q}}$ as the sum of the weights of the set of fillings of kT of L-tuples of permutations $\Sigma = (\sigma^{(1)}, \ldots, \sigma^{(L)})$ such that for each i, the elements of $\sigma^{(i)}$ are increasing within each brick of kT and where the weight of such a filling is $\mathbf{Q}^{inv(\Sigma)} \mathbf{P}^{coinv(\Sigma)}$. For example, if $T = (2, 1, 2) \in \mathcal{B}_{(1,2^2),5}$, $k = 3$, and $L = 3$, then such a filling of kT is pictured below.

			−1			1			1		x			1
$\sigma^{(1)}$	5	8	9	11	13	15	4	7	10	1	2	3	6	12	14
$\sigma^{(2)}$	1	2	7	10	11	12	4	9	14	3	5	6	8	13	15
$\sigma^{(3)}$	3	5	9	10	13	14	1	4	11	2	6	7	8	12	15

We order the cells of such a filled brick tabloid from left to right. We can interpret the term $(x-1)^{n-\ell(\mu)}$ as taking such a filling and labeling the cells of the form sk which are not at the end of a brick with either an x or -1 and labeling each cell at the end of a brick with 1. This was done in the previous figure. Objects $\mathcal{O}$ constructed in this way will be called *labeled filled brick tabloids*. We define the weight of $\mathcal{O}$, $W(\mathcal{O})$, to be the product over all the labels of the cells times $\mathbf{Q}^{inv(\Sigma)}\mathbf{P}^{coinv(\Sigma)}$ if T is filled with permutations $\Sigma = (\sigma^{(1)}, \ldots \sigma^{(L)})$. Thus for the object above,

$$W(\mathcal{O}) = (-1)xq_1^{inv(\sigma^{(1)})}q_2^{inv(\sigma^{(2)})}q_3^{inv(\sigma^{(3)})}p_1^{coinv(\sigma^{(1)})}$$
$$\times p_2^{coinv(\sigma^{(2)})}p_3^{coinv(\sigma^{(3)})}.$$

Let $\mathcal{L}F^{(k)}(kn)$ denote the set of all objects that can be created in this way. It follows that

$$[kn]_{\mathbf{P},\mathbf{Q}}!\xi(h_{kn}) = \sum_{\mathcal{O}\in\mathcal{L}F^{(k)}(kn)} W(\mathcal{O}).$$

To finish the proof, let us define an involution $I : \mathcal{L}F^{(k)}(kn) \rightarrow \mathcal{L}F^{(k)}(kn)$. Given $\mathcal{O} \in \mathcal{L}F^{(k)}(kn)$, read the cells of $\mathcal{O}$ from left to right looking for the first cell kc for which either

(i) kc is labeled with -1, or

(ii) kc is at the end of end of brick b, the cell $kc+1$ is immediately to the right of kc and starts another brick b', and each permutation $\sigma^{(i)}$ increases as we go from cell kc to $kc+1$.

If we are in case (i), then $I(\mathcal{O})$ is the labeled filled brick tabloid which is obtained from $\mathcal{O}$ by taking the brick b that contains kc and splitting b into two bricks b_1 and b_2 where b_1 contains the cells of b up to and including the cell kc and b_2 contains the remaining cells of b and the label on kc is changed from -1 to 1. If in case (ii), $I(\mathcal{O})$ is the labeled filled brick tabloid which is obtained from $\mathcal{O}$ by combining the two bricks b and b' into a single brick and changing the label on cell kc from 1 to -1. If neither case (i) or case (ii) applies, then we let $I(\mathcal{O}) = \mathcal{O}$. For example, the image of the brick tabloid depicted above under I is below:

			1			1			1			x			1
$\sigma^{(1)}$	5	8	9	11	13	15	4	7	10	1	2	3	6	12	14
$\sigma^{(2)}$	1	2	7	10	11	12	4	9	14	3	5	6	8	13	15
$\sigma^{(3)}$	3	5	9	10	13	14	1	4	11	2	6	7	8	12	15

If $I(\mathcal{O}) \neq \mathcal{O}$, then $W(I(\mathcal{O})) = -W(\mathcal{O})$ since we change the label on cell c from 1 to -1 or vice versa. Moreover, I is an involution. Thus, I

shows that

$$[kn]_{\mathbf{P},\mathbf{Q}}!\xi_k(h_{kn}) = \sum_{\mathcal{O}\in\mathcal{LF}^{(k)}(kn)} W(\mathcal{O}) = \sum_{\mathcal{O}\in\mathcal{LF}^{(k)}(kn),I(\mathcal{O})=\mathcal{O}} W(\mathcal{O}).$$

We are therefore led to examine the fixed points under I. If $I(\mathcal{O}) = \mathcal{O}$, then $\mathcal{O}$ can have no cells which are labeled with -1. Also it must be the case that between any two consecutive bricks at least one of the underlying permutations $\sigma^{(i)}$ must decrease. Each cell kc which is not at the end of the brick in $\mathcal{O}$ is labeled with x and each of the permutations $\sigma^{(i)}$ has a rise at kc so that $kc \in Comris_{0,k}(\Sigma)$. All the other cells of the form kc in $\mathcal{O}$ except the last cell are at the end of a brick which has another brick to its right in which case $kc \notin Comris_{0,k}(\Sigma)$. All such cells have label 1 so that $W(\mathcal{O}) = x^{comris_{0,k}(\Sigma)}\mathbf{Q}^{inv(\Sigma)}\mathbf{P}^{coinv(\Sigma)}$.

Now if we are given $\Sigma = (\sigma^{(1)},\dots,\sigma^{(L)}) \in (\mathcal{C}_{kn}^{(0,k,k)})^L$, we can construct a fixed point of I from Σ by using $(\sigma^{(1)},\dots,\sigma^{(L)})$ to fill a tabloid of shape (kn), then drawing the bricks so that the cells kc which end bricks are precisely the cells kc where one of the permutations $\sigma^{(i)}$ decreases from kc to $kc+1$. This shows that

$$\sum_{\mathcal{O}\in\mathcal{LF}^{(k)}(kn),I(\mathcal{O})=\mathcal{O}} W(\mathcal{O}) = \sum_{\Sigma\in(\mathcal{C}_{kn}^{0,k,k})^L} x^{comris_{0,k}(\Sigma)}\mathbf{Q}^{inv(\Sigma)}\mathbf{P}^{coinv(\Sigma)}.$$

This completes the proof of the theorem. $\qquad\qquad\qquad\qquad\Box$

To find the generating function in equation (3), we apply the homomorphism ξ_k to both sides of the identity

$$1 + \sum_{n\geq 1} h_n t^n = H(t) = \frac{1}{E(-t)}$$

to find that

$$1 + \sum_{n\geq 1} \frac{t^{kn}}{[kn]_{\mathbf{P},\mathbf{Q}}!} \sum_{\Sigma\in(\mathcal{C}_{kn}^{0,k,k})^L} x^{comris_{0,k}(\Sigma)}\mathbf{Q}^{inv(\Sigma)}\mathbf{P}^{coinv(\Sigma)} = \frac{1}{\xi_k(E(-t))}.$$

$$(9)$$

Now

$$\xi_k(E(-t)) = 1 + \sum_{n \geq 1}(-t)^{kn}(-1)^{kn-1}\frac{(x-1)^{n-1}\mathbf{P}^{\binom{kn}{2}}}{[kn]_{\mathbf{P},\mathbf{Q}}!}$$

$$= \frac{1}{1-x}\left(1 - x + \sum_{n \geq 1}\frac{t^{kn}(x-1)^n\mathbf{P}^{\binom{kn}{2}}}{[kn]_{\mathbf{P},\mathbf{Q}}!}\right)$$

$$= \frac{1}{1-x}\left(-x + e_{\mathbf{P},\mathbf{Q},k}(t(x-1)^{1/k})\right). \tag{10}$$

Combining (9) and (10) gives the following theorem.

Theorem 3.3. *For $k \geq 2$,*

$$1 + \sum_{n \geq 1}\frac{t^{kn}}{[kn]_{\mathbf{P},\mathbf{Q}}!}\sum_{\boldsymbol{\Sigma} \in (\mathcal{C}_{kn}^{0,k,k})^L} x^{comris_{0,k}(\boldsymbol{\Sigma})}\mathbf{Q}^{inv(\boldsymbol{\Sigma})}\mathbf{P}^{coinv(\boldsymbol{\Sigma})}$$

$$= \frac{1-x}{-x + e_{\mathbf{P},\mathbf{Q},k}(t(x-1)^{1/k})}$$

and

$$1 + \sum_{n \geq 1}\frac{t^{kn}}{[kn]_{\mathbf{P},\mathbf{Q}}!}\sum_{\boldsymbol{\Sigma} \in (\mathcal{E}_{kn}^{0,k,k})^L} \mathbf{Q}^{inv(\boldsymbol{\Sigma})}\mathbf{P}^{coinv(\boldsymbol{\Sigma})} = \frac{1}{e_{\mathbf{P},\mathbf{Q},k}(t(-1)^{1/k})}.$$

Next, suppose that $k \geq 2$ and $1 \leq j \leq k-1$. Define a function ν on the positive integers by setting $\nu(j) = 0$ if $j \not\equiv 0 \mod k$ and

$$\nu(kn) = \frac{[kn]_{\mathbf{P},\mathbf{Q}}\downarrow_{k-j}}{\mathbf{P}^{(k-j)kn - \binom{k-j+1}{2}}}$$

where for any $1 \leq s \leq n$, $[n]_{\mathbf{P},\mathbf{Q}}\downarrow_s = [n]_{\mathbf{P},\mathbf{Q}}[n-1]_{\mathbf{P},\mathbf{Q}}\cdots[n-s+1]_{\mathbf{P},\mathbf{Q}}$. This definition of $\nu(kn)$ is designed so that

$$\nu(kn)\xi_k(e_{nk}) = (-1)^{kn-1}\frac{(x-1)^{n-1}\mathbf{P}^{\binom{kn}{2}}}{[kn]_{\mathbf{P},\mathbf{Q}}!}\frac{[kn]_{\mathbf{P},\mathbf{Q}}\downarrow_{k-j}}{\mathbf{P}^{\sum_{s=1}^{k-j}kn-j}}$$

$$= (-1)^{kn-s}\frac{(x-1)^{n-1}\mathbf{P}^{\binom{k(n-1)}{2}+j}}{[k(n-1)+j]_{\mathbf{P},\mathbf{Q}}!}.$$

Theorem 3.4. *For any $k \geq 2$ and $1 \leq j \leq k-1$, $\xi_k(p_{j,\nu}) = 0$ if $j \not\equiv 0$ mod k and for all $n \geq 1$,*

$$[k(n-1)+j]_{\mathbf{P},\mathbf{Q}}!p_{kn,\nu} = \sum_{\boldsymbol{\Sigma} \in (\mathcal{C}_{k(n-1)+j}^{0,j,k})^L} x^{comris_{0,k}(\boldsymbol{\Sigma})}\mathbf{Q}^{inv(\boldsymbol{\Sigma})}\mathbf{P}^{coinv(\boldsymbol{\Sigma})}.$$

Proof. The relationship between $p_{n,\nu}$ and the elementary symmetric functions gives

$$\xi_k(p_{n,\nu}) = \sum_{\mu \vdash n} (-1)^{n-\ell(\mu)} \nu(B_{\mu,n}) \xi(e_\mu). \tag{11}$$

If $n \not\equiv 0 \mod k$, then every $\mu = (\mu_1, \ldots, \mu_{\ell(\mu)})$ on the right-hand side of (11) must contain a part $\mu_i \not\equiv 0 \mod k$ and hence $\xi_k(e_\mu) = 0$. Thus $\xi_k(p_{n,\nu}) = 0$ in this case. Similarly, if μ is a partition of kn where $n \geq 1$, then $\xi(e_\mu) = 0$ unless μ consists entirely of parts which are multiples of k. Thus, in the expansion of $\xi_k(h_{nk})$, we can restrict ourselves to partitions μ of the form $k\lambda$ where λ is a partition of n. Therefore,

$$[k(n-1)+j]_{\mathbf{P},\mathbf{Q}}! \xi_k(p_{kn,\nu})$$

$$= [k(n-1)+j]_{\mathbf{P},\mathbf{Q}}! \sum_{\mu \vdash n} (-1)^{kn-\ell(\mu)} \nu(B_{k\mu,kn}) \xi(e_{k\mu})$$

$$= [k(n-1)+j]_{\mathbf{P},\mathbf{Q}}! \sum_{\mu \vdash n} (-1)^{kn-\ell(\mu)}$$

$$\times \sum_{(b_1,\ldots,b_{\ell(\mu)}) \in \mathcal{B}_{\mu,n}} \nu(kb_{\ell(\mu)}) \xi_k(e_{kb_{\ell(\mu)}})$$

$$\times \prod_{i=1}^{\ell(\mu)-1} \frac{(-1)^{kb_i-1}(x-1)^{b_i-1} \mathbf{P}^{\binom{kb_i}{2}}}{[kb_i]_{\mathbf{P},\mathbf{Q}}!}$$

$$= [k(n-1)+j]_{\mathbf{P},\mathbf{Q}}! \sum_{\mu \vdash n} (-1)^{kn-\ell(\mu)}$$

$$\times \sum_{(b_1,\ldots,b_{\ell(\mu)}) \in \mathcal{B}_{\mu,n}} (-1)^{kb_{\ell(\mu)}-1} \frac{(x-1)^{b_{\ell(\mu)}-1} \mathbf{P}^{\binom{k(b_{\ell(\mu)}-1)+j}{2}}}{[k(b_{\ell(\mu)}-1)+j]_{\mathbf{P},\mathbf{Q}}!}$$

$$\times \prod_{i=1}^{\ell(\mu)-1} \frac{(-1)^{kb_i-1}(x-1)^{b_i-1} \mathbf{P}^{\binom{kb_i}{2}}}{[kb_i]_{\mathbf{P},\mathbf{Q}}!}$$

$$= \sum_{\mu \vdash n} \sum_{(b_1,\ldots,b_{\ell(\mu)}) \in \mathcal{B}_{\mu,n}} (x-1)^{n-\ell(\mu)}$$

$$\times \begin{bmatrix} kn-k+j \\ kb_1,\ldots,kb_{\ell(\mu)-1},k(b_{\ell(\mu)}-1)+j \end{bmatrix}_{\mathbf{P},\mathbf{Q}}$$

$$\times \mathbf{P}^{\binom{kb_{\ell(\mu)}-k+j}{2}+\sum_{i=1}^{\ell(\mu)-1}\binom{kb_i}{2}}.$$

By Lemma 3.2, we can interpret

$$\begin{bmatrix} k(n-1)+j \\ kb_1,\ldots,kb_{\ell(\mu)-1},k(b_{\ell(\mu)}-1)+j) \end{bmatrix}_{\mathbf{P},\mathbf{Q}} \mathbf{P}^{\binom{k(b_{\ell(\mu)}-1)+j}{2}} \mathbf{P}^{\sum_{i=1}^{\ell(\mu)-1}\binom{kb_i}{2}}$$

as the set of fillings of the brick tabloid $U = (kb_1, \ldots, kb_{\ell(\mu)-1}, k(b_{\ell(\mu)} - 1) + j)$ of L-tuples of permutations $\Sigma = (\sigma^{(1)}, \ldots, \sigma^{(L)})$ such that for each i, the elements of $\sigma^{(i)}$ are increasing within each brick of U and we weight such a filling with $\mathbf{Q}^{inv(\Sigma)}\mathbf{P}^{coinv(\Sigma)}$. In fact, we shall think of U as the brick tabloid $kT = (kb_1, \ldots, kb_{\ell(\mu)})$ where the last $k - j$ cells of the last brick are blank.

Again, order the cells of such a filled brick tabloid from left to right. Interpret the term $(x-1)^{n-\ell(\mu)}$ as taking such a filling and labeling the cells of the form sk which are not at the end of a brick with either an x or -1 and labeling each cell at the end of a brick with 1. Below is an example of such a tabloid.

		-1			1			1		x			1
$\sigma^{(1)}$	5	8	9	11	12	13	4	7	10	1	2	3	6	
$\sigma^{(2)}$	1	2	7	10	11	12	4	9	13	3	5	6	8	
$\sigma^{(3)}$	3	5	9	10	12	13	1	4	11	2	6	7	8	

Call such an object $\mathcal{O}$ a labeled filled brick tabloid and define the weight of $\mathcal{O}$, $W(\mathcal{O})$, to be product over all the labels of the cells times $\mathbf{Q}^{inv(\Sigma)}\mathbf{P}^{coinv(\Sigma)}$ if T is filled with permutations $\Sigma = (\sigma^{(1)}, \ldots \sigma^{(L)})$. Thus for the object pictured above,

$$W(\mathcal{O}) = (-1)xq_1^{inv(\sigma^{(1)})}q_2^{inv(\sigma^{(2)})}q_3^{inv(\sigma^{(3)})}p_1^{coinv(\sigma^{(1)})}$$
$$p_2^{coinv(\sigma^{(2)})}p_3^{coinv(\sigma^{(3)})}.$$

We let $\mathcal{L}F^{(k,j)}(kn)$ denote the set of all objects that can be created in this way from brick tabloids T in $\mathcal{B}_n$. Then it follows that

$$[k(n-1)+j]_{\mathbf{P},\mathbf{Q}}!\xi_k(p_{kn,\nu}) = \sum_{\mathcal{O} \in \mathcal{L}F^{(k,j)}(kn)} W(\mathcal{O}).$$

We now define an involution $I : \mathcal{L}F^{(k,j)}(kn) \to \mathcal{L}F^{(k,j)}(kn)$ exactly as we did in the proof of Theorem 3.1. That is, given $\mathcal{O} \in \mathcal{L}F^{(k,j)}(kn)$, read the cells of $\mathcal{O}$ from left to right looking for the first instance of either:

(i) kc is labeled with -1, or

(ii) kc is at the end of end of brick b, the cell $kc+1$ is immediately to the right of kc and starts another brick b', and each permutation $\sigma^{(i)}$ increases as we go from kc to $kc+1$.

If in case (i), then $I(\mathcal{O})$ is the labeled filled brick tabloid which is obtained from $\mathcal{O}$ by taking the brick b that contains kc and splitting b into two bricks b_1 and b_2 where b_1 contains the cells of b up to and including the cell kc and b_2 contains the remaining cells of b and changing the label on kc from -1 to 1. If in case (ii), $I(\mathcal{O})$ is the labeled filled brick tabloid which is obtained from $\mathcal{O}$ by combining the two bricks b and b' into a single brick and changing the label on cell kc from 1 to -1. If neither case (i) or case (ii) applies, take $I(\mathcal{O}) = \mathcal{O}$. As an example, the image of the above figure under this map is below:

			1			1			1			x				1
$\sigma^{(1)}$	5	8	9	11	12	13	4	7	10	1	2	3	6			
$\sigma^{(2)}$	1	2	7	10	11	12	4	9	13	3	5	6	8			
$\sigma^{(3)}$	3	5	9	10	12	13	1	4	11	2	6	7	8			

Then we argue exactly as in Theorem 3.1 that

$$\sum_{\mathcal{O}\in\mathcal{LF}^{(k,j)}(kn),\,I(\mathcal{O})=\mathcal{O}} W(\mathcal{O})$$

$$= \sum_{\Sigma\in(\mathcal{C}^{0,j,k}_{k(n-1)+j})^{L}} x^{comris_{0,k}(\Sigma)} \mathbf{Q}^{inv(\Sigma)} \mathbf{P}^{coinv(\Sigma)},$$

as desired. $\qquad\square$

We can now apply ξ_k to the relationship between $p_{n,\nu}$ and the elementary symmetric functions to arrive at (4). We have

$$\xi_k\left(\sum_{n\geq 1}(-1)^{n-1}\nu(n)e_n t^n\right)$$

$$= \sum_{n\geq 1}(-1)^{kn-1}\nu(kn)\xi_k(e_{kn})t^{kn}$$

$$= \sum_{n\geq 1}(-1)^{kn-1}(-1)^{kn-1}\frac{(x-1)^{n-1}\mathbf{P}^{\left(\frac{k(n-1)+j}{2}\right)}}{[k(n-1)+j]_{\mathbf{P},\mathbf{Q}}!}t^{kn}$$

$$= \frac{1}{x-1}\sum_{n\geq 1}\frac{(x-1)^{n}\mathbf{P}^{\left(\frac{k(n-1)+j}{2}\right)}}{[k(n-1)+j]_{\mathbf{P},\mathbf{Q}}!}t^{kn}$$

$$= \frac{1}{x-1}e^{(j)}_{\mathbf{P},\mathbf{Q},k}\left(t(x-1)^{1/k}\right).$$

Therefore, we have that

$$\sum_{n\geq 1}\xi_k(p_{kn,\nu})t^{kn} = \sum_{n\geq 1}\frac{t^{kn}}{[k(n-1)+j]_{\mathbf{P},\mathbf{Q}}!}$$
$$\times \sum_{\mathbf{\Sigma}\in(\mathcal{C}^{0,j,k}_{k(n-1)+j})^L} x^{comris_{0,k}(\mathbf{\Sigma})}\mathbf{Q}^{inv(\mathbf{\Sigma})}\mathbf{P}^{coinv(\mathbf{\Sigma})}$$
$$= \frac{\frac{1}{x-1}e^{(j)}_{\mathbf{P},\mathbf{Q},k}(t(x-1)^{1/k})}{\frac{1}{1-x}\left(-x+e_{\mathbf{P},\mathbf{Q},k}(t(x-1)^{1/k})\right)}$$
$$= \frac{-e^{(j)}_{\mathbf{P},\mathbf{Q},k}(t(x-1)^{1/k})}{-x+e_{\mathbf{P},\mathbf{Q},k}(t(x-1)^{1/k})},$$

and we have arrived at the following theorem.

Theorem 3.5. *For any* $k \geq 2$ *and* $1 \leq j \leq k-1$,

$$\sum_{n\geq 1}\frac{t^{kn}}{[k(n-1)+j]_{\mathbf{P},\mathbf{Q}}!}\sum_{\mathbf{\Sigma}\in(\mathcal{C}^{0,j,k}_{k(n-1)+j})^L} x^{comris_{0,k}(\mathbf{\Sigma})}\mathbf{Q}^{inv(\mathbf{\Sigma})}\mathbf{P}^{coinv(\mathbf{\Sigma})}$$
$$= \frac{-e^{(j)}_{\mathbf{P},\mathbf{Q},k}(t(x-1)^{1/k})}{-x+e_{\mathbf{P},\mathbf{Q},k}(t(x-1)^{1/k})}$$

and

$$\sum_{n\geq 1}\frac{t^{kn}}{[k(n-1)+j]_{\mathbf{P},\mathbf{Q}}!}\sum_{\mathbf{\Sigma}\in(\mathcal{E}^{0,j,k}_{k(n-1)+j})^L} \mathbf{Q}^{inv(\mathbf{\Sigma})}\mathbf{P}^{coinv(\mathbf{\Sigma})}$$
$$= \frac{-e^{(j)}_{\mathbf{P},\mathbf{Q},k}(t(-1)^{1/k})}{e_{\mathbf{P},\mathbf{Q},k}(t(-1)^{1/k})}.$$

Theorem 3.5 also gives us the generating functions for permutations in $\mathcal{C}^{j,k,k}_{k(n-1)+j}$ where $1 \leq j \leq k-1$. That is, given $\sigma = \sigma_1\cdots\sigma_n \in S_n$, let

$$\sigma^c = (n+1-\sigma_1)\cdots(n+1-\sigma_n) \qquad \text{and} \qquad \sigma^r = \sigma_n\cdots\sigma_1.$$

Given $\mathbf{\Sigma} = (\sigma^{(1)},\ldots,\sigma^{(L)}) \in S_n^L$, we let $(\mathbf{\Sigma}^c)^r = (\tau^{(1)},\ldots,\tau^{(L)}) \in S_n^L$ where for each i, $\tau^{(i)} = ((\sigma^{(i)})^c)^r$. Then $\mathbf{\Sigma} \in (\mathcal{C}^{0,j,k}_{k(n-1)+j})^L$ if and only if $(\mathbf{\Sigma}^c)^r \in (\mathcal{C}^{j,k,k}_{k(n-1)+j})^L$ and $comris_{0,k}(\mathbf{\Sigma}) = comris_{j,k}(((\mathbf{\Sigma})^c)^r)$. Thus the only cases that we have left are the generating functions for for permutations in $\mathcal{C}^{i,j,k}_{i+k(n-1)+j}$ where $1 \leq i,j \leq k-1$. Such generating functions are considered in the next section.

4   The case $i \neq 0$ and $j < k$

The goal of this section is to prove the identity in (5). To do this, we will use the symmetric functions p_{n,β_1,β_2} where $\beta_1(j) = \beta_2(j) = 0$ if $j \neq 0$ mod k for $n \geq 2$, and otherwise

$$\beta_1(kn) = \frac{[kn]_{\mathbf{P},\mathbf{Q}} \downarrow_{k-i}}{\mathbf{P}^{(k-i)kn - \binom{k-i+1}{2}}} \quad \text{and} \quad \beta_2(kn) = \frac{[kn]_{\mathbf{P},\mathbf{Q}} \downarrow_{k-j}}{\mathbf{P}^{(k-j)kn - \binom{k-j+1}{2}}}.$$

We have defined $\beta_1(kn)$ and $\beta_2(kn)$ so that for $n \geq 2$

$$\beta_1(kn)\xi_k(e_{nk}) = (-1)^{kn-1}\frac{(x-1)^{n-1}\mathbf{P}^{\binom{kn}{2}}}{[kn]_{\mathbf{P},\mathbf{Q}}!}\frac{[kn]_{\mathbf{P},\mathbf{Q}}\downarrow_{k-i}}{\mathbf{P}^{\sum_{s=1}^{k-i}kn-s}}$$

$$= (-1)^{kn-1}\frac{(x-1)^{n-1}\mathbf{P}^{\binom{k(n-1)+i}{2}}}{[k(n-1)+i]_{\mathbf{P},\mathbf{Q}}!}. \tag{12}$$

and

$$\beta_2(kn)\xi_k(e_{nk}) = (-1)^{kn-1}\frac{(x-1)^{n-1}\mathbf{P}^{\binom{kn}{2}}}{[kn]_{\mathbf{P},\mathbf{Q}}!}\frac{[kn]_{\mathbf{P},\mathbf{Q}}\downarrow_{k-j}}{\mathbf{P}^{\sum_{s=1}^{k-j}kn-s}}$$

$$= (-1)^{kn-1}\frac{(x-1)^{n-1}\mathbf{P}^{\binom{k(n-1)+j}{2}}}{[k(n-1)+j]_{\mathbf{P},\mathbf{Q}}!}. \tag{13}$$

Theorem 4.1. *For $k \geq 2$ and $1 \leq i,j \leq k-1$, $\xi_k(p_{j,\beta_1,\beta_2}) = 0$ if $j \not\equiv 0$ mod k or $j < 2k$. Otherwise, for $n \geq 2$,*

$$[i + k(n-2) + j]_{\mathbf{P},\mathbf{Q}}!\xi_k(p_{kn,\beta_1,\beta_2})$$

$$= ((n-1)x^{n-2} - x^{n-1})\mathbf{P}^{\binom{i+(k(n-2)+j}{2}}$$

$$+ \sum_{\Sigma \in (C^{i,j,k}_{i+k(n-2)+j})^L} x^{comris_{i,k}(\Sigma)}\mathbf{Q}^{inv(\Sigma)}\mathbf{P}^{coinv(\Sigma)}.$$

Proof. We have

$$\xi_k(p_{m,\beta_1,\beta_2}) = \sum_{\substack{T=(b_1,\ldots,b_s) \\ b_1+\cdots+b_s=m, s \geq 2}} (-1)^{\sum_{r=1}^{s} b_r - 1}\beta_1(b_1)\beta_2(b_2)\prod_{r=1}^{s} \xi_k(e_{b_r}).$$

Thus if $m \not\equiv 0 \mod k$, then at least one $b_r \not\equiv 0$ mod k and hence $\xi_k(b_r) = 0$. Thus $\xi_k(p_{m,\beta_1,\beta_2}) = 0$. Similarly if $m < 2k$, then either $b_1 < k$ in which case $\beta_1(b_1) = 0$ or $b_2 < k$ in which case $\beta_2(b_2) = 0$ so again we can conclude that $\xi_k(p_{m,\beta_1,\beta_2}) = 0$. Similarly, if μ is a partition of $m = kn$ where $n \geq 2$, then all the b_r's must be multiples of

k. It follows that for $n \geq 2$,

$$[i + k(n-2) + j]_{\mathbf{P},\mathbf{Q}}! \xi_k(p_{kn,\beta_1,\beta_2})$$
$$= [i + k(n-2) + j]_{\mathbf{P},\mathbf{Q}}!$$

$$\times \sum_{\substack{T=(b_1,\ldots,b_s) \\ b_1+\cdots+b_s=n, s\geq 2}} (-1)^{\sum_{r=1}^s kb_r - 1} \beta_1(kb_1)\beta_2(kb_2) \prod_{r=1}^s \xi_k(e_{kb_r}).$$

For each $T = (b_1, b_2, b_3 \ldots, b_s)$ appearing on the right-hand side of the above equation, we let $\bar{T} = (b_1, b_3, \ldots, b_s, b_2) = (c_1, \ldots, c_s)$. Then isolating a portion of the above equation,

$$\beta_1(kb_1)\beta_2(kb_2) \prod_{r=1}^s \xi_k(e_{kb_r})$$

$$= (-1)^{kb_1-1} \frac{(x-1)^{b_1-1} \mathbf{P}^{\binom{k(b_1\frac{-1}{2})+i}{2}}}{[k(b_1-1)+i]_{\mathbf{P},\mathbf{Q}}!}$$

$$\times (-1)^{kb_2-1} \frac{(x-1)^{b_2-1} \mathbf{P}^{\binom{k(b_2\frac{-1}{2})+j}{2}}}{[k(b_2-1)+j]_{\mathbf{P},\mathbf{Q}}!}$$

$$\times \prod_{r=3}^s (-1)^{kb_r-1} \frac{(x-1)^{b_r-1} \mathbf{P}^{\binom{kb_r}{2}}}{[kb_r]_{\mathbf{P},\mathbf{Q}}!}.$$

The factors of (-1) above will cancel the $(-1)^{\sum_{r=1}^s kb_i - 1}$ term. We can combine the denominators that appear in the right-hand side of the above equation with the $[i + k(n-2) + j]_{\mathbf{P},\mathbf{Q}}!$ term to form the multinomial coefficient

$$\begin{bmatrix} i + k(n-2) + j \\ i + k(c_1-1), kc_2, \ldots, kc_{s-1}, k(c_s-1) + j \end{bmatrix}_{\mathbf{P},\mathbf{Q}}.$$

It follows that

$$[i + k(n-2) + j]_{\mathbf{P},\mathbf{Q}}! \xi_k(p_{kn,\beta_1,\beta_2})$$

$$= \sum_{\substack{T=(c_1,\ldots,c_s) \\ c_1+\cdots+c_s=n, s\geq 2}} \begin{bmatrix} i + k(n-2) + j \\ i + k(c_1-1), kc_2, \ldots, kc_{s-1}, k(c_s-1) + j \end{bmatrix}_{\mathbf{P},\mathbf{Q}}$$

$$\times \prod_{r=1}^s (x-1)^{c_r-1}.$$

By Lemma 3.2, we can interpret

$$\begin{bmatrix} i + k(n-2) + j \\ i + k(c_1-1), kc_2, \ldots, kc_{s-1}, k(c_s-1) + j \end{bmatrix}_{\mathbf{P},\mathbf{Q}}$$

as the set of filling of the brick tabloid

$$U = (i + k(c_1 - 1), kc_2, \ldots, kc_{s-1}, , k(c_s - 1) + j)$$

of L-tuples of permutations $\boldsymbol{\Sigma} = (\sigma^{(1)}, \ldots, \sigma^{(L)})$ such that for each i, the elements of $\sigma^{(i)}$ are increasing within each brick of U and we weight such a filling with $\mathbf{Q}^{inv(\boldsymbol{\Sigma})}\mathbf{P}^{coinv(\boldsymbol{\Sigma})}$. In fact, we shall think of U as the brick tabloid $kT = (kc_1, \ldots, kc_s)$ with the first $k - i$ cells of the first brick blank and the last $k - j$ cells of the last brick blank.

Again, order the cells of such a filled brick tabloid from left to right and interpret the term $\prod_{r=1}^{s}(x - 1)^{c_r - 1}$ as taking such a filling and labeling the cells of the form sk which are not at the end of a brick with either an x or -1 and labeling each cell at the end of a brick with 1. An example is pictured below:

		-1			1			1		x			1
$\sigma^{(1)}$	5	8	9	11	12	4	7	10	1	2	3	6	
$\sigma^{(2)}$	1	2	7	10	11	4	9	12	3	5	6	8	
$\sigma^{(3)}$	3	5	9	10	12	1	4	11	2	6	7	8	

Again, $\mathcal{O}$ is a labeled filled brick tabloid. We define the weight of $\mathcal{O}$, $W(\mathcal{O})$, to be product over all the labels of the cells times $\mathbf{Q}^{inv(\boldsymbol{\Sigma})}\mathbf{P}^{coinv(\boldsymbol{\Sigma})}$ if T is filled with permutations $\boldsymbol{\Sigma} = (\sigma^{(1)}, \ldots, \sigma^{(L)})$.

Let $\mathcal{L}F^{(k,i,j)}(kn)$ denote the set of all objects that can be created in this way from brick tabloids $T = (c_1, \ldots, c_s)$ where $s \geq 2$ and $c_1 + \cdots + c_s = n$. Then it follows that

$$[i + k(n - 2) + j]_{\mathbf{P},\mathbf{Q}}! \xi_k(p_{kn, \beta_1, \beta_2}) = \sum_{\mathcal{O} \in \mathcal{L}F^{(k,i,j)}(kn)} W(\mathcal{O}).$$

Next we define an involution $I : \mathcal{L}F^{(k,i,j)}(kn) \to \mathcal{L}F^{(k,i,j)}(kn)$ which is a slight variation of our previous two involutions. That is, given $\mathcal{O} \in \mathcal{L}F^{(k,i,j)}(kn)$, read the cells of $\mathcal{O}$ in the same order that we read the underlying permutations and look for the first cell kc such that either:

(i) kc is labeled with -1 or
(ii) kc is at the end of end of brick b, the cell $kc + 1$ is immediately to the right of kc and starts another brick b', and each permutation $\sigma^{(i)}$ increases as we go from kc to $kc + 1$.

If we are in case (i), then $I(\mathcal{O})$ is the labeled filled brick tabloid which is obtained from $\mathcal{O}$ by taking the brick b that contains kc and splitting b into two bricks b_1 and b_2 where b_1 contains the cells of b up to and including the cell kc and b_2 contains the remaining cells of b and changing the label on kc from -1 to 1. In case (ii), if $\mathcal{O}$ has at least three bricks, then $I(\mathcal{O})$ is the labeled filled brick tabloid which is obtained from $\mathcal{O}$ by combining the two bricks b and b' into a single brick and changing the label on cell kc from 1 to -1. However, if we are in case (ii) and $\mathcal{O}$ has exactly 2 bricks, then $I(\mathcal{O}) = \mathcal{O}$. Finally, if neither case (i) or case (ii) applies, then we let $I(\mathcal{O}) = \mathcal{O}$. For example, the image of the above figure under I is

		1			1			1		x				1
$\sigma^{(1)}$	5	8	9	11	12	4	7	10	1	2	3	6		
$\sigma^{(2)}$	1	2	7	10	11	4	9	12	3	5	6	8		
$\sigma^{(3)}$	3	5	9	10	12	1	4	11	2	6	7	8		

This sign-reversing weight-preserving involution I shows

$$[i + k(n-2) + j]_{\mathbf{P},\mathbf{Q}}! \xi_k(p_{n,\beta_1,\beta_2})$$

$$= \sum_{\mathcal{O} \in \mathcal{L}F^{(k,i,j)}(kn)} W(\mathcal{O})$$

$$= \sum_{\mathcal{O} \in \mathcal{L}F^{(k,i,j)}(kn), I(\mathcal{O})=\mathcal{O}} W(\mathcal{O}). \tag{14}$$

If $I(\mathcal{O}) = \mathcal{O}$, then $\mathcal{O}$ can have no cells which are labeled with -1. If $\mathcal{O}$ has at least 3 bricks, then if must be the case that between any two consecutive bricks of $\mathcal{O}$, at least one of the underlying permutations $\sigma^{(i)}$ must decrease. It follows that each cell kc which is not at the end of the brick in $\mathcal{O}$ is labeled with x and each of the permutations $\sigma^{(i)}$ has a rise at kc so that $kc \in Comris_{i,k}(\Sigma)$. All the other cells of the form kc in $\mathcal{O}$ other than the last cell are at the end of brick which has another brick to its right in which case $kc \notin Comris_{i,k}(\Sigma)$. All such cells have label 1 so that $W(\mathcal{O}) = x^{comris_{i,k}(\Sigma)} \mathbf{Q}^{inv(\Sigma)} \mathbf{P}^{coinv(\Sigma)}$.

Next consider the fixed points $\mathcal{O}$ of I which have two bricks. Again $\mathcal{O}$ can have no cells labeled with -1. There are two cases here. If the last cell of first brick is kc and at least one of the underlying permutations $\sigma^{(i)}$ decreases form kc to $kc + 1$, then again it will be the case that $W(\mathcal{O}) = x^{comris_{i,k}(\Sigma)} \mathbf{Q}^{inv(\Sigma)} \mathbf{P}^{coinv(\Sigma)}$. However, if each of the underlying permutations $\sigma^{(i)}$ increases from kc to $kc + 1$, then it must be the case that each of the $\sigma^{(i)}$'s is the identity permutation. In this case, $W(\mathcal{O}) = x^{n-2}\mathbf{P}^{\left(i+k\binom{n-2}{2}+j\right)}$. Moreover, since kc can be either $1, \ldots, n-1$, it follows that the contribution of $\mathbf{Id} = (\sigma^{(1)}, \ldots, \sigma^{(L)})$ where each $\sigma^{(i)}$ is the identity permutation to (14) is $(n-1)x^{n-2}\mathbf{P}^{\left(i+k\binom{n-2}{2}+j\right)}$. However the contribution of $\mathbf{Id}$ to

$$\sum_{\Sigma \in (C^{i,j,k}_{i+k(n-2)+j})^L} x^{comris_{i,k}(\Sigma)} \mathbf{Q}^{inv(\Sigma)} \mathbf{P}^{coinv(\Sigma)}$$

is $x^{n-1}\mathbf{P}^{\left(i+k\binom{n-2}{2}+j\right)}$. It thus follows that $\sum_{\mathcal{O} \in \mathcal{L}F^{(k,i,j)}(kn), I(\mathcal{O})=\mathcal{O}} W(\mathcal{O})$ is equal to

$$(n-1)x^{n-2}\mathbf{P}^{\left(i+k\binom{n-2}{2}+j\right)} - x^{n-1}\mathbf{P}^{\left(i+k\binom{n-2}{2}+j\right)}$$
$$+ \sum_{\Sigma \in (C^{i,j,k}_{kn})^L} x^{comris_{i,k}(\Sigma)} \mathbf{Q}^{inv(\Sigma)} \mathbf{P}^{coinv(\Sigma)}$$

as desired. $\qquad\qquad\qquad\qquad\qquad\qquad\qquad\qquad\qquad\qquad\qquad\qquad\qquad\square$

We can now apply ξ_k to the identity

$$\sum_{n \geq 1} p_{n,\beta_1,\beta_2} t^n = \frac{\left(\sum_{n \geq 1}(-1)^{n-1}\beta_1(n)e_n t^n\right)\left(\sum_{n \geq 1}(-1)^{n-1}\beta_2(n)e_n t^n\right)}{E(-t)}.$$

Using (12), we can find that

$$\xi\left(\sum_{n \geq 1}(-1)^{n-1}\beta_1(n)e_n t^n\right) = \sum_{n \geq 1}(-1)^{kn-1}\beta_1(kn)\xi_k(e_{kn})t^{kn}$$

$$= \sum_{n \geq 1} \frac{(x-1)^{n-1}\mathbf{P}^{\left(k\binom{n-1}{2}+i\right)}t^{kn}}{[k(n-1)+i]_{\mathbf{P},\mathbf{Q}}!}$$

$$= \frac{1}{x-1}\sum_{n \geq 1} \frac{(x-1)^n t^{kn}\mathbf{P}^{\left(k\binom{n-1}{2}+i\right)}}{[k(n-1)+i]_{\mathbf{P},\mathbf{Q}}!}$$

$$= \frac{1}{x-1}e^{(i)}_{\mathbf{P},\mathbf{Q},k}(t(x-1)^{1/k}) \qquad (15)$$

Similarly,

$$\xi\left(\sum_{n \geq 1}(-1)^{n-1}\beta_2(n)e_n t^n\right) = \frac{1}{x-1}e^{(j)}_{\mathbf{P},\mathbf{Q},k}(t(x-1)^{1/k}). \qquad (16)$$

Combining (10), (15), and (16), we see that

$$
\sum_{n \geq 1} \xi_k(p_{n,\beta_1,\beta_2}) t^n = \sum_{n \geq 2} \frac{(n-1) x^{n-2} \mathbf{P}^{\binom{i+k\binom{n-2}{2}+j}{2}} t^{kn}}{[i+k(n-2)+j]_{\mathbf{P},\mathbf{Q}}!}
$$

$$
- \sum_{n \geq 2} \frac{x^{n-1} \mathbf{P}^{\binom{i+k\binom{n-2}{2}+j}{2}} t^{kn}}{[i+k(n-2)+j]_{\mathbf{P},\mathbf{Q}}!}
$$

$$
+ \sum_{n \geq 2} \frac{t^{kn}}{[i+k(n-2)+j]_{\mathbf{P},\mathbf{Q}}!}
$$

$$
\times \sum_{\boldsymbol{\Sigma} \in (\mathcal{C}^{i,j,k}_{k(n-1)+j})^L} x^{comris_{i,k}(\boldsymbol{\Sigma})} \mathbf{Q}^{inv(\boldsymbol{\Sigma})} \mathbf{P}^{coinv(\boldsymbol{\Sigma})}
$$

$$
= \frac{\frac{1}{x-1} e^{(i)}_{\mathbf{P},\mathbf{Q},k}(t(x-1)^{1/k}) \frac{1}{x-1} e^{(j)}_{\mathbf{P},\mathbf{Q},k}(t(x-1)^{1/k})}{\frac{1}{1-x}\left(-x + e_{\mathbf{P},\mathbf{Q},k}(t(x-1)^{1/k})\right)}
$$

$$
= \frac{e^{(i)}_{\mathbf{P},\mathbf{Q},k}(t(x-1)^{1/k}) e^{(j)}_{\mathbf{P},\mathbf{Q},k}(t(x-1)^{1/k})}{(1-x)\left(-x + e_{\mathbf{P},\mathbf{Q},k}(t(x-1)^{1/k})\right)},
$$

and we have therefore proved the following theorem, achieving the final goal set out in the introduction.

Theorem 4.2. *For any* $k \geq 2$ *and* $1 \leq i,j \leq k-1$,

$$
\sum_{n \geq 2} \frac{t^{kn}}{[i+k(n-2)+j]_{\mathbf{P},\mathbf{Q}}!}
$$

$$
\times \sum_{\boldsymbol{\Sigma} \in (\mathcal{C}^{i,j,k}_{i+k(n-2)+j})^L} x^{comris_{i,k}(\boldsymbol{\Sigma})} \mathbf{Q}^{inv(\boldsymbol{\Sigma})} \mathbf{P}^{coinv(\boldsymbol{\Sigma})}
$$

$$
= \sum_{n \geq 2} \frac{x^{n-1} \mathbf{P}^{\binom{i+k\binom{n-2}{2}+j}{2}} t^{kn}}{[i+k(n-2)+j]_{\mathbf{P},\mathbf{Q}}!} - \sum_{n \geq 2} \frac{(n-1) x^{n-2} \mathbf{P}^{\binom{i+k\binom{n-2}{2}+j}{2}} t^{kn}}{[i+k(n-2)+j]_{\mathbf{P},\mathbf{Q}}!}
$$

$$
+ \frac{e^{(i)}_{\mathbf{P},\mathbf{Q},k}(t(x-1)^{1/k}) e^{(j)}_{\mathbf{P},\mathbf{Q},k}(t(x-1)^{1/k})}{(1-x)((-x + e_{\mathbf{P},\mathbf{Q},k}(t(x-1)^{1/k}))}
$$

and

$$
\sum_{n \geq 2} \frac{t^{kn}}{[i+k(n-2)+j]_{\mathbf{P},\mathbf{Q}}!} \sum_{\boldsymbol{\Sigma} \in (\mathcal{E}^{i,j,k}_{i+k(n-2)+j})^L} \mathbf{Q}^{inv(\boldsymbol{\Sigma})} \mathbf{P}^{coinv(\boldsymbol{\Sigma})}
$$

$$
= \frac{e^{(i)}_{\mathbf{P},\mathbf{Q},k}(t(-1)^{1/k}) e^{(j)}_{\mathbf{P},\mathbf{Q},k}(t(-1)^{1/k})}{e_{\mathbf{P},\mathbf{Q},k}(t(-1)^{1/k})} - \frac{\mathbf{P}^{\binom{i+j}{2}} t^{2k}}{[i+j]_{\mathbf{P},\mathbf{Q}}!}.
$$

A few remarks are in order. The Salié numbers count permutations $\sigma \in S_n$ such that $Des(\sigma) = \{2, 4, \ldots, 2k\}$ where $2k \leq n$. Carlitz [6] showed that the generating function of the Salié numbers is given by $\cosh(t)/\cos(t)$. More recently, Prodinger [20] and Guo and Zeng [12] studied q-analogues of the Salié numbers. The methods of this paper can also be used to prove analogues of (3), (4), and (5) for the set of permutations $\sigma \in S_n$ such that $Des(\sigma) \subseteq \{i + ks, i + k(s+1), \ldots, i + kt\}$ for some $0 \leq s < t$ where $kt < n$. We will show in a forthcoming paper that we can obtain such generating functions by varying the weight functions described in this paper in an appropriate manner.

In [21], Remmel and Riehl have shown how to find the generating functions for permutations which contain a given descent set by applying ring homomorphisms to symmetric function identities involving ribbon Schur functions. Their methods give a systematic way to find the following generating functions for any $S \subset \{1, 2, \ldots\}$:

(i) $\displaystyle\sum_{n=0}^{\infty} \frac{u^n}{n!} \sum_{\sigma \in S_n,\, S \subseteq Des(\sigma)} x^{des(\sigma)}$

(ii) $\displaystyle\sum_{n=0}^{\infty} \frac{u^n}{(n!)^2} \sum_{(\sigma,\tau) \in S_n \times S_n,\, S \subseteq Comdes(\sigma,\tau)} x^{comdes(\sigma,\tau)}$

(iii) $\displaystyle\sum_{n=0}^{\infty} \frac{u^n}{[n]_q!} \sum_{\sigma \in S_n,\, S \subseteq Des(\sigma)} x^{des(\sigma)} q^{inv(\sigma)}$

(iv) $\displaystyle\sum_{n=0}^{\infty} \frac{u^n}{[n]_{q,p}!} \sum_{\sigma \in S_n,\, S \subseteq Des(\sigma)} x^{des(\sigma)} q^{inv(\sigma)} p^{coinv(\sigma)}$

(v) $\displaystyle\sum_{n=0}^{\infty} \frac{u^n}{[n]_q![n]_p!} \sum_{(\sigma,\tau) \in S_n \times S_n,\, S \subseteq Comdes(\sigma,\tau)} x^{comdes(\sigma,\tau)} q^{inv(\sigma)} p^{inv(\tau)}$

It is also possible to derive similar analogues of (3), (4), and (5). That is, we can find generating function for L-tuples of permutations in $C_{i+nk+j}^{i,j,k}$ such that $Comris_{i,k}(\Sigma)$ contains a given finite set of the form $\{s_1 k, s_2 k, \ldots, s_r k\}$. One can also get other regular patterns of descents in permutations by replacing the increasing fillings within bricks by more complicated patterns. Such ideas will be developed more fully in upcoming papers.

References

[1] D. André. Développements de sec x et de tang x. *C. R. Math. Acad. Sci. Paris*, 88:965–967, 1879.

[2] D. André. Sur les permutations alternées. *J. Math. Pures Appl.*, 7:167–184, 1881.

[3] G. E. Andrews and D. Foata. Congruences for the q-secant numbers. *European J. Combin.*, 1(4):283–287, 1980.

[4] G. E. Andrews and I. Gessel. Divisibility properties of the q-tangent numbers. *Proc. Amer. Math. Soc.*, 68(3):380–384, 1978.

[5] F. Brenti. Permutation enumeration symmetric functions, and unimodality. *Pacific J. Math.*, 157(1):1–28, 1993.

[6] L. Carlitz. The coefficients of $\cosh x/\cos x$. *Monatsh. Math.*, 69:129–135, 1965.

[7] L. Carlitz. Sequences and inversions. *Duke Math. J.*, 37:193–198, 1970.

[8] Ö. Eğecioğlu and J. B. Remmel. Brick tabloids and the connection matrices between bases of symmetric functions. *Discrete Appl. Math.*, 34(1-3):107–120, 1991.

[9] J.-M. Fédou and D. Rawlings. Statistics on pairs of permutations. *Discrete Math.*, 143(1-3):31–45, 1995.

[10] D. Foata. Further divisibility properties of the q-tangent numbers. *Proc. Amer. Math. Soc.*, 81(1):143–148, 1981.

[11] I. M. Gessel. Some congruences for generalized Euler numbers. *Canad. J. Math.*, 35(4):687–709, 1983.

[12] V. J. W. Guo and J. Zeng. Some arithmetic properties of the q-Euler numbers and q-Salié numbers. *European J. Combin.*, 27(6):884–895, 2006.

[13] T. M. Langley and J. B. Remmel. Enumeration of m-tuples of permutations and a new class of power bases for the space of symmetric functions. *Adv. in Appl. Math.*, 36(1):30–66, 2006.

[14] D. J. Leeming and R. A. MacLeod. Some properties of generalized Euler numbers. *Canad. J. Math.*, 33(3):606–617, 1981.

[15] I. G. Macdonald. *Symmetric functions and Hall polynomials*. Oxford Mathematical Monographs. The Clarendon Press Oxford University Press, New York, second edition, 1995.

[16] A. Mendes. *Building generating functions brick by brick*. PhD thesis, University of California, San Diego, 2004.

[17] A. Mendes and J. B. Remmel. Generating functions for statistics on $C_k \wr S_n$. *Sém. Lothar. Combin.*, 54A:Art. B54At, 40 pp., 2005/07.

[18] A. Mendes and J. B. Remmel. Permutations and words counted by consecutive patterns. *Adv. in Appl. Math.*, 37(4):443–480, 2006.

[19] H. Prodinger. Combinatorics of geometrically distributed random variables: new q-tangent and q-secant numbers. *Int. J. Math. Math. Sci.*, 24(12):825–838, 2000.

[20] H. Prodinger. q-enumeration of Salié permutations. *Ann. Comb.*, 11(2):213–225, 2007.

[21] J. B. Remmel and A. Riehl. Generating functions for permutations which contain a given descent set. *Electron. J. Combin.*, 17(1):Research Paper 27, 33 pp., 2010.

[22] B. E. Sagan and P. Zhang. Arithmetic properties of generalized Euler numbers. *Southeast Asian Bull. Math.*, 21(1):73–78, 1997.

[23] R. P. Stanley. Binomial posets, Möbius inversion, and permutation enumeration. *J. Combinatorial Theory Ser. A*, 20(3):336–356, 1976.

[24] R. P. Stanley. *Enumerative combinatorics. Vol. 1*, volume 49 of *Cambridge Studies in Advanced Mathematics*. Cambridge University Press, Cambridge, 1997.

[25] J. D. Wagner. The permutation enumeration of wreath products $C_k \wr S_n$ of cyclic and symmetric groups. *Adv. in Appl. Math.*, 30(1-2):343–368, 2003.

Packing rates of measures and a conjecture for the packing density of 2413

Cathleen Battiste Presutti

Department of Mathematics
Ohio University - Lancaster

Lancaster, Ohio 43130 USA

Walter Stromquist

Department of Mathematics and Statistics
Swarthmore College

Swarthmore, Pennsylvania 19081 USA

Abstract

We give a new lower bound of 0.10472422757673209041 for the packing density of 2413, justify it by a construction, and conjecture that this value is actually equal to the packing density. Along the way we define the packing rate of a permutation with respect to a measure, and show that maximizing the packing rate of a pattern over all measures gives the packing density of the pattern.

In this paper we consider the packing density of the pattern 2413. This pattern is significant because it is not layered, and because up to

Fig. 1. The conjecture is based on this measure, μ_2

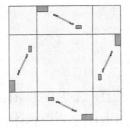

Fig. 2. The measure μ_∞

symmetry it is the smallest nontrivial pattern that is simple in the sense of [3]. We conjecture that its packing density is given by

$$\delta(2413) = 0.10472422757673209041\ldots,$$

and we show by a construction that this value is a lower bound. It is slightly larger than the lower bound of $0.10425\ldots$ given in [4].

We leap ahead briefly to describe the construction. Figure 1 describes a probability distribution on the unit square. Probability is concentrated on the dark shaded rectangles and the dark shaded segments. The distribution is described below. It is not uniform along the segments or in the rectangles; in fact, the rectangles are "recursion bubbles," meaning that each of them is a scaled-down replica of the entire figure. To construct a permutation of size n (for n large) with a large number of occurrences of the pattern 2413, we select n points independently from this distribution, and treat them as the graph of a permutation. In the limit of large n, with probability one, the packing density of 2413 in the resulting permutations approaches the value given above for $\delta(2413)$. (Figure 1 is not drawn to scale. If it were, the smaller recursion boxes would be too small to see.)

Permutations constructed in this way tend to consist of an initial increasing sequence, then two interleaved increasing sequences (one of high values, one of low values), then a final increasing sequence. Figure 9, below, shows the graph of one such permutation (with $n = 8$) and Figure 11, a prototype of Figure 1, suggests the more general pattern.

The basic definitions related to packing densities are reviewed in Section 1.

Our principal technique is to reinterpret packing densities in the language of measures. By a measure we mean a probability distribution on the unit square. In Section 2 we define the packing rate of a pattern with respect to a measure, and define the packing rate $\delta'(\pi)$ of a pattern π as the supremum of its packing rates over all measures. Our main result, in Section 4, is that the packing rate of a pattern is equal to its packing density, $\delta'(\pi) = \delta(\pi)$. Finding the packing density of a pattern is then a matter of finding an optimal measure for the pattern.

We return to the packing density of 2413 in Section 5. The language of measures allows us to bring to bear the techniques of analysis, including the calculus of variations and extensive calculations involving integrals of probability distribution functions. In Sections 6 to 8 we define four-segment measures, and by extensive calculation we find the optimal measure for 2413 within this class. In Sections 9 to 11 we improve the measure slightly by the use of recursion bubbles. We conjecture that the optimal measure is the one we call μ_2, which is the measure that is illustrated in Figure 1, on which the conjecture is based.

Figure 2 illustrates an attractive alternative called μ_∞, in which there is an infinite sequence of recursion bubbles at each end of each segment. We do not believe that μ_∞ is optimal, but we cannot rule it out.

1 Packing densities

Let $\pi \in S_m$. A sequence $x_1, \ldots, x_m$ *has the order type of* π if, for all i and j, $x_i < x_j \Leftrightarrow \pi_i < \pi_j$. This requires at least that the terms x_i be distinct. If $\sigma \in S_n$ then an *occurrence of* π *in* σ is an m-term subsequence of σ that has the order type of π. The number $\nu(\pi, \sigma)$ of such occurrences is called the *packing number of* π *in* σ, and the ratio

$$\delta(\pi, \sigma) = \frac{\nu(\pi, \sigma)}{\binom{n}{m}} \tag{1}$$

is called the *packing density of* π *in* σ. Clearly $0 \le \delta(\pi, \sigma) \le 1$. In this context π is called a *pattern*. We always assume that $\pi \in S_m$, $\sigma \in S_n$, and $n \ge m \ge 1$.

For a fixed pattern π we are concerned with finding permutations $\sigma \in S_n$ that maximize the packing density, especially in the limit as $n \to \infty$. Write

$$\delta(\pi, n) = \max_{\sigma \in S_n} \delta(\pi, \sigma). \tag{2}$$

If σ realizes this maximum—that is, if $\delta(\pi, \sigma) = \delta(\pi, n)$—then σ is called an *optimizer* (or "optimizing permutation") of size n for π. The *packing density of* π is

$$\delta(\pi) = \lim_{n \to \infty} \delta(\pi, n). \tag{3}$$

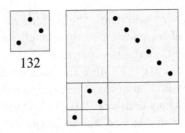

132

Fig. 3. $\nu(132, 132987654) = 46$

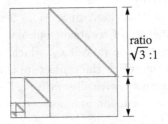

ratio
$\sqrt{3}$:1

Fig. 4. A measure for 132

Galvin showed that the sequence $\{\delta(\pi, n)\}$ is non-increasing, so its limit always exists.

Equivalently we could define the packing density as the largest number D for which there is a sequence of permutations $\sigma_1, \sigma_2, \ldots$ of increasing size with

$$D = \lim_{i \to \infty} \delta(\pi, \sigma_i). \qquad (4)$$

Such a sequence with $D = \delta(\pi)$ is called an *optimizing sequence* for π, and the permutations σ_i are called (collectively) *near-optimizers*. They do not need actually to be optimizers; they need only be close enough to give the right limit.

As an example consider the pattern $\pi = 132$. If $\sigma = 132987654$ then $\nu(\pi, \sigma) = 46$ and $\delta(\pi, \sigma) = 46/84 \approx 0.548$. This turns out to be the unique optimizer of size 9 for π, so $\delta(\pi, 9) = 46/84$ as well. These permutations are illustrated in Figure 3. The shape of σ suggests a recursive construction of near-optimizers for larger n. In fact, as is well known, this construction does produce an optimizing sequence for 132, whose packing density turns out to be $\delta(132) = 2\sqrt{3} - 3 \approx 0.464$. (Even for $\pi = 132$ it is not so easy to find optimizers for particular values of *n.* Rounding issues arise and many possibilities need to be considered. Near-optimizers are easier.)

A simpler illustration of the near-optimizers for 132 is in Figure 4. The points in the graph of σ line up along the diagonal lines in the figure, and are distributed uniformly by length (to the extent possible for any particular value of n).

Figure 4 could be understood as simply a guide to the imagination. We prefer to give it a more formal meaning: we interpret pictures like this as defining probability measures on the unit square. In the next sections we will clarify this interpretation and show how it relates to packing densities.

2 Packing rates for measures

We consider probability measures μ on the unit square $S = [0,1] \times [0,1] \subseteq \mathbf{R}^2$. In this section we define $\delta'(\pi, \mu)$, the packing rate of a pattern π with respect to a measure μ. The packing rate of the pattern, $\delta'(\pi)$, is the supremum of the rates $\delta'(\pi, \mu)$ over all measures μ.

Recall that a *measure* μ on S assigns a non-negative value $\mu(A)$ to each Borel set $A \subseteq S$ in such a way that $\mu(\cup A_i) = \sum (\mu(A_i))$ whenever $\{A_i\}$ is a finite or countable sequence of pairwise disjoint sets. It is a *probability measure* if $\mu(S) = 1$. Borel sets are subsets of S that can obtained from closed rectangles in finitely many steps, each step being a complementation, a union or intersection of finitely many sets, or a union or intersection of countably many sets. In this paper all of the sets we encounter are Borel sets and "measure" always means probability measure.

A measure can be interpreted as a guide for selecting points randomly from S. When we say that a point is selected "according to μ" we mean that the probability that the point is in any set A is $\mu(A)$. Our plan is to pick m points independently according to μ and look at the order type of the resulting configuration.

Suppose that an m-tuple of points in S has no repeated x coordinates and no repeated y coordinates. We say that it *has the order type of* π if, when the points are arranged in order of increasing x coordinates, their y coordinates form a sequence with the order type of π. See Figure 5. (The order of the points in the m-tuple does not affect the order type.) An m-tuple that has a repeated x coordinate or a repeated y coordinate is called *degenerate* and has no order type.

An *order-preserving transformation* of S is a map of the form $(x, y) \mapsto$

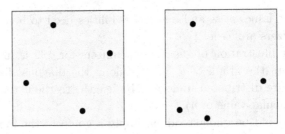

Fig. 5. Graph of 2413, and a 4-tuple with the order type of 2413

$(f(x), g(y))$ where each of f and g is an order-preserving bijection of $[0,1]$. Transformations of this type preserve the order type of any m-tuple. Given any two m-tuples with the same order type, we can find an order-preserving transformation of S that maps one onto the other.

Let $\pi \in S_m$ be a pattern and let μ be a measure. The *packing rate* of π with respect to μ is the probability, if m points are selected independently according to μ, that they have the order type of π. We denote the packing rate by $\delta'(\pi, \mu)$.

More precisely: Let $\mu^m = \mu \times \cdots \times \mu$ be the product measure on S^m, and let $C_\pi \subseteq S^m$ contain all m-tuples of points that have order type π. Then the packing rate is

$$\delta'(\pi, \mu) = \mu^m(C_\pi). \tag{5}$$

The notation δ' for packing rates is temporary. After we relate packing rates to packing densities in Theorem 4.1 we will replace δ' with δ everywhere.

Examples of packing rates.

(i) Let μ be the uniform measure on S (Figure 6). Then all order types are equally likely. We have $\delta'(123, \mu) = 1/6$ and in general

$$\delta'(\pi, \mu) = \frac{1}{m!} \tag{6}$$

if π has size m.

(ii) Let μ be concentrated on the main diagonal of S (Figure 7). Then

$$\delta'(123, \mu) = 1 \tag{7}$$

and $\delta'(\pi, \mu) = 0$ for any other pattern π of size 3. It isn't necessary that μ be uniform on the diagonal as long as single points have zero probability.

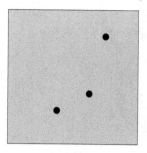

Fig. 6. uniform   Fig. 7. diagonal

(iii) Let μ be concentrated along countably many diagonal segments as shown in Figure 4, with probability being proportional to length. Then

$$\delta'(132, \mu) = 2\sqrt{3} - 3 \approx 0.464. \tag{8}$$

This is equal to the packing density of 132. We call this an "optimal measure" for 132, because no other measure gives a higher packing rate.

(iv) **A challenge.** Let μ be uniform on a disk in S (Figure 8). Then what is $\delta'(123, \mu)$? (We don't know!)

(v) **Template measures.** Let $\tau \in S_k$ be a permutation and form a measure μ_τ as shown in Figure 9. The measure is concentrated uniformly on the union of k small squares arranged like the graph of τ. Then μ_τ is called the *template measure* corresponding to τ. We will use template measures in the proof of Theorem 4.1 and we will have more to say about them in Section 5. For now, consider the packing rate of a pattern π with respect to its own template measure μ_π. If m points are selected from m different cells in the template—an event that occurs with probability $m!/m^m$—then they are guaranteed to form an occurrence of τ. Therefore,

$$\delta'(\pi, \mu_\pi) \geq \frac{m!}{m^m}. \tag{9}$$

The *packing rate of π* is the supremum of $\delta'(\pi, \mu)$:

$$\delta'(\pi) = \sup_\mu \delta'(\pi, \mu), \tag{10}$$

the supremum being taken over all probability measures μ on S. An

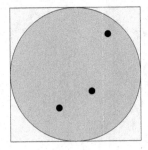

Fig. 8. disk

optimal measure for π (or "optimizer" when we are considering only measures) is a measure μ that achieves the supremum.

The example of the template measure shows that $\delta'(\pi) \geq m!/m^m$ for any $\pi \in S_m$.

3 Limits of measures

Is there an optimal measure for every pattern? That is, is the supremum in (10) really a maximum? The answer is yes, and we can prove it by forming a limit of of a sequence of measures whose packing rates approach the supremum. But the proof requires care for two reasons:

- We need a suitable definition for the limit of a sequence of measures; and
- Limits of measures do not always respect packing rates.

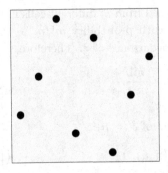
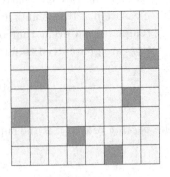

Fig. 9. $\tau = 35827146$ and the template measure μ_τ

As a cautionary example consider the measures μ_j defined as in Figure 10. Each μ_j is concentrated (uniformly) on the square $[1/(j+1), 1/j]^2 \subseteq S$ for $j = 1, 2, \ldots$. The only candidate for a limiting measure is the measure μ defined by a point mass at the origin. Then for each j we have $\delta'(123, \mu_j) = 1/6$, but $\delta'(123, \mu) = 0$.

The trouble, of course, is that the limit measure allows degenerate m-tuples. We need to identify circumstances in which this does not occur.

First we define limits of measures. We say that μ is the limit of a sequence of measures $\{\mu_j\}$,

$$\mu = \lim \mu_j,$$

if

$$\int_S f(p)d\mu(p) = \lim_{j \to \infty} \int_S f(p)d\mu_j(p) \tag{11}$$

for every continuous function $f : S \to \mathbf{R}$. With this definition the limit of a sequence is unique (if it exists) and the probability measures on S form a compact topological space. This means that from any sequence of measures $\{\nu_i\}$ we can select a sequence $\{\mu_j\}$ that has a limit measure.

It is not generally true that $\mu(A) = \lim \mu_j(A)$ for an arbitrary set A. (Consider $A = \{(0,0)\}$ in the cautionary example.) But it can be shown from the definition that if A is a closed set,

$$\mu(A) \geq \limsup_{j \to \infty} \mu_j(A). \tag{12}$$

and if B is an open set,

$$\mu(B) \leq \liminf_{j \to \infty} \mu_j(B). \tag{13}$$

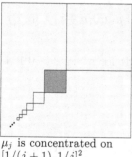

μ_j is concentrated on $[1/(j+1), 1/j]^2$.

μ is a point mass at the origin.

Fig. 10. Limiting measures do not always respect packing rates

Well-behaved measures. We now identify some special classes of measures. For this paper, a measure μ is *smooth* if $\mu(A) = 0$ whenever A is a vertical or horizontal line. If points are selected independently according to a smooth measure, the probability of their forming a degenerate configuration is zero.

A measure μ is *normalized* if its projection onto each axis is the uniform measure; that is, if

$$\mu([0, a] \times [0, 1]) = \mu([0, 1] \times [0, a]) = a \tag{14}$$

for every $a \in [0, 1]$. Every normalized measure is smooth. Better, as we now show, the limit of a sequence of normalized measures is necessarily normalized, and limits of normalized measures respect packing rates.

Lemma 3.1. *If $\mu = \lim_{j \to \infty} \mu_j$ and each μ_j is normalized, then μ is also normalized and for any pattern π,*

$$\delta'(\pi, \mu) = \lim_{j \to \infty} \delta'(\pi, \mu_j).$$

Proof. To see that μ is normalized note that

$$\mu([0, a] \times [0, 1]) \geq \limsup_{j \to \infty} \mu_j([0, a] \times [0, 1]) \tag{15}$$

because this set is closed. Every term in the right-hand sequence is a, so $\mu([0, a] \times [0, 1]) \geq a$. If $\epsilon > 0$ then

$$\mu([0, a] \times [0, 1]) \leq \mu([0, a + \epsilon) \times [0, 1])$$
$$\leq \liminf_{j \to \infty} \mu_j([0, a + \epsilon) \times [0, 1]) \tag{16}$$

so $\mu([0, a] \times [0, 1]) < a + \epsilon$. Since this holds for every $\epsilon > 0$ it follows that

$$\mu([0, a] \times [0, 1]) = a. \tag{17}$$

The same is true in the other dimension, so μ is normalized.

To see that these limits respect packing rates, note that according to any of the normalized measures μ or μ_j, the boundary of each C_π has measure zero. Therefore

$$
\begin{aligned}
\delta'(\pi, \mu) &= \mu^m(C_\pi) \\
&= \mu^m(\text{int } C_\pi) \\
&\leq \liminf \mu_j^m(\text{int } C_\pi) \\
&= \liminf \delta'(\pi, \mu_j) \\
&\leq \limsup \delta'(\pi, \mu_j) \\
&= \limsup \mu_j^m(\text{cl } C_\pi) \\
&\leq \mu^m(\text{cl } C_\pi) \\
&= \mu^m(C_\pi) \\
&= \delta'(\pi, \mu).
\end{aligned}
\tag{18}
$$

This is enough to force $\lim \delta'(\pi, \mu_j) = \delta'(\pi, \mu)$. $\qquad\square$

Lemma 3.2. *If μ is any measure on S then there is a normalized measure $\tilde{\mu}$ such that $\delta'(\pi, \tilde{\mu}) \geq \delta'(\pi, \mu)$ for every pattern π.*

We call $\tilde{\mu}$ a *normalization* of μ. It is unique if μ is smooth. We conclude from Lemma 3.2 that if we want to maximize $\delta'(\pi, \nu)$ it suffices to look among normalized measures ν.

Proof. If μ is smooth we can find an order-preserving transformation of S that maps μ to a normalized measure. More precisely, we can define $\tilde{\mu}$ by

$$
\tilde{\mu}([0, a] \times [0, b]) = \mu([0, x] \times [0, y])
\tag{19}
$$

for every pair (a, b), where x is the least value for which

$$
\mu([0, x] \times [0, 1]) = a
\tag{20}
$$

and y is the least value for which

$$
\mu([0, 1] \times [0, y]) = b.
\tag{21}
$$

The values specified in (19) are enough to determine $\tilde{\mu}$ on all Borel sets. Now $\delta'(\pi, \tilde{\mu}) = \delta'(\pi, \mu)$ for all patterns π.

If μ is not smooth, we apply (19) whenever x and y are uniquely determined, and then extend $\tilde{\mu}$ arbitrarily to a normalized measure on S. The implied mapping is not a bijection, and some m-tuples which have no order type at all (because an x coordinate is repeated or a y coordinate is repeated) may be mapped to m-tuples that do have order types. So, the probability of an order type π arising may be greater under $\tilde{\mu}$ than under μ, and we may have $\delta'(\pi, \tilde{\mu}) > \delta'(\pi, \mu)$.   □

The next theorem says that every pattern has an optimal measure.

Theorem 3.3. *For every pattern π there is a normalized measure μ for which $\delta'(\pi, \mu) = \sup_\nu \delta'(\pi, \nu)$, the supremum being taken over all measures ν on S.*

Proof. Let $D = \sup_\nu \delta'(\pi, \nu)$. Find a sequence $\{\nu_i\}$ of measures such that

$$\lim_{i \to \infty} \delta'(\pi, \nu_i) = D. \tag{22}$$

Replacing each ν_i with its normalization does not change the limit (it might increase some values of $\delta'(\pi, \nu_i)$, but not above D) so we might as well assume that each ν_i is normalized. We can find among them a subsequence $\{\mu_j\}$ of $\{\nu_i\}$ that converges to a limit measure μ. Then μ is necessarily normalized and we have

$$\delta'(\pi, \mu) = \lim_{j \to \infty} \delta'(\pi, \mu_j) = D \tag{23}$$

as required.   □

4 Packing rates are packing densities

Theorem 4.1. *For every pattern π,*

$$\delta'(\pi) = \delta(\pi).$$

Proof. First we show that $\delta(\pi) \geq \delta'(\pi)$.

More generally, if μ is any measure we show that there exists a sequence of permutations $\{\sigma_i\}$ such that $\lim_{i \to \infty} \delta(\pi, \sigma_i) \geq \delta'(\pi, \mu)$. Since this result holds for every measure μ, including the optimal measure, it follows that $\delta(\pi) \geq \delta'(\pi)$.

First suppose that a permutation σ of length n is chosen randomly according to μ—that is, n points are selected independently according to μ, and σ is their order type. Suppose then that we select an m-element subsequence from σ, also randomly. Then we might has well have chosen m-element subsequence directly according to μ, so the probability that it is an occurrence of π is exactly $\delta'(\pi, \mu)$. This means that the expected value of $\delta(\pi, \sigma)$ is at least $\delta'(\pi, \mu)$.

It follows that (for each n) there exists at least one specific permutation σ such that $\delta(\pi, \sigma) \geq \delta'(\pi, \mu)$. From the sequence of these permutations select a subsequence $\{\sigma_i\}$ for which the limit $\lim_{i \to \infty} \delta(\pi, \sigma_i) = D$ exists; then $\delta(\pi) \geq D \geq \delta'(\pi, \mu)$. Since this result holds for every measure μ, including the optimal measure, it follows that $\delta(\pi) \geq \delta'(\pi)$.

Next we show that $\delta'(\pi) \geq \delta(\pi)$.

Let $\{\sigma_i\}$ be a sequence of permutations of increasing size satisfying $\lim \delta(\pi, \sigma_i) = \delta(\pi)$. From each σ_i construct a template measure ν_i as defined above. Suppose that each σ_i has size n_i.

Select an m-tuple according to ν_i. For it to have the order type of π it suffices that

- the points come from m different boxes, and
- the boxes correspond to an occurrence of π in σ_i.

The probability of the first event is $\frac{n_i!/(n_i-m)!}{n_i^m}$ and then the conditional probability of the second event is $\delta(\pi, \sigma_i)$. Therefore

$$\delta'(\pi, \nu_i) \geq \frac{n_i!/(n_i - m)!}{n_i^m} \delta(\pi, \sigma_i). \tag{24}$$

Each ν_i is normalized, so there is a subsequence $\{\mu_j\}$ with a limit μ that is also normalized, and

$$
\begin{aligned}
\delta'(\pi, \mu) &= \lim_{j \to \infty} \delta'(\pi, \mu_j) \\
&= \lim_{i \to \infty} \left(\frac{n_i!/(n_i - m)!}{n_i^m} \right) \lim_{i \to \infty} \delta(\pi, \sigma_i) \\
&= \delta(\pi). \tag{25}
\end{aligned}
$$

It follows that $\delta'(\pi) \geq \delta(\pi)$. $\qquad\square$

We now abandon the notation δ' in favor of δ in all uses. The packing rate of π with respect to a measure μ is $\delta(\pi, \mu)$, and the packing density is $\delta(\pi)$ whether it arises from a sequence of permutations or an optimal measure.

Open question: Is the normalized optimal measure for π unique?

5 The packing density of 2413

With the language of measures now firmly in place, we come to the packing density of 2413. In this section we summarize the existing lower bounds on $\delta(2413)$, all of which have been obtained using template measures in one form or another.

Recursive template measures. Let $\tau \in S_k$. In Section 2 we defined the template measure μ_τ to be uniform on k small squares arranged like the graph of τ (Figure 9). One way for a pattern π to occur in μ_τ is for m points to be chosen from different squares which happen to correspond to an occurrence of π in τ. The probability of this event is given in equation (24); in the current context it is

$$\frac{k!/(k-m)!}{k^m}\delta(\pi,\tau).$$

If we substitute $\delta(\pi,\tau) = \nu(\pi,\tau)/\binom{k}{m}$ we get the equivalent form

$$\frac{m!}{k^m}\nu(\pi,\tau). \tag{26}$$

This construction can be refined by modifying the measure within each small square to be a reduced-scale copy of the measure on S itself. We call the result the *recursive template measure* corresponding to τ. (From now on we will reserve the notation μ_τ for the recursive template.) In a recursive template there is another good way for occurrences of a pattern π to occur: if m points are drawn from the same square (probability k/k^m) then they form an occurrence of π with probability $\delta(\pi,\mu_\tau)$. Combining these two ways gives

$$\delta(\pi,\mu_\tau) \geq \frac{m!}{k^m}\nu(\pi,\tau) + \frac{k}{k^m}\delta(\pi,\mu_\tau) \tag{27}$$

which can be solved to give

$$\delta(\pi,\mu_\tau) \geq \frac{(m!)\nu(\pi,\tau)}{k^m - k}. \tag{28}$$

This formula appears in [4]. It is an inequality because there may be yet other ways for a pattern to occur (although this is not the case when $\pi = 2413$).

For any pattern π we can use this construction with $\tau = \pi$ (whence $\nu(\pi,\tau) = 1$) to obtain

$$\delta(\pi,\mu_\pi) \geq \frac{m!}{m^m - m} \tag{29}$$

which gives a lower bound for the packing density of any pattern of size m. In the case of $\pi = 2413$, we have $m = 4$ and the bound is

$$\delta(2413) \geq 2/21 \approx 0.0952. \tag{30}$$

In [2] the same construction was applied using $\tau = 35827146$ (Figure 9) to obtain

$$\delta(2413) \geq 51/511 \approx 0.099804. \tag{31}$$

This follows from equation (28) with $k = 8$, $m = 4$, and $\nu(2413, 35827146) = 17$.

Warren [5] used this construction with $k = 12$ and $\tau = 5\ 4\ 7\ 12\ 11\ 3\ 10\ 2\ 1\ 6\ 9\ 8$ to obtain

$$\delta(2413) \geq 16/157 \approx 0.101911. \tag{32}$$

Weighted templates. The above construction can be improved in another way. We can alter the probabilities allocated to the small squares in the template. The probabilities (weights) can be assigned arbitrarily, as long as they add to 1.

Presutti [4] uses a template based on the permutation

$$579(11)(16)4(15)3(14)2(13)168(10)(12)$$

(with $m = 16$) and optimizes weights using Mathematica to obtain

$$\delta(2413) \geq 0.104250980068974874, \tag{33}$$

which is the best lower bound that has appeared.

Empirical results. Other researchers have used empirical methods to find optimizing permutations σ for 2413, including cases with large n. Michael Albert, Nik Ruskuc, and Imre Leader found optimizers and near-optimizers for large values of n, and were able to use them to establish lower bounds greater than 51/511. Albert and Vince Vatter (separately) used simulated annealing to find additional examples [1]. The optimizers seem to have a consistent form. If σ is one of these optimizers, then generally σ consists of...

- An initial, increasing segment, consisting of middle-range values;
- A segment with two interleaved decreasing sequences, one with high values and one with low values; and
- A terminal, increasing segment, with middle-range values overlapping those of the initial segment.

In effect, the points in the graph of σ seem to be lining up along the segments illustrated in Figure 11, below. The optimizers also have some local complications corresponding to the "recursion bubbles" that we introduce in Section 9.

The template permutations used above are all of this form. For example, 35827146 (illustrated in Figure 9) consists of an initial increasing segment 35, two interleaved decreasing sequences 8_7 (with high values) and 2_1 (with low values) and a terminal increasing segment 46. The templates used by Warren (size 12) and Presutti (size 16) are also of this form. (Well, actually, Warren's template is not of this form, and does not even have four-fold symmetry. But the slightly modified template $\tau = 457(12)(11)3(10)21689$ is of the above form and has exactly the same 2413-occurrences as Warren's template. None of these examples is large enough to show recursion bubbles.)

This form of the optimizers motivates the definition of a "four-segment measure" in the next section.

6 Four-segment measures

We define a class of measures that offer good packing rates for 2413.

A *symmetrical four-segment measure* (SFS measure) is a measure on S which is

- symmetrical with respect to four-fold rotations of S, and
- concentrated on the line segment from $(1/4, 1/4)$ to $(3/4, 0)$ and the three segments obtained from it by rotations of S.

That means that the measure is concentrated on the four segments illustrated in Figure 11.

There is no requirement that the measures be uniform on the segments. In fact, the measures in this class differ precisely in their distributions along the segments. Because of symmetry, each is determined by the distribution along the bottom segment.

We give a name to this distribution. Let μ be a four-segment measure and let $t \in [0, 1]$. Then let $F(t)$ be the probability, given that a point is on the bottom segment, that it is in the leftmost fraction t of the segment.

One way to make this definition precise is to write

$$F(t) = 4\mu([1/4, 1/4 + t/2] \times [0, 1/4]) \tag{34}$$

for $t \in [0, 1]$.

(This is an awkward formula, mainly because—for convenience in later calculations—we have chosen to make F have domain $[0, 1]$ and range $[0, 1]$. This forces a mismatch of coordinates. While the argument t

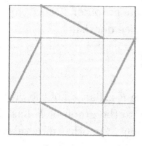 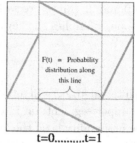

t=0.........t=1

F(t) =
fraction of this
segment's probability
that is in the
leftmost fraction t of
the segment

Fig. 11. The segments on which four-segment measures are concentrated

runs from 0 to 1, the coordinate x runs from $1/4$ to $3/4$. One unit of t corresponding to $1/2$ unit of x.)

Now any SFS measure is completely determined by the function F. Like any cumulative distribution function on $[0, 1]$, F is non-decreasing and satisfies $F(1) = 1$. It is not necessarily differentiable.

An SFS measure is not *a priori* smooth or normalized. Since we are concerned with maximizing packing rates we can limit our attention to smooth measures, but we can't usually normalize an SFS measure without bending the segments. We can, however, partially normalize the measure by requiring that, projected onto the x axis, the measure be uniform on $[1/4, 3/4]$. To preserve symmetry we then do the same thing for the y axis. The geometry of the four segments is such that the two operations do not interfere with each other. This process does not alter the packing rate for 2413 or any other pattern, so we might as well limit our attention to SFS measures that are partially normalized in this sense.

This assumption gives us the formula

$$F(t) + (1 - F(1 - t)) = 2t \tag{35}$$

for every $t \in [0, 1]$. We call this the *normalization identity,* and we assume that this relationship holds for every SFS measure we consider. This relationship forces the measure to be smooth and forces the distribution F to be continuous. It also implies that the slope of F is bounded between 0 and 2. (More precisely, since the graph of F need not always have a slope, it implies that every difference quotient

$$\frac{F(b) - F(a)}{b - a}$$

is in the interval $[0, 2]$.) On intervals where F is flat, all of the probability is in the upper segment; on intervals where the graph of F has slope 2, all of the probability is on the lower segment. On other intervals there is probability on both segments (which accounts for interleaved sequences in the empirical optimizing permutations).

We now have a class of measures with which to proceed. We make the following non-conjecture:

Non-Conjecture 6.1. The optimal measure for 2413 is a symmetrical four-segment measure determined by a distribution function F satisfying (35).

We call this a non-conjecture because we will prove in Section 9 that it is false. In that section we will show that adding recursion bubbles to the best SFS measure increases the packing rate; hence, the best SFS measure isn't optimal. Still, it's a starting point. Our plan in the next two sections is to find, with proof, the SFS measure that optimizes the packing density of 2413 among SFS measures. Then we can begin to improve that measure with recursion bubbles.

7 The packing rate for an SFS measure

In this section we give a formula for the packing rate of 2413 with respect to a symmetrical four-segment measure. Most of the rest of the section consists of the proof, which involves heavy calculation. Theorem 7.2, at the end of the section, gives an alternative formula.

Theorem 7.1. *Let μ be the symmetrical four-segment measure determined by a distribution function F satisfying the normalization identity (35). Then the packing rate of 2413 with respect to μ is given by*

$$\delta(2413, \mu) = \frac{5}{32} + \frac{3}{4} \left(\int_{t=0}^{1} F(t)\, dt \right)^2$$
$$+ \int_{t=0}^{1} \left(\left(\frac{3}{4}t - \frac{9}{8} \right) F(t)^2 - \frac{1}{4} F(t)^3 \right) dt. \quad (36)$$

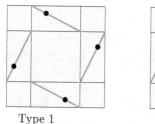
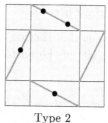
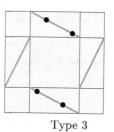

Type 1   Type 2   Type 3

Fig. 12. Three types of 2413-occurrences

Proof. For full generality we will use the notation of Stieltjes integrals. Recall that the integral

$$\int_{t=a}^{b} H(t)\,dF(t) \tag{37}$$

gives the mean of $H(t)$ when t is drawn from the probability distribution defined by F. If F has a derivative f then the integral can be understood as

$$\int_{t=a}^{b} H(t)\,dF(t) = \int_{t=a}^{b} H(t)f(t)\,dt. \tag{38}$$

More generally the integral is defined using Riemann-like sums:

$$\int_{t=a}^{b} H(t)\,dF(t) = \lim \sum_{i=0}^{n} H(x_i^*)\left(F(x_i) - F(x_{i-1})\right) \tag{39}$$

where x_i^* is an arbitrary point in $[x_{i-1}, x_i]$ and the limit is over partitions $a \le x_0 < x_1 < \cdots < x_n = b$ with decreasing mesh size. The formula for integration by parts is

$$\int H(t)\,dF(t) = H(t)F(t) - \int F(t)\,dH(t) \tag{40}$$

or

$$\int_{t=a}^{b} H(t)\,dF(t) = H(t)F(t)\big|_{t=a}^{b} - \int_{t=a}^{b} F(t)\,dH(t). \tag{41}$$

With these tools in hand we begin the evaluation of $\delta(2413, \mu)$. There are three ways for a 4-tuple of points selected according to μ to be an occurrence of 2413: Types 1, 2, and 3, each illustrated in Figure 12.

The 4-tuple is an occurrence of Type 1 if

- One point is chosen from each segment (probability 3/32);
- The top point is to the left of the bottom point; and

- The right point is above the left point.

Let X be the probability (given that there is one point on each segment) that the top point is to the left of the bottom point. Then X is given by

$$X = \int_{t=0}^{1} (1 - F(1-t))dF(t). \tag{42}$$

Using the identity (35) and integration by parts we obtain

$$
\begin{aligned}
X &= \int_{t=0}^{1} (2t - F(t))dF(t) \\
&= 2\int_{t=0}^{1} tdF(t) - \int_{t=0}^{1} F(t)dF(t) \\
&= 2\left(1 - \int_{t=0}^{1} F(t)dt\right) - \frac{1}{2}F(t)^2\big|_{t=0}^{1} \\
&= 2\left(1 - \int_{t=0}^{1} F(t)dt\right) - \frac{1}{2} \\
&= \frac{3}{2} - 2\int_{t=0}^{1} F(t)dt. \tag{43}
\end{aligned}
$$

As an example, consider the measure μ determined by a uniform distribution along the four segments; that is, consider the case of $F(t) = t$. In this case we would expect to find $X = 1/2$, and that is indeed the value given by equation (43).

By symmetry the probability that the right point is above the left point is also X, so the probability of a Type 1 occurrence is

$$\frac{3}{32}X^2 = \frac{27}{128} + \frac{3}{8}\left(\int_{t=0}^{1} F(t)dt\right)^2 - \frac{9}{16}\int_{t=0}^{1} F(t)dt. \tag{44}$$

The 4-tuple is an occurrence of Type 2 if

- Two points are on the top segment and one each are on the left and bottom segments (probability 3/64 before allowing for rotations);
- The point on the bottom segment is horizontally between the points on the top segment.

Let Y be the probability, given that the points are on the correct segments, that the bottom point is horizontally between the top points. Then Y is given by

$$Y = 2\int_{t=0}^{1} F(1-t)(1 - F(1-t))dF(t). \tag{45}$$

The initial factor 2 appears because the two points on the top segment can occur in either order. Substituting $1 - t$ for t makes this

$$Y = 2 \int_{t=0}^{1} F(t) \left(1 - F(t)\right) dF(1 - t). \tag{46}$$

Differentiating (35) gives

$$dF(1 - t) = 2dt - dF(t), \tag{47}$$

and substituting this into our expression gives

$$Y = 4 \int_{t=0}^{1} F(t) \left(1 - F(t)\right) dt - 2 \int_{t=0}^{1} F(t) \left(1 - F(t)\right) dF(t). \tag{48}$$

The second integral is $\left(\frac{1}{2}F(t)^2 - \frac{1}{3}F(t)^3\right)\big|_{t=0}^{1} = \frac{1}{6}$, so we can rewrite (48) as

$$Y = -\frac{1}{3} + \int_{t=0}^{1} \left(4F(t)dt - 4F(t)^2\right) dt. \tag{49}$$

For example, if $F(t) = t$, then $Y = \frac{1}{3}$ as we would expect. Now the probability of a Type 2 occurrence (including a factor of 4 to account for the rotations) is

$$4 \cdot \frac{3}{64} Y = -\frac{1}{16} + \int_{t=0}^{1} \left(\frac{3}{4}F(t)dt - \frac{3}{4}F(t)^2\right) dt. \tag{50}$$

A 4-tuple is a Type 3 occurrence if

- Two points are from the top segment and two from the bottom segment (probability 3/128 before allowing for rotations); and
- From the left, the four points have the order bottom, top, bottom, top.

Let Z be the probability, given that the points are on the correct segments, that they have the correct order. Then Z is given by

$$Z = 4 \int_{t=0}^{1} \int_{s=1-t}^{1} F(1 - t)F(1 - s)dF(s)dF(t). \tag{51}$$

In this formulation t represents the distance from the left end of the bottom segment to the rightmost bottom point, and s represents the distance from the right end of the top segment to the leftmost of the top points. These must satisfy $s > 1 - t$, hence the limits of the integrals. The order requirement is that one bottom point is left of s, one top point is at s, one bottom point is at t, and one top point is right of t; hence factors of $F(1-t)$, $dF(s)$, $F(1-s)$, $dF(t)$ respectively. The initial factor

of 4 is there because the top points can exchange roles and the bottom
points can exchange roles.

Write this expression as an iterated integral:

$$Z = 4 \int_{t=0}^{1} F(1-t) \left[\int_{s=1-t}^{1} F(1-s)dF(s) \right] dF(t). \qquad (52)$$

To evaluate the inner integral we first substitute $1-s$ for s, again using
the identity $dF(1-s) = 2ds - dF(s)$:

$$
\begin{aligned}
\int_{s=1-t}^{1} F(1-s)dF(s) &= \int_{s=0}^{t} F(s)dF(1-s) \\
&= \int_{s=0}^{t} F(s)(2ds - dF(s)) \\
&= 2\int_{s=0}^{t} F(s)ds - \int_{s=0}^{t} F(s)dF(s) \\
&= 2\int_{s=0}^{t} F(s)ds - \left(\frac{1}{2}F(s)^2 \right)\Big|_{s=0}^{t} \\
&= 2G(t) - \frac{1}{2}F(t)^2
\end{aligned}
$$

where $G(t) = \int_{s=0}^{t} F(s)ds$. Substituting this into the main integral gives

$$
\begin{aligned}
Z &= 4\int_{t=0}^{1} F(1-t) \left(2G(t) - \frac{1}{2}F(t)^2 \right) dF(t) \\
&= 4\int_{t=0}^{1} (1 - 2t + F(t)) \left(2G(t) - \frac{1}{2}F(t)^2 \right) dF(t).
\end{aligned}
$$

Expanding this integral into six terms and integrating (usually by parts) gives

$$
\begin{aligned}
Z &= 8 \int_{t=0}^{1} G(t)dF(t) - 16 \int_{t=0}^{1} tG(t)dF(t) + 8 \int_{t=0}^{1} F(t)G(t)dF(t) \\
&\quad -2 \int_{t=0}^{1} F(t)^2 dF(t) + 4 \int_{t=0}^{1} tF(t)^2 dF(t) - 2 \int_{t=0}^{1} F(t)^3 dF(t) \\
&= 8 \left(\int_{t=0}^{1} F(t)dt - \int_{t=0}^{1} F(t)^2 dt \right) \\
&\quad -16 \left(\int_{t=0}^{1} F(t)dt - \frac{1}{2} \left(\int_{t=0}^{1} F(t)dt \right)^2 - \int_{t=0}^{1} tF(t)^2 dt \right) \\
&\quad +8 \left(\frac{1}{2} \int_{t=0}^{1} F(t)dt - \frac{1}{2} \int_{t=0}^{1} F(t)^3 dt \right) \\
&\quad -2 \left(\frac{1}{3} \right) + 4 \left(\frac{1}{3} - \frac{1}{3} \int_{t=0}^{1} F(t)^3 dt \right) - 2 \left(\frac{1}{4} \right) \\
&= \frac{1}{6} + 8 \left(\int_{t=0}^{1} F(t)dt \right)^2 \\
&\quad + \int_{t=0}^{1} \left(-4F(t) - 8F(t)^2 - \frac{16}{3}F(t)^3 + 16tF(t)^2 \right) dt.
\end{aligned}
$$

When $F(t) = t$ this is $1/6$ as we would expect. The probability of a Type-3 occurrence (multiplying by 2 to account for the rotation) is

$$
\begin{aligned}
2 \cdot \frac{3}{128} Z &= \frac{1}{128} + \frac{3}{8} \left(\int_{t=0}^{1} F(t)dt \right)^2 \\
&\quad + \int_{t=0}^{1} \left(-\frac{3}{16}F(t) - \frac{3}{8}F(t)^2 - \frac{1}{4}F(t)^3 + \frac{3}{4}tF(t)^2 \right) dt. \quad (53)
\end{aligned}
$$

Combining the probabilities for the three types, we have

$$
\begin{aligned}
\delta(2413, \mu) &= \frac{5}{32} + \frac{3}{4} \left(\int_{t=0}^{1} F(t)dt \right)^2 \\
&\quad + \int_{t=0}^{1} \left(\left(\frac{3}{4}t - \frac{9}{8} \right) F(t)^2 - \frac{1}{4}F(t)^3 \right) dt \quad (54)
\end{aligned}
$$

as required. $\square$

We aren't free to choose $F(t)$ arbitrarily for all $t \in [0, 1]$. We may choose $F(t)$ on $[0, 1/2]$ subject to certain constraints, but then the values on $[1/2, 1]$ are forced on us by the normalization identity (35). It is helpful, therefore, to have an alternative to Theorem 7.1 in which the integrals are limited to the interval $[0, 1/2]$.

Theorem 7.2. *Let μ be the symmetrical four-segment measure deter-mined by a distribution function F satisfying (35). Then the packing rate of 2413 with respect to μ is given by*

$$\delta(2413, \mu) = \frac{3}{32} + 3 \left(\int_{t=0}^{1/2} F(t)dt \right)^2$$

$$+ \int_{t=0}^{1/2} \left[\left(3t - \frac{3}{4} \right) F(t) + \left(\frac{3}{2}t - \frac{9}{4} \right) F(t)^2 - \frac{1}{2}F(t)^3 \right] dt. \quad (55)$$

Proof. Divide each of the integrals in Theorem 7.1 into two integrals, substitute $1 - t$ for t in the second integral, simplify using the normal-ization identity, and recombine the integrals. For example:

$$\int_{t=0}^{1} F(t)dt = \int_{t=0}^{1/2} F(t)dt + \int_{t=1/2}^{1} F(t)dt$$

$$= \int_{t=0}^{1/2} F(t)dt + \int_{t=0}^{1/2} F(1-t)dt$$

$$= \int_{t=0}^{1/2} F(t)dt + \int_{t=0}^{1/2} (1 - 2t + F(t))dt$$

$$= 2\int_{t=0}^{1/2} F(t)dt + \int_{t=0}^{1/2} (1 - 2t)dt$$

$$= 2\int_{t=0}^{1/2} F(t)dt + \frac{1}{4} \quad (56)$$

The other integral in (36) can be restated in the same way, and the results can be combined to give (55). We leave the calculation to the reader. (Actually we invite the reader to leave the calculation to us. This is a good time to thank the volunteer referees who make mathematical publication possible.) $\square$

8 The optimal SFS measure

In this section we use the calculus of variations to find a measure that maximizes $\delta(2413, \mu)$ among symmetrical four-segment measures μ.

Define a functional Φ by

$$\Phi[F] = \frac{3}{32} + 3 \left(\int_{t=0}^{1/2} F(t)dt \right)^2$$

$$+ \int_{t=0}^{1/2} \left(\left(3t - \frac{3}{4} \right) F(t) + \left(\frac{3}{2}t - \frac{9}{4} \right) F(t)^2 - \frac{1}{2}F(t)^3 \right) dt \quad (57)$$

when F is defined on the interval $[0, 1/2]$. This is the formula from Theorem 7.2, which says that $\delta(\pi, \mu) = \Phi[F]$ when μ is the SFS measure determined by F. To find the optimal SFS measure, we need to maximize $\Phi[F]$ subject to certain constraints on F.

We are free to choose any distribution F provided that $F(0) = 0$, F is non-decreasing, and F satisfies the normalization identity (35). Equivalently: We can choose $F(t)$ arbitrarily on the interval $0 \le t \le 1/2$ subject to two constraints:

- $F(0) = 0$, and
- The difference quotients of F satisfy

$$0 \le \frac{F(t) - F(s)}{t - s} \le 2 \quad (58)$$

whenever $0 \le s < t \le 1/2$.

Then F can be extended to all of $[0, 1]$ using equation (35), and equation (58) is automatically satisfied on the entire interval. These requirements also force F to be continuous and nondecreasing on $[0, 1]$ and to satisfy $F(1) = 1$.

We say that F is *unconstrained* at t if, in some neighborhood of t, the difference quotients are bounded away from 0 and 2. Otherwise, F is *constrained* at t. The easiest way for F to be constrained at t is for the graph of F to have slope 0 or 2 on an interval containing t, but F can also be constrained at t (for example) if t is a limit point of such intervals. If F is unconstrained at t, we are free to make positive or negative adjustments to F in a neighborhood of t in an attempt to maximize $\delta(2413, \mu)$.

(When F is constrained at t, the corresponding measure has probability only on one segment—on top when $F(t) = 0$, and on the bottom when F has slope 2. When F is unconstrained, there is probability on both segments.)

Theorem 8.1. *Let* $J = \int_0^{1/2} F(t)dt$. *If* F *maximizes* $\delta(2413, \mu)$ *subject* *to the above requirements, then F(t) must be given by*

$$F(t) = \sqrt{\left(t - \frac{1}{2}\right)^2 + \frac{3}{2} + 4J} + \left(t - \frac{3}{2}\right) \qquad (59)$$

whenever F is unconstrained at t.

This is a local requirement. We will prove it first, then extend it to a global description of F in the next theorem.

Proof. Let H be any function with a continuous derivative on $[0, 1/2]$ and satisfying $H(0) = 0$ and $H(t) = 0$ whenever F is constrained at t. Then F may be altered by adding or subtracting a small multiple of H. It follows that the derivative of

$$\Phi[F + \epsilon H]$$

with respect to ϵ must be zero at $\epsilon = 0$. Compute:

$$\lim_{\epsilon \to 0} \frac{\Phi[F + \epsilon H] - \Phi[F]}{\epsilon}$$
$$= \int_{t=0}^{1} \left(\left(6J + 3t - \frac{3}{4}\right) + \left(3t - \frac{9}{2}\right) F(t) - \frac{3}{2} F(t)^2\right) H(t)\, dt. \qquad (60)$$

This expression must be zero for F to be optimal, for any suitable H. When F is unconstrained at t we can choose H to be positive in a small neighborhood of t. Therefore we must have

$$\left(6J + 3t - \frac{3}{4}\right) + \left(3t - \frac{9}{2}\right) F(t) - \frac{3}{2} F(t)^2 = 0 \qquad (61)$$

whenever $F(t)$ is unconstrained.

This can be solved uniquely for $F(t)$ (since $F(t) \geq 0$) giving

$$F(t) = \sqrt{\left(t - \frac{1}{2}\right)^2 + \frac{3}{2} + 4J} + \left(t - \frac{3}{2}\right). \qquad (62)$$

as required.   □

That is a local result. To understand the behavior of F globally, we must know when F is unconstrained. If $J < 1/8$ (which is the case for all plausible F) we can check that equation (59) never gives a slope greater than 2, but that it sometimes does give negative values for F.

In fact, equation (59) gives $F(t) \leq 0$ whenever $t \leq 1/4 - 2J$. Write $t^* = 1/4 - 2J$. This means that for the optimal F we must have $F(t) = 0$

when $t \leq t^*$, and $F(t)$ given by the formula when $t^* \leq t \leq 1/2$. To summarize:

$$F(t) = \begin{cases} 0 & \text{when } t \leq t^* \\ \sqrt{\left(t - \frac{1}{2}\right)^2 + \frac{3}{2} + 4J} + \left(t - \frac{3}{2}\right) & \text{when } t^* \leq t \leq \frac{1}{2}. \end{cases} \tag{63}$$

This formula for F is circular because it makes F depend on its own integral J. In fact, there is only one value of J that makes the formula consistent, which we will call J^*, and only one corresponding value of t^*. We turn to Mathematica for a numerical integral and solution:

$$J^* \approx 0.05110454191162339225 \tag{64}$$

$$t^* = \frac{1}{4} - 2J^* \approx 0.14779091617675321550 \tag{65}$$

(This calculation is almost analytic. If K is the smallest positive solution of $K \ln K = K - 5/2$ then the above values are given by $J^* = (K - 3/2)/4$ and $t^* = 1 - K/2$.)

Extending F to $[0, 1]$ using (35) leaves the formula (59) unchanged. For $t > 1 - t^*$ it gives $F(t) = 2t - 1$. We have proved:

Theorem 8.2. *There is a unique distribution F that maximizes $\Phi[F]$ for four-segment measures. If J^* and t^* are chosen as above with approximate values given by (64) and (65), then $J^* = \int_0^{1/2} F(t)dt$ and F is given on $[0, 1]$ by*

$$F(t) = \begin{cases} 0 & \text{when } 0 \leq t \leq t^* \\ \sqrt{\left(t - \frac{1}{2}\right)^2 + \frac{3}{2} + 4J} + \left(t - \frac{3}{2}\right) & \text{when } t^* \leq t \leq 1 - t^* \\ 2t - 1 & \text{when } 1 - t^* \leq t \leq 1. \end{cases} \tag{66}$$

$\square$

Figure 13 is a graph of F. The function is convex, has $F(1/2) \approx 0.30553$ and $F'(1/2) = 1$, and has no derivative at t^* or $1 - t^*$. We call the corresponding measure μ. Its packing density is calculated from (59):

$$\delta(2413, \mu) = \Phi[F] \approx 0.10472339512772223636. \tag{67}$$

This number is a new lower bound for the packing density $\delta(2413)$, and we have proven that it is the best packing rate possible using a symmetrical four-segment measure.

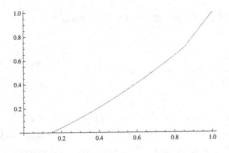

Fig. 13. Graph of F for μ or μ_1

9 The first recursion bubble

Having found the optimal four-segment measure, we now improve it using recursion.

The measure μ defined at the end of the last section is determined by a function F with $F(t) = 0$ for $t \in [0, t^*]$. This corresponds to the part of S with $x \in [1/4, 1/4 + t^*/2]$. In this region all of the probability is concentrated on the upper segment, in the rectangle $R = [1/4, 1/4 + t^*/2] \times [1 - t^*/4, 1]$. The probability itself is $\mu(R) = t^*/2$.

There is no probability above, below, or to the left or right of this rectangle, so it is easy to check that no occurrence of 2413 includes more than one point from this rectangle. Therefore nothing is lost by rearranging probability within the rectangle.

Define μ_1 recursively by $\mu_1 = \mu$ except on the rectangle R and its rotated images, in which μ_1 is a reduced-scale image of μ itself. (It makes no difference that the transformation between S and R is not aspect-preserving.) We call the four altered rectangles *recursion bubbles*. (We have seen recursion bubbles before, in recursive templates and in Figure 3.)

Now we have the same 2413 occurrences as before, plus additional occurrences when all four points fall within one of the recursion bubbles (probability $4(t^*/2)^4$) and happen to form a 2413 occurrence within the bubble. Therefore

$$\delta(2413, \mu_1) = \Phi[F] + 4\left(\frac{t^*}{2}\right)^4 \delta(2413, \mu_1) \qquad (68)$$

or, solving,

$$\delta(2413, \mu_1) = \frac{\Phi[F]}{1 - 4\left(\frac{t^*}{2}\right)^4}. \qquad (69)$$

Using the values of $\Phi[F]$ and t^* from the previous section we obtain

$$\delta(2413, \mu_1) \approx 0.10473588696991414716\ldots, \tag{70}$$

a new lower bound for $\delta(2413)$. The increase due to the recursion bubble appears in the fifth decimal place.

10 The second recursion bubble

Shouldn't the recursion bubble be bigger?

The measure μ was optimal in the absence of recursion. With recursion, there is a greater advantage to selecting points in the recursion box than there was before. At the margin, shouldn't that shift the optimum configuration in the direction of a larger bubble?

So, let's increase the size of the bubble. We can't do that in isolation, because it would cause F to be inconsistent with (59) immediately to the right of the bubble. We must allow F to increase with slope 2 until it catches up with the formula. This creates a small region in which all of the probability is on the lower segment, so we might as well turn it into a second recursion bubble.

The resulting measure is the same as an SFS with this distribution:

$$F(t) = \begin{cases} 0 & \text{when } t \leq t_1 \\ 2(t - t_1) & \text{when } t_1 \leq t \leq t_2 \\ \sqrt{\left(t - \frac{1}{2}\right)^2 + \frac{3}{2} + 4J} + \left(t - \frac{3}{2}\right) & \text{when } t_2 \leq t \leq 1 - t_2 \\ 1 - 2t_1 & \text{when } 1 - t_2 \leq t \leq 1 - t_1 \\ 2t - 1 & \text{when } 1 - t_1 \leq t. \end{cases}$$

We have extended F to $[0,1]$ using the normalization identity. There are now two recursion bubbles on each segment, one corresponding to the interval $[0, t_1]$ (probability $(t_1/2)^2$ for each box) and one corresponding to the interval $[t_1, t_2]$ (probability $((t_2 - t_1)/2)^2$ for each box). Both t_1 and t_2 are parameters that we can choose, along with J, subject to the requirement that F be continuous at t_2 and have integral J on $[0, 1/2]$.

We optimize t_1 and t_2 by naked calculation:

$$t_1 = 0.14861089461296151506\ldots$$
$$t_2 = 0.14909030676438411460\ldots$$

and if μ_2 is the measure with these revisions, we get

$$\delta(2413, \mu_2) = \frac{\Phi[F]}{1 - 4\left(\frac{t_1}{2}\right)^4 - 4\left(\frac{t_2 - t_1}{2}\right)^4} \approx 0.10473602526603545023\ldots.$$

The improvement over μ_1 is about 10^{-7}. This is the best lower bound we have found for $\delta(2413)$.

Conjecture 10.1. The measure μ_2 is optimal for 2413 and the packing density of 2413 is $\delta(2413) = 0.10473602526603545023\ldots$.

The measure μ_2 is illustrated in Figure 1.

How much is proof and how much is conjecture? We conjectured that the optimal measure would be related to a four-segment measure, and we proved that the optimal four-segment measure is given by Theorem 8.2. We conjectured that adding two recursion bubbles would make this optimal, and calculated the best location of the recursion bubbles by brute force. Hence, gaps remain before we can be sure that μ_2 is optimal.

11 More bubbles

An alternative possibility is that the recursion bubbles continue to multiply, alternating between the top and bottom segment and reaching a limit point before the center of the segment. We cannot calculate a positive contribution even for the third box, which may just mean that it is too small to be found by our methods. A measure μ_∞ with an infinite sequence of recursion blocks is illustrated in Figure 2.

References

[1] M. H. Albert. personal communication. 2007.

[2] M. H. Albert, M. D. Atkinson, C. C. Handley, D. A. Holton, and W. Stromquist. On packing densities of permutations. *Electron. J. Combin.*, 9(1):Research Paper 5, 20 pp., 2002.

[3] R. Brignall. A survey of simple permutations. In *this volume*, pages 41–65.

[4] C. B. Presutti. Determining lower bounds for packing densities of non-layered patterns using weighted templates. *Electron. J. Combin.*, 15(1):Research paper 50, 10, 2008.

[5] D. Warren. *Optimizing the packing behavior of layered permutation patterns.* PhD thesis, University of Florida, 2005.

On the permutational power of token passing networks

Michael Albert†
Department of Computer Science
University of Otago
Dunedin New, Zealand

Steve Linton
School of Computer Science
University of St Andrews
St Andrews, Fife, Scotland

Nik Ruškuc
School of Mathematics and Statistics
University of St Andrews
St Andrews, Fife, Scotland

Abstract

A token passing network is a directed graph with one or more specified input vertices and one or more specified output vertices. A vertex of the graph may be occupied by at most one token, and tokens are passed through the graph. The reorderings of tokens that can arise as a result of this process are called the language of the token passing network. It was known that these languages correspond through a natural encoding to certain regular languages. We show that the collection of such languages is relatively restricted, in particular that only finitely many occur over each fixed alphabet.

1 Introduction

The study of graphs whose vertices can be occupied by tokens, or pebbles, which are moved along the edges has ranged from recreational

† Supported by EPSRC grant GR/S41074/01.

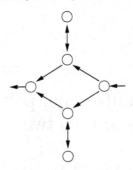

Fig. 1. $S_{2,2}$, a network of two stacks of capacity two in parallel

mathematics [8, 10] to motion planning and related topics [3, 4, 6]. In most of these papers the problem is restricted to moving a fixed set of pebbles within a given graph, generally aiming to obtain a specific configuration. On the other hand, early works such as [5, 7, 9] dealt in a similar way with moving tokens, now thought of as items of data, within a network (represented as a directed graph) with the aim of producing specified outputs from a fixed input, or sorted output from a variable input. There is no standard name to describe such networks and their operation. To emphasise their abstract and general nature we have chosen to call them *token passing networks*, a name derived from the title of [2].

The problem of identifying permutations which could be produced when the network was restricted to a fixed size was considered in [2]. They showed that, under a natural encoding scheme, the collection of permutations generated by a token passing network is always a regular language. The principal aim of this paper is to extend the analysis of these collections of permutations and establish in Theorem 2.1 and Theorem 2.2 that, in effect, for each alphabet size, there are only finitely many such languages.

In Section 6 we provide a complete catalog of these languages over the three letter alphabet, along with networks producing them. We also provide some examples to show that certain natural conjectures about the behaviour of these networks are not correct. The results of Section 6 are obtained by explicit implementation in GAP of some of the implicit computational methods introduced in [1] and [2].

Formal definitions of token passing networks, and the languages associated with them will be found in the next section. We conclude this introduction with an informal, but illustrative example drawn from [2] (Figure 8 in that article). Consider the network $S_{2,2}$ shown in Figure 1. This network consists essentially of two stacks in parallel, each capable of containing up to two items. Input arrives at the rightmost node (shown by an inward pointing arrow without a source), and output occurs at the leftmost node (an outward pointing arrow without a target). This network operates non-deterministically, with the following basic operations:

- If the input node is empty, then a new token may be added to the network at the input node. The input source is considered to be a (potentially infinite) queue containing tokens labelled 1, 2, ...
- If the source of an edge is occupied by a token, and its target is not, then the token may be moved along that edge.
- If the output node is occupied, then the token on it may be removed from the network.
- If the network is empty, operation may halt.

The output of a particular run of the network is the permutation $\pi = p_1 p_2 \cdots p_n$ where p_i is the label of the ith token removed from the network. The set of all such permutations will be denoted $\mathrm{Out}(S_{2,2})$.

Once four tokens are present in the network $S_{2,2}$ the next output symbol will be one of those four. Therefore, each element p_i in an output permutation π is one of the four smallest of the remaining elements. So, the output of the network can be encoded as a string over the alphabet $\{1, 2, 3, 4\}$ where p_i is encoded by its rank in the set $\{p_j : j \geq i\}$ (with 1 denoting the smallest element). Theorem 2 of [2] then implies that under this encoding $\mathrm{Out}(S_{2,2})$ forms a regular language. Of perhaps more interest are the basis elements of $\mathrm{Out}(S_{2,2})$. These are permutations not in $\mathrm{Out}(S_{2,2})$ but with the property that if any single symbol is deleted from them, then the resulting permutation (that is the output permutation of the remaining input items) is in $\mathrm{Out}(S_{2,2})$. It was noted in [2] that this set is infinite, containing at least the permutations:

$$4, 1, 6, 3, 8, 5, 10, 7, \ldots, 4n, 4n - 3, 2, 4n - 1$$

for any n. As a consequence of the results of [1] and our explicit computations outlined in Section 6 below we can report that these permutations together with all the permutations of length 5 beginning with a 5 are the complete set of basis elements of $\mathrm{Out}(S_{2,2})$.

2 Definitions and basic results

In a token passing network as defined informally above, we move tokens from vertex to vertex along directed edges, sometimes adding new tokens to the network at specified input vertices, and sometimes removing them from specified output vertices. Operation is to halt at any point when there are no tokens in the network. This informal approach is important in *understanding* how token passing networks operate. However, we must formalise these definitions, and will do so now. Our definitions and notation are slightly (but not materially) different from those of [2], and we will discuss these differences and the reasons for them after presenting our definitions.

A *token passing network*, $T(G, I, O)$, consists of a directed graph G together with non-empty subsets I and O of the vertices of G. When clear from context, we suppress the parameters. Elements of the set I are called *input vertices*, while those of O are called *output vertices*. As is normal, we make no distinction notationally between a graph and its underlying set of vertices.

Token passing networks are to be thought of as devices, which accept input tokens in some order and produce a permutation of the input tokens as their output. Since the nature of the tokens is not significant in analysing the behaviour of token passing networks, the tokens are represented by the numbers $1, 2, \ldots, n$ according to the order in which they are added to the network. At any stage in the operation of these devices, each vertex of G will either be *occupied* by some single token k, or be *unoccupied*.

The operation of these devices consists of a sequence of *primitive operations*. A primitive operation is one of the following:

Input The next input token can be placed on any unoccupied input vertex. Thereby, the input vertex becomes occupied by the token.

Movement If a token k occupies a vertex v, there is an edge $v \to w$ and w is unoccupied, then the token can be moved from v to w. Thereby, v becomes unoccupied, and k occupies w.

Output If an output vertex is occupied by a token k, then k can be removed from the network. Thereby, the output vertex becomes unoccupied, and k is appended to the output sequence.

A *run* of T begins with G completely unoccupied and follows a sequence of primitive operations, concluding at some point when G is again completely unoccupied. The *output class*, Out($\mathcal{T}$) consists of all the permutations π that can be produced as output sequences by some run of $\mathcal{T}$. It is well-known (and implicit in all of [2, 5, 7, 9]) that this output class is closed under deletion. That is, if $\pi \in$ Out($\mathcal{T}$) and we delete a symbol of π, then re-index the remaining symbols, preserving their relative order, to produce a permutation π' (for example, after deleting the symbol 3 from 25341 we obtain 2541 which is re-indexed to 2431) then it is also the case that $\pi' \in$ Out($\mathcal{T}$). This is easily seen as we can produce the permutation π' by following the run of $\mathcal{T}$ which produced π but ignoring any operation on the token representing the element we wish to delete. In other words, the set of all permutations that can be produced by runs of $\mathcal{T}$ forms a *pattern class*.

The style of argument of the preceding paragraph, will be repeated in a number of different contexts. When we use it, we will generally refer to the tokens we are ignoring in some modified run of $\mathcal{T}$ as *ghost tokens*.

A key observation made in [2] is that, in considering the operation of $\mathcal{T}$ only the relative ranks of the tokens remaining in the network are important. This leads to the notion of *rank encoding*. Formally, for the permutation $\pi = p_1 p_2 \cdots p_n$ its rank encoding $r(\pi)$ is the sequence $r_1 r_2 \cdots r_n$ where

$$r_i = |\{j \, : \, j \geq i \,\text{and}\, p_j \leq p_i\}|\,.$$

For instance, the rank encoding of 25341 is 24221. As noted in [2] a sequence $r_1 r_2 \cdots r_n$ is the rank encoding of some permutation π if and only if $r_{n-i} \leq i + 1$ for all i, and if this is satisfied, then π is uniquely determined. Henceforth, we make no distinction between a permutation and its rank encoding.

Since the graph underlying $\mathcal{T}$ is finite, the rank of any output element is at most the size of the underlying graph. Hence (the rank encodings of) Out($\mathcal{T}$) is a language over a finite alphabet. The main result of [2] (Theorem 1) is that this language is regular. Basically, the reason this is true is that we can associate a finite automaton to $\mathcal{T}$ whose states are represented by maps f from G to $\{0, 1, 2, \ldots, |G|\}$, where $f(v) = 0$ if v is unoccupied, and $f(v) = k > 0$ if v is occupied by the kth largest token currently in the network. An equivalent representation is to represent a state in which there are m tokens currently in the network by the sequence $v_1 v_2 \cdots v_m$ where v_i is the vertex occupied by the ith largest token. With respect to this representation, the *primitive transitions* corresponding to the primitive operation of $\mathcal{T}$ can be easily described. From a state s:

Input If $i \in I$ and i does not occur in s, then there is a transition
$$s \to si.$$

Movement If $s = avb$ and $v \to w$ is an edge of G and w does not occur in s then there is a transition $s \to awb$.

Output If $s = aob$ with $o \in O$ then there is a transition $s \to ab$

Note that the input transitions always add an element of largest (current) rank, while after an output transition, the ranks of any remaining larger elements decreases by 1 as a consequence of the deletion in s. Though there is no formal distinction between this representation of states and that used in [2], the simple form of the primitive transitions in this representation is particularly convenient in actual implementations.

Again, as shown in [2], the runs of $\mathcal{T}$ can be identified with the accepting computations of a non-deterministic finite state automaton (on the states defined above) by considering input and movement transitions as ϵ-transitions, and output transitions as transitions on the symbol k where k is the index of the output symbol in s. The language accepted by this automaton, which we denote $L(\mathcal{T})$ is precisely the rank encoding of the set of permutations in $\text{Out}(\mathcal{T})$.

It is probably prudent to pause at this point and illustrate the preceding definitions with an example. Consider a graph G consisting of three vertices u, v and w, together with a single edge $v \to w$. Designate u and v as input vertices and u and w as output vertices. The corresponding token passing network can produce the permutation 31524 whose rank encoding is 31311 as shown in the following table, where each row corresponds to a state, moving from one row to the next represents a primitive transition, and we show both the *actual* tokens located at the vertices, and the corresponding state.

u	v	w	State	Output
			ϵ	
	1		v	
		1	w	
	2	1	wv	
3	2	1	wvu	
	2	1	wv	3
	2		v	31
		2	w	31
	4	2	wv	31
5	4	2	wvu	31
	4	2	wv	313
	4		v	3131
		4	w	3131
			ϵ	31311

The main formal differences between the definitions above and those of [2] are: we permit multiple input and output vertices, there is no distinction between internal nodes and the input and output nodes, and tokens are added to and removed from the network, rather than beginning at the input node and finishing at the output node. From a practical standpoint these differences are immaterial. Any token passing network as we have defined it can be converted to one of the type defined in [2] simply by adding two new nodes namely the input and output nodes, together with edges directed from the input node to each vertex in I and from each vertex of O to the output node. The reverse process converts the networks considered in [2] to ones as we have defined them. Corresponding networks of these two types produce the same set of permutations. We have formulated the definitions as above, because it facilitates considering subgraphs of G as token passing networks in their own right.

When we consider token passing networks as a whole it will occasionally be useful to consider token passing networks that have only one input and one output vertex, which are distinct. In this case, a vertex may only be occupied if if lies on a directed path from the input vertex to the output vertex. Such vertices will be called *useful*. We will refer to networks with unique and distinct input and output vertices in which every vertex is useful as *standard* token passing networks. The same construction as above, followed by a pruning of useless vertices, shows that for any token passing network $\mathcal{T}$ there exists a standard token passing network $\mathcal{T}'$ having at most two more vertices than $\mathcal{T}$ such that $\text{Out}(\mathcal{T}) = \text{Out}(\mathcal{T}')$.

Let k be a positive integer, and let K be a set of k elements considered as a graph with no edges. The token passing network $\mathcal{B}_k = \mathcal{T}(K, K, K)$ will be called a *buffer* of size k. Since we can freely add or remove any element of rank at most k during the operaton of $\mathcal{B}_k$, the language $B(k) = L(\mathcal{B}_k)$ consists of the rank encodings of all permutations where the rank of any element is bounded by k. These permutations are referred to as *k-bounded*. Many other token passing networks produce this same language, for instance a directed cycle of k vertices with a single vertex serving as both input and output vertex†.

The language accepted by a token passing network, $\mathcal{T}(G, I, O)$, is a sublanguage of $B(|G|)$ since at most $|G|$ tokens can occupy the graph. We define the *boundedness* of $\mathcal{T}$ to be the minimum k such that $L(\mathcal{T}) \subseteq B(k)$.

We will also consider token passing networks restricted to operate with a total of at most c tokens in the network at any one time. In terms of the states introduced above, we restrict the operation of $\mathcal{T}$ to states represented by sequences of length at most c. We refer to such networks as *capacity restricted token passing networks* and denote the token passing network $\mathcal{T}$ restricted in this way by $\mathcal{T}_c$. Obviously $L(\mathcal{T}_c) \subseteq B(c) \cap L(\mathcal{T})$. We will see in Section 6 that, in general, this inclusion is proper. Specifically in Figure 6 we illustrate a token passing network which can produce certain 4-bounded permutations, but not without containing at least 5 tokens at some point in its operation.

The main results of this paper are the following:

Theorem 2.1. *Let c be a fixed positive integer. Then:*

$$\{L(\mathcal{T}) : \mathcal{T} \text{ a token passing network of boundedness } c\}$$

is finite.

---

† To see this, consider operating the cycle as follows: add $k - 1$ tokens (or all the remaining tokens if there are fewer than $k-1$ remaining) to the cycle. Move them so that the input/output vertex is vacant. If the next symbol to be output is k, add it to the cycle and output it immediately. If it is smaller than k, continue moving elements around the cycle until the desired element is on the input/output vertex, then output it. Add one more element to the cycle (if necessary) and continue in the same fashion.

Theorem 2.2. *Let c be a fixed positive integer. Then:*

$$\{L(\mathcal{T}_c) \, : \, \mathcal{T} \ a \ token \ passing \ network\}$$

is finite.

These two results indicate that the permutational power of token passing networks is relatively restricted, in that there are infinitely many pattern classes represented by regular sublanguages of $B(c)$. From another viewpoint, they say that, for a fixed boundedness or capacity bound, there is a finite test set, T, of permutations such that, if two token passing networks satisfying the boundedness conditions generate the same subset of T then they generate the same language. Using the catalogues provided in Section 6, by inspection of the possible classes it follows that for boundedness (or capacity bound) 2, the set $T = \{21\}$ suffices, while for boundedness 3 we may take:

$$T = \{21, 321, 312, 31542, 324651\}.$$

3 Strongly cycle connected graphs

We say that a directed graph is *strongly cycle connected* if it has a strongly connected spanning subgraph in which every edge belongs to a directed cycle of length at least three. For instance any directed graph on at least three vertices containing a directed Hamilton cycle is strongly cycle connected and the bi-directed orientation of an undirected graph is strongly cycle connected if and only if it is biconnected, that is, contains no bridge edges.

Recall that a k buffer is a token passing network on k vertices whose language is all of $B(k)$. The next lemma shows that strongly cycle connected graphs are, in some sense, almost like buffers. This will be used in the proof of the main result, in that it implies that a token passing network of boundedness c cannot contain many vertices belonging to strongly cycle connected subgraphs.

Lemma 3.1. *Let C be a strongly cycle connected graph containing m vertices and I and O be non empty subsets of C. The language of $\mathcal{T}(C, I, O)$ contains $B(m-2)$. If $m \in \{3, 4\}$ then it contains $B(m-1)$.*

Proof. We will show that if $m - 2$ or fewer tokens are present in C, and if v is any vertex of C occupied by some token a, and $v \to w$ is an edge of C, then we can move a to w. In fact we will prove that this is possible

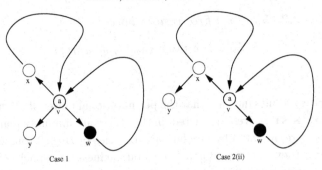

Fig. 2. Critical cases for the argument concerning buffering capacity of strongly cycle connected graphs

after we have deleted any edges from C which do not belong to proper cycles.

This will suffice to prove the first part of the lemma. For, if fewer than $m - 2$ tokens are in C then treating one of the unoccupied vertices as a "ghost token" and applying the above operation repeatedly, we can arrange for an input vertex to be unoccupied, allowing further input. Likewise, we can output any token from the network if it is occupied by $m - 2$ or fewer tokens. So, any word in $B(m - 2)$ can be produced by the network.

We refer to unoccupied vertices as *holes*. When we refer to the *distance from u to z* we mean the length of the shortest directed path from u to z. We will frequently make use of the fact that if a cycle C contains a hole, then we may advance the tokens in C along it by successively moving the hole backwards. When we wish to move a token t to a vertex v on the cycle by this method we will indicate this by the phrase *cycle t along C to v* or something similar.

Our assumption is that there are at least two holes. By moving holes backwards along paths we may assume that there are two holes at x and y either (Case 1) both at distance 1 from v, or (Case 2) x at distance 1, and y at distance 2 (along a path $v \rightarrow x \rightarrow y$). If either x or y is equal to w there is no problem (just move the token immediately). Otherwise, choose a cycle A containing the edge $v \rightarrow w$. If A contains a hole we're done, after moving it backwards along the cycle to w. So, suppose that A contains no hole. The critical situations in these two cases are illustrated in Figure 2 which may aid in visualising the details of the following arguments.

In case 1, choose a cycle A_x containing $v \rightarrow x$. If w belongs to A_x we can just use A_x in place of A in the preceding paragraph. So assume there is a vertex u of A not on A_x. Move the token a from v to x and move the hole from y backwards along the path from u to y, preserving the hole at v. Now cycle a back to v along A_x, reaching a state where A contains a hole.

Now consider case 2. Consider any cycle C_x containing $v \to x$. Then either (i) $y \in C_x$ or (ii) $y \notin C_x$.

(sub-case i) Again if every vertex of A belongs to this cycle we're done. If not, move a hole from this cycle to v, and then move this hole back along A to a vertex u not on C_x. Cycle C_x again until a returns to its original position, and A now contains a hole.

(sub-case ii) Cycle along C_x until there is a hole at v, but a does not occupy x. Move the token at x to y. This creates two holes at v and x. Move the hole at x onto A. Move a back onto A along C_x. Now A has a hole and we're done.

The final sentence of the lemma follows simply by considering the possible cases. A strongly cycle connected graph with 3 vertices contains a directed triangle, and trivially can produce any element of $B(2)$. If there are 4 vertices then either there is a directed 4-cycle, or two 3-cycles containing a common edge. In either case it is easy to check that all elements of $B(3)$ can be produced. □

The 5 vertex strongly connected graph consisting of two directed 3-cycles sharing a common vertex does not produce all of $B(4)$. So in general, the bound $m - 2$ in the lemma above cannot be improved. However, it is easy to see that, if G is a graph which has the property that for every edge $v \to w$, there is a path in $G \setminus v$ from w to every other vertex, then the language generated by a token passing network based on G contains $B(|G| - 1)$ since we can guarantee the movement of a token along any edge $v \to w$ provided there is at least one hole in the graph.

Corollary 3.2. *Let T be a standard token passing network of boundedness c. If t is the number of vertices of the underlying graph G belonging to directed cycles of length 3 or more, then $t < 3c$.*

Proof. Suppose for the sake of contradiction that $3c$ or more vertices of G belong to directed cycles of length at least 3. Consider the subgraph of G consisting only of edges which belong to cycles of length at least 3 (and their endpoints). Each component of this graph is strongly cycle connected. Let the components be $C_1, C_2, \ldots, C_k$ and note that $|C_1| +$

$|C_2| + \cdots + |C_k| \geq 3c$. Define

$$f(m) = \begin{cases} m - 2 & (m = 3, 4) \\ m - 3 & (m > 4). \end{cases}$$

Using the lemma above, for $1 \leq i \leq k$ we can place $f(|C_i|)$ tokens into C_i, and still be able to move the next token from input to output. However, as $f(m) \geq m/3$ for all m, it follows that $f(|C_1|) + f(|C_2|) + \cdots + f(|C_k|) \geq c$ so this contradicts the assumption that $\mathcal{T}$ is c-bounded. $\qquad \square$

Effectively, a strongly cycle connected subgraph of size m can safely "store" $f(m)$ tokens without interfering with the movement of tokens from input to output.

4 Token passing networks with fixed boundedness

The aim of this section is to prove Theorem 2.1. So, let a fixed positive real number c be given, and consider a language $L \subseteq B(c)$ produced by at least one token passing network of boundedness c. Among the networks which produce this language choose a standard one, $\mathcal{T}$, with the least possible number of vertices.

We will establish an upper bound, independent of L, on the number of vertices of $\mathcal{T}$. Of course this suffices to prove the theorem since there are only finitely many token passing networks of such sizes.

Choose a shortest directed path S:

$$i = v_0 \rightarrow v_1 \rightarrow \cdots \rightarrow v_m = o$$

from the input to the output vertex. We will call this the *spine* of $\mathcal{T}$. The complement of the spine will be called the *body*, its set of vertices will be denoted B, and the number of vertices in the body will be denoted b.

Since the boundedness of $\mathcal{T}$ is c, it follows that $b < c$. For, we can fill all the vertices of the body with tokens, working in reverse order of distance from i to avoid any possible blockages and then move the next token along the spine from input to output. If $b \geq c$ this would contradict the c-boundedness of $\mathcal{T}$.

We say that a vertex $v \in S$ *is spanned* if there is some vertex $w \in B$, and vertices $v_p \in S$ not following v and $v_f \in S$ not preceding v such that there is a directed path from v_p to w not meeting S except at v_p and one from w to v_f not meeting S except at v_f. Note that either v_p or v_f (or even both) might equal v.

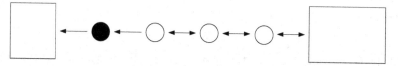

Fig. 3. Schematic of a token passing network containing two internal queues. The shaded node is an internal queue of length one. The boxes may contain additional structure.

For each vertex $w \in B$, there is an earliest vertex $v_i \in S$ which allows a directed path from v_i to w not meeting S except at v_i, and a latest vertex $v_t \in S$ allowing a directed path from w to v_t not meeting S except at v_t. Since the sum of the lengths of these paths is at most $2b$ the distance on S between v_i and v_t must be at most $2b$. In particular the number of elements of S which are spanned is less than $2c^2$.

Furthermore by the Corollary 3.2, fewer than $3c$ of the vertices of the spine can lie on cycles of length 3 or more.

Consider any vertex $v \in S$ which is neither spanned, nor belongs to any cycle of length 3 or more. We call such a vertex *queuelike* because the only possible edges having v as an endpoint are of the form:

$$u \leftrightarrow v \leftrightarrow w$$

where $u, w \in S$ are the predecessor and successor of v in S.

A maximal interval v_i through v_j of S consisting entirely of queuelike vertices with the additional property that for $i \leq k < j$ both the edges $v_i \rightarrow v_{i+1}$ and $v_{i+1} \rightarrow v_i$ are present in $\mathcal{T}$ will be called an *internal queue*. The structure of a token passing network containing an internal queue is illustrated in Figure 3 which also illustrates the fact that it is possible for an internal queue to consist of a single vertex. Note that every queuelike vertex belongs to a unique internal queue.

We now define the *backwards mobility* of an internal queue Q. This is the largest number k such that if tokens 1 through k are placed in order in Q (with 1 closest to the output vertex) then it is possible to rearrange them in Q in such a way that 1 is no longer the first element, using only the vertices of $\mathcal{T}$ between i and the end of Q.

Observation 4.1. The backwards mobility of an internal queue cannot exceed c.

Proof. Suppose that we had an internal queue Q of backwards mobility at least $c + 1$. Then certainly (by using virtual or ghost tokens if necessary) we could rearrange the initial contents 1 through $c + 1$ of Q into a sequence:

$$a, \ldots, 1, \ldots$$

where 1 does not occur in the first position. By leaving the elements before 1 fixed we could now move 1 further down the sequence. So, after a series of such movements we could rearrange the contents of Q into some permutation π ending with 1. By repeating this same series of movements we could generate all the powers of π, including π^{-1}. However, in π^{-1}, token $c + 1$ has been moved to the front of the queue. It could now be output from $\mathcal{T}$, contradicting c-boundedness. □

There is an analogous notion of *forwards mobility*, the largest number k such that if tokens 1 through k are placed in order in Q then it is possible to rearrange them in Q in such a way that k is no longer the last element, using only the vertices of $\mathcal{T}$ between the end of Q and o. By similar reasoning to the above, the forwards mobility of an internal queue is also at most c.

Observation 4.2. If the length of an internal queue is greater than the sum of its forwards and backwards mobilities, then it is possible to produce a token passing network generating the same language as $\mathcal{T}$ but having fewer vertices.

Proof. The key idea in proving this observation is that in such an internal queue, we can safely delete a vertex from the middle. It is instructive to begin with an extreme case of this observation, namely an internal queue whose forward and backward mobilities are both 0. Then, as soon as an object is added to the queue, because the backward mobility is 0, it will necessarily be the first to eventually leave the queue. So, we can reschedule any action involving addition of further elements to the queue until this element has finally left the queue. In fact, we can postpone its addition to the queue until the situation following the queue has been adjusted to the state at which this element leaves the queue for the final time. This then allows the element to simply be pushed directly through the queue. Thus the queue actually performs no useful function and can be short-circuited.

In the general case, suppose that we are about to add a token, a, to Q which already contains as many tokens as the sum of its forward and backward mobilities. This addition would create a token, k, in the middle of the queue whose order, relative to its successors and predecessors in the queue would have to remain fixed until some token had been removed permanently from the queue. The token k functions as a barrier between operations of T between i and itself, and operations between itself and o. So any operations of the latter type that precede the permanent removal of some element from Q can be advanced to occur before the addition of a to Q ensuring that at no point does Q ever contain more tokens than the sum of its forward and backward mobilities. Now the observation follows since vertices of Q in excess of this number are now superfluous to its effective operation. $\qquad\square$

It follows from these two observations that, under the conditions imposed on T, no internal queue has length greater than $2c$. In fact, no internal queue can be bounded at both ends by one way edges, for such a queue has zero mobility in either direction. Thus there can be at most two internal queues in any block of queuelike elements. Since there are at most $2c^2$ non-queuelike vertices on S and at most two internal queues each of size at most $2c$ between non-queuelike vertices, there can be at most $8c^3$ queuelike vertices in T, establishing Theorem 2.1.

5 Capacity restricted token passing networks

We now turn to the proof of Theorem 2.2. In this case the proof relies on a decomposition of the directed graph G underlying T. If we can identify part of the graph which can be operated in such a way as to simulate the effect of a buffer of size c, then $L(T_c)$ (respectively $L(T)$) must equal $B(c)$. Conversely, we will argue that if we cannot find such parts of the graph, then the structure of the graph as a whole must be very simple, and, if the graph is sufficiently large, parts could be pruned away without affecting its permutational power.

Our graph decomposition occurs on two levels. The first is fairly standard. Let a directed graph G be given. Define an equivalence relation on G by relating v and w if there is a walk from v to w and also one from w to v. The induced subgraphs on the equivalence classes of this relation are strongly connected, and are called the *strongly connected components* of G. Moreover, we can define an acyclic directed graph A on the quotient set, by connecting two equivalence classes V and W if there is an edge from some element of V to some element of W. Similarly, a strongly connected graph S (i.e. one having only a single strongly connected component) has a maximal quotient T which is a bi-directed tree (the only type of strongly connected graph which has no cycles of length at least 3). The equivalence classes of this quotient map are strongly cycle connected, and we will call them the strongly cycle connected components of S.

The proof of the theorem now follows a standard inductive approach as in the previous section. We begin with a language of the form $L(\mathcal{T}_c)$. Among all the token passing networks whose capacity c restrictions generate this language we choose a standard one $\mathcal{T}$ with underlying graph G of minimal size. We then argue that the sizes of its strongly cycle connected components, and the two quotients A and T can be bounded, establishing a bound on $|G|$. The result then follows.

For convenience suppose that $L(\mathcal{T}_c) \neq B(c)$. The following lemma is directed towards limiting the size of the directed acyclic quotient A of G.

Lemma 5.1. *Let D be a directed acyclic graph, and let i and o be specified vertices of D. Suppose that for every vertex v of D there is a directed path from i to v and also one from v to o. If D has a directed spanning tree from i which has c or more leaves, then $L(\mathcal{T}_c(D, \{i\}, \{o\})) = B(c)$.*

Proof. We prove this result by describing an algorithm for actually running this network to produce any given word $\omega \in B(c)$. Take an inward directed spanning tree T leading to the output vertex o, and identify in it c vertices which are the leaves of some outward directed spanning tree, S, from i. We will use these vertices as storage, and we shall arrange matters so that the following invariant properties are maintained:

- If a token is stored at v in T and w is a storage vertex such that the path in T from w to o passes through v then a token is also stored at w.
- If tokens are stored at v and w in T, and the path in T from v to o passes through w, then the token stored at w occurs earlier in ω than the one stored at v.

These properties certainly hold at the beginning of the run, since no tokens are stored. Suppose that we have reached some intermediate step of the run, and the next token of ω which we have not yet produced is ω_i. If ω_i is in storage, then by the second invariant property, the path from it to o is unblocked by other stored elements, and so we can move it to o, remove it from the network, and maintain both invariant properties.

Suppose now that ω_i has not yet been placed in storage. Consider the next input token, α, (which may not represent ω_i). Choose a currently unoccupied storage vertex, v, at the maximum available distance from o in T. If placing α at v maintains both invariants, then do so. If not, then consider the subtree of T rooted at v. All the storage vertices nearest the root are occupied. Among the occupants is one which occurs earliest in ω. Move this token from its current location v_1 to v. Now, if placing α at v_1 maintains both invariants then do so. Otherwise, consider the subtree of T rooted at v_1. Proceed as in the preceding step. Eventually, we must free a storage location at which α can be placed, since at worst, when we eventually must arrive at a leaf of T then we can place α there. Thus α can be added to storage while maintaining the invariants.

Continue this procedure for as long as necessary. In each phase we either store a new element, or output an element required next in ω. The invariants ensure that we never get blocked in any way, so eventually we will have succeeded in producing all of ω. □

Note that a minor modification of the proof establishes the same result when D has an inward directed spanning tree to o having c or more leaves.

Consider the graph A which is the directed acyclic quotient of G. By the lemma above and the remark following it we may assume the in-degree and out-degree of any vertex of A is at most c.

Choose an outward directed spanning tree T from i in A with the maximum possible number of leaves. By Lemma 5.1 this tree has at most $c-1$ leaves. Thus it also has at most $c-2$ branch vertices (vertices which do not have degree 2 consisting of an in-edge and an out-edge). Consider any segment, S, between a branch vertex and a leaf, or between two branch vertices. There cannot be incoming edges to this segment from off the segment at as many as c vertices. If there were, we could use these c vertices as a form of storage to produce all of $B(c)$ (the argument is much the same as, but simpler than, that for Lemma 5.1). Moreover, the in-degree of any vertex in A is at most c, thus there are fewer than c^2 incoming edges to S from off S. However, there cannot be any edges internal to S other than the ones belonging to S, for the existence of such an edge would create either a cycle (impossible as A is acyclic), or the possibility of creating a spanning tree with more leaves.

The total number of segments is at most $2c-4$, hence the total number of edges of A which are not edges of T is bounded above by $(2c-4)c^2$. Suppose that there were a segment longer than $(2c-4)c^2M$ (for a value of M to be chosen later). Then this segment would contain a sequence of M consecutive vertices, each of degree exactly two in A. These vertices represent strongly connected components of G, so their actual structure may be somewhat more complex. However, each such component which is anything other than a simple 2-way path, connected only at its two ends within G is capable of storing at least one element. Thus, there cannot be as many as c vertices of this segment which represent components not of this type. Choosing $M > c^2$ we find more than c consecutive vertices of the segment which represent either individual vertices of G, or two way paths connected only at their endpoints. Any manipulation of c or fewer tokens along such a path can be carried out equally well when one of the vertices of the path is deleted, yielding a contradiction.

Now it remains only to show that the strongly connected components of G which are not simple two way paths connected only at their endpoints, can also contribute only a bounded amount to the size of G.

Consider a bi-directed tree T which is the quotient of such a strongly connected component of G by identifying vertices belonging to a common strongly cycle connected components. If T contains $c+1$ or more leaves, then it can store c items, and so produces $B(c)$. On the other hand if it contains at most c leaves, then it can contain at most $c-1$ branch vertices. Consider once more, a segment S of T.

As in the previous argument, there cannot be as many as c vertices of S which have an incoming edge from off S. Finally, by Lemma 3.1 there can not be as many as c vertices of S that represent non-trivial strongly cycle connected components. So, if S has more than $2c^2$ elements, then there is a block of more than c consecutive elements of S which are actually vertices of G and whose only adjacencies in G are the edges involving them in S. One of these can be deleted without affecting the language produced by $\mathcal{T}_c$, giving a contradiction.

Since the total size of the strongly cycle connected components of G is already known to be bounded, we have succeeded in establishing Theorem 2.2.

6 Examples

We illustrate some of the preceding results with some example networks. In all cases the illustrated networks are standard ones. This section makes certain claims of exhaustiveness, and of the forms of permutations provided by various networks without including any proofs. This is because the proofs for exhaustiveness consist of considering the arguments of the previous two sections in the cases $c = 2$ and $c = 3$ and performing some obvious simplifications to reduce the number of networks that need be considered. Then GAP was used to generate these networks exhaustively, and construct their corresponding automata, which

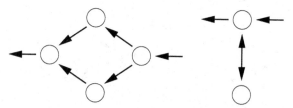

Fig. 4. Simple token passing networks producing all 2-bounded permutations.

were then minimised. A representative of each type was then chosen for illustrative purposes. Likewise, the non-obvious claims about the basis elements are also obtained through implementation in GAP of the methods of [1].

The 2-bounded pattern classes produced by token passing networks are not very interesting. Aside from the 1-bounded class, there is only one, which is the class of all 2-bounded permutations. Two simple networks producing it are shown in Figure 4.

Following the proofs of Theorems 2.1 and 2.2 for the case of 3-bounded classes shows that the underlying networks cannot be very complex. There are in fact precisely five 3-bounded classes that can be produced by token passing networks (and considering capacity boundedness does not provide any others). The five classes are:

(A) All 3-bounded permutations. These can be produced by adding an extra vertex in the middle section of the left hand network in Figure 4, or by combining two copies of the right hand network in series, as well as by many other networks.

(B) The 3-bounded permutations avoiding the pattern 321. These can be produced by two queues in parallel.

(C) The 3-bounded permutations avoiding the pattern 312. These can be produced by a stack.

(D) The 3-bounded permutations avoiding both the pattern 31542 and 32541.

(E) A class whose basis is infinite, given in the language of the rank encoding by:

$$322321, \ 3213(31)^*321.$$

Networks producing the latter four classes are shown in Figure 5.

It is tempting to suppose that if the next token required by the output sequence is already in the network, then it can be output before any further tokens are added. However, a notable feature of the bottom

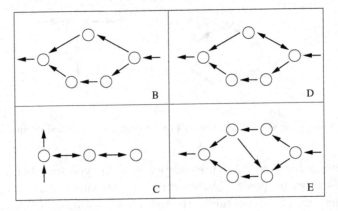

Fig. 5. Examples of networks producing each possible non-universal 3-bounded class.

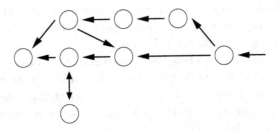

Fig. 6. A network which produces the 4-bounded permutation with encoding 4222434111 but not without holding at least five tokens at some point.

right network in Figure 5 is that it produces the permutation 32541, but in doing so, the element 2 cannot be output before the element 4 is added to the network.

Our final example is a token passing network $\mathcal{T}$ which has the property that:

$$L(\mathcal{T}_4) \neq L(\mathcal{T}) \bigcap B(4).$$

The network shown in Figure 6 can produce the permutation

$$4\,2\,3\,5\,8\,7\,10\,1\,6\,9$$

whose rank encoding is:

$$4\,2\,2\,2\,4\,3\,4\,1\,1\,1.$$

However, it cannot do so without at some point having five tokens in the network, namely token 10 must be added before token 7 is output. As this network has a boundedness of 5 this still leaves open the question of whether it is ever necessary to add more tokens to the network than its boundedness.

7 Summary and conclusions

We have shown that the permutational power of token passing networks is relatively limited, at least in the variety of classes of permutations that they can produce. For the sake of simplicity, the arguments we used in proving Theorems 2.1 and 2.2 were extremely conservative in the numerical bounds derived. In many cases multiplicative bounds could have been replaced by additive ones. So the actual size of the smallest network producing any given c-bounded language (if such a network exists) is quite small. This allowed us to determine the complete catalogue of such languages for $c = 3$. A similar catalogue for $c = 4$ would probably be feasible, though the collection of bases is already much richer in this case.

We have skirted the issue of the complexity of determining $L(\mathcal{T})$ given $\mathcal{T}$. The underlying non deterministic automaton has, potentially, $|G|!$ states, although in practice many of these are unreachable and so need never be considered. The equivalent minimal deterministic automaton often exhibits a very straightforward structure. Determining the basis elements using the methods of [1] requires several applications of non-deterministic transducers, complementation, and re-determinization and so is of exponential complexity in the worst case, and empirically also in practice. However, the final automaton produced by this procedure is generally quite small. It has been possible, by various ad hoc methods, to extend the practical range in which these computations can be carried out. Still, a general theoretical understanding of why the deterministic automata for the language and its basis are so simple is lacking as is a method to exploit this apparent phenomenon in constructing one from the other.

The proof of Lemma 5.1 contains an algorithm for the efficient solution of the following problem:

Given A directed acyclic graph G with a specified input vertex i, and output vertex o and such that for any vertex v there is a directed path from i to v and from v to o, together with a positive integer c, not greater than the number of leaves of some directed spanning tree of G rooted at i.

Problem Use this graph to sort an incoming sequence of packets provided only with the guarantee that no packet will be preceded by c or more packets which should follow it.

This problem is the inverse view of the problem addressed in Lemma 5.1 and the algorithm provided in the proof of the lemma provide an online linear time solution of it in which no packet ever moves more than $2|G|$ times.

We have seen that the relationship between $L(\mathcal{T}_c)$ and $L(\mathcal{T}) \cap B(c)$ is not entirely simple. However, a range of questions of a similar character remain open. For example, given a token passing network of boundedness c what is the minimum number of tokens we must allow in the network to guarantee producing every permutation in its language?

References

[1] M. H. Albert, M. D. Atkinson, and N. Ruškuc. Regular closed sets of permutations. *Theoret. Comput. Sci.*, 306(1-3):85–100, 2003.

[2] M. D. Atkinson, M. J. Livesey, and D. Tulley. Permutations generated by token passing in graphs. *Theoret. Comput. Sci.*, 178(1-2):103–118, 1997.

[3] V. Auletta, A. Monti, M. Parente, and P. Persiano. A linear-time algorithm for the feasibility of pebble motion on trees. *Algorithmica*, 23(3):223–245, 1999.

[4] V. Auletta and P. Persiano. Optimal pebble motion on a tree. *Inform. and Comput.*, 165(1):42–68, 2001.

[5] D. E. Knuth. *The art of computer programming. Vol. 1: Fundamental algorithms*. Addison-Wesley Publishing Co., Reading, Mass., 1969.

[6] C. H. Papadimitriou, P. Raghavan, M. Sudan, and H. Tamaki. Motion planning on a graph (extended abstract). In S. Goldwasser, editor, *35th Annual Symposium on Foundations of Computer Science*, pages 511–520. IEEE, 1994.

[7] V. R. Pratt. Computing permutations with double-ended queues, parallel stacks and parallel queues. In *STOC '73: Proceedings of the fifth annual ACM symposium on Theory of computing*, pages 268–277, New York, NY, USA, 1973. ACM Press.

[8] D. Ratner and M. Warmuth. The $(n^2 - 1)$-puzzle and related relocation problems. *J. Symbolic Comput.*, 10(2):111–137, 1990.

[9] R. Tarjan. Sorting using networks of queues and stacks. *J. Assoc. Comput. Mach.*, 19:341–346, 1972.

[10] R. M. Wilson. Graph puzzles, homotopy, and the alternating group. *J. Combinatorial Theory Ser. B*, 16:86–96, 1974.

Problems and conjectures presented at the problem session

assembled by Vincent Vatter

1 A very brief introduction to permutation patterns

We say a permutation π *contains* or *involves* the permutation σ if deleting some of the entries of π gives a permutation that is order isomorphic to σ, and we write $\sigma \le \pi$. For example, 534162 contains 321 (delete the values 4, 6, and 2). A permutation *avoids* a permutation if it does not contain it.

This notion of containment defines a partial order on the set of all finite permutations, and the downsets of this order are called *permutation classes*. For a set of permutations B define $\mathrm{Av}(B)$ to be the set of permutations that avoid all of the permutations in B. Clearly $\mathrm{Av}(B)$ is a permutation class for every set B, and conversely, every permutation class can be expressed in the form $\mathrm{Av}(B)$.

For the problems we need one more bit of notation. Given permutations π and σ of lengths m and n, respectively, their *direct sum*, $\pi \oplus \sigma$, is the permutation of length $m+n$ in which the first m entries are equal to π and the last n entries are order isomorphic to σ while their *skew sum*, $\pi \ominus \sigma$, is the permutation of length $m+n$ in which the first m entries are order isomorphic to π while the last n entries are equal to π. For example, $231 \oplus 321 = 231654$ and $231 \ominus 321 = 564321$.

2 Growth rates

We define the *growth rate* of a permutation class $\mathrm{Av}(B)$ as

$$\mathrm{gr}(\mathrm{Av}(B)) = \limsup_{n \to \infty} \sqrt[n]{|\mathrm{Av}(B) \cap S_n|},$$

where S_n denotes the set of all permutations of length n. (This limit supremum is not known to be a limit except in special cases (e.g., when $|B| = 1$) where a supermultiplicativity argument applies, see Arratia [5];

339

it is, however, known to be finite so long as $B \neq \emptyset$ by the Marcus-Tardos Theorem [16].)

Question 2.1 (contributed by Mike Atkinson). For all permutations β and $k \geq 1$, do we have $\mathrm{gr}(\mathrm{Av}(k \cdots 21, \beta)) = \mathrm{gr}(\mathrm{Av}(k \cdots 21, 1 \oplus \beta))$?

Subsequent research conducted at the University of Otago (and communicated to this author by Albert) provides a generalization of this question for $k = 3$: they proved that $\mathrm{gr}(\mathrm{Av}(321, \alpha \oplus \beta)) = \mathrm{gr}(\mathrm{Av}(321, \alpha \oplus 1 \oplus \beta))$ for all permutations α and β (to get the $k = 3$ case of Problem 2.1, take α to be the empty permutation)†. It is also not hard to see that

$$\mathrm{gr}(\mathrm{Av}(k \cdots 21, \alpha \oplus 1 \oplus \beta)) = \mathrm{gr}(\mathrm{Av}(k \cdots 21, \alpha \oplus 12 \oplus \beta))$$

for all k, α, and β. (Communicated by Atkinson.)

For *principally based classes*, i.e., classes of the form $\mathrm{Av}(\beta)$ for a single permutation β, there have been a series of conjectures which we briefly recap. For any $\pi \in S_3$, the growth rate of $\mathrm{Av}(\pi)$ is 4, because the π-avoiding permutations are counted by the Catalan numbers. The growth rate of $\mathrm{Av}(12 \ldots k)$ is $(k - 1)^2$ by Regev [19]. It was an old conjecture that the growth rate of $\mathrm{Av}(\pi)$ is $(k - 1)^2$ for all $\pi \in S_k$, but this was disproved by Bóna [10], who gave an exact enumeration of $\mathrm{Av}(1342)$ which shows that its growth rate is 8. It is also tempting from this data to conjecture that the growth rate of $\mathrm{Av}(\pi)$ is always an integer, or at least rational; Bóna [12] disproved this by showing that the growth rate of $\mathrm{Av}(12453)$ is $9 + 4\sqrt{2}$.

Question 2.2 (contributed by Miklós Bóna). Are all growth rates of principally based classes algebraic integers?

Note that Vatter [23], building on the work of Albert and Linton [4], has constructed permutation classes of every growth rate at least 2.48188, so there are permutation classes with non-algebraic growth rates (though they needn't be principally based). The obvious candidate to provide a negative answer to Question 2.2 is the class $\mathrm{Av}(1324)$, for which we know only that $\mathrm{gr}(\mathrm{Av}(1324)) > 9.47$ (proved by Albert, Elder, Rechnitzer, Westcott, Zabrocki [3], this bound disproved Arratia's earlier conjecture [5] that $\mathrm{gr}(\mathrm{Av}(\beta)) \leq (|\beta| - 1)^2$.)

---

† A generalization of this result appears in Albert et al. [2].

3 Sorting

A stack is a last-in first-out linear sorting device with push and pop operations. The greedy algorithm for stack sorting a permutation $\pi = \pi(1)\pi(2)\ldots\pi(n)$ goes as follows. First we push $\pi(1)$ onto the stack. Now suppose at some later stage that the letters $\pi(1),\ldots,\pi(i-1)$ have all been either output or pushed on the stack, so we are reading $\pi(i)$. We push $\pi(i)$ onto the stack if and only if $\pi(i)$ is lesser than any element on the stack. Otherwise we pop elements off the stack until $\pi(i)$ is less than any remaining stack element and we push $\pi(i)$ onto the stack. For sorting with one stack this *greedy algorithm* is optimal, and its analysis quickly leads one to the conclusion that a single stack can sort precisely those permutations in Av(231).

West [25] considered the permutations that can be sorted by using the above greedy algorithm twice. He proved that the permutations sortable by this algorithm are those that avoid 2341 and in which every copy of 3241 and be extended by a single entry to a copy of 35241. However, this algorithm is *not* optimal, i.e., there are permutations that can be sorted by two stacks in series but cannot be sorted in this manner. In fact, the following question remains open:

Question 3.1. Is it decidable in polynomial time (in n) if the permutation π of length n can be sorted by two stacks in series?

Atkinson, Murphy, and Ruškuc [6] considered a sorting machine consisting of two stacks in series, subject to the restriction that the entries in each of the stacks must remain ordered. They presented an optimal algorithm for sorting with such a machine, showed that the class of sortable permutations is

$$\text{Av}(\{2, 2k-1, 4, 1, 6, 3, 8, 5, \ldots, 2k, 2k-3 : k \geq 2\}),$$

and constructed a bijection between this set and Av(1342). This suggests the following question.

Question 3.2 (contributed by Miklós Bóna). Is there a natural sorting machine / algorithm which can sort precisely the class Av(1342)?

For more information on stack sorting we refer the reader to Bóna's survey [11], and for more open problems we refer to the problems presented at Permutation Patterns 2005 [14].

4 Wilf-equivalence

We say that the sets B_1 and B_2 are *Wilf-equivalent* if $|\text{Av}_n(B_1)| = |\text{Av}_n(B_2)|$ for all natural numbers n, that is, if B_1 and B_2 are equally avoided. Clearly natural symmetries of permutation classes give Wilf-equivalences, but many nontrivial Wilf-equivalences have been found. For example, it is a classic result that every permutation in S_3 is Wilf-equivalent to every other permutation in S_3. Another example is given above: $\{2, 2k-1, 4, 1, 6, 3, 8, 5, \ldots, 2k, 2k-3 : k \geq 2\}$ is Wilf-equivalent to 1342. To date, the following problem remains open:

Problem 4.1. Find necessary and sufficient conditions for two permutations to be Wilf-equivalent.

There has been considerable work on the sufficient conditions front:

- Stankova [22] constructed a bijection between the generating trees of Av(4132) and Av(3142), establishing that 4132 and 3142 are Wilf-equivalent,
- Stankova and West [21] proved that $231 \ominus \beta$ and $132 \ominus \beta$ are Wilf-equivalent for all permutations β,
- Backelin, West, and Xin [9] proved that, for all k and β, $12 \cdots k \oplus \beta$ and $k \cdots 21 \oplus \beta$ are Wilf-equivalent (thus generalizing the results of West [24] and Babson and West [8]).

These results, together with computer calculations, complete the classification of singleton Wilf-equivalences up to and including permutations of length 7.

Necessary conditions, on the other hand, have so far been lacking, and the only general way to show that α and β are *not* Wilf-equivalent is to compute $|\text{Av}_n(\alpha)|$ and $|\text{Av}_n(\beta)|$ until they disagree.

More general necessary conditions would require results on permutations of small "codimension" that contain α and β. More precisely, consider then function

$$g_k(\beta) = |\{\text{permutations } \pi \text{ of length } |\beta| + k \text{ which contain } \beta\}|.$$

We have

$$|\text{Av}_n(\beta)| = n! - g_{n-|\beta|}(\beta),$$

so α and β are not Wilf-equivalent if and only if $g_k(\alpha)$ and $g_k(\beta)$ differ for some k.

For a permutation β of length k, Pratt [17] seems to have been the first to observe that

$$g_1(\beta) = k^2 + 1$$

while Ray and West [18] show that

$$g_2(\beta) = \left(k^4 + 2k^3 + k^2 + 4k + 4 - 2j\right)/2$$

for some $0 \le j \le k - 1$.

Problem 4.2 (contributed by Vince Vatter). Express the quantity j above in terms of statistics of β.

Problem 4.3 (contributed by Vince Vatter). Find a formula for $g_3(\beta)$.

5 Long subsequences

The longest increasing subsequence (LIS) problem asks (in our language) for the greatest k such that a given permutation π of length n contains $12 \cdots k$. The fastest algorithm for computing the LIS is $O(n \log n)$, due to Schensted [20], and this bound is essentially best possible. Albert et al. [1] studied the longest X-subsequence (LXS) problem, which asks, for a set X of permutations, for the longest member of X that a permutation π of length n contains. They presented $O(n^2 \log n)$ algorithms to compute the LXS for all cases where $X = \mathrm{Av}(B)$ and $B \subset S_3$ except for the case $X = \mathrm{Av}(231)$, where they gave a dynamic programming algorithm with runtime $O(n^5)$.

Problem 5.1 (contributed by Michael Albert). Give a faster algorithm for the $X = \mathrm{Av}(231)$ case of the LXS problem.

6 Generalized patterns

A *generalized* (also known as *blocked*, *gapped*, or *Babson-Steingrímsson*, after their inventors [7]) pattern is one including dashes indicating the entries that need not occur consecutively (recall that no entries need occur consecutively in the normal pattern-containment order). For example, 24135 contains only one copy of 1-23, namely 235; the entries 245 do not form a copy of 1-23 because 4 and 5 are not adjacent.

In some cases, e.g., 2-1 and 21 or 2-31 and 2-3-1, avoiding a generalized pattern is equivalent to avoiding the underlying classical pattern. This leads us to:

Question 6.1 (contributed by Einar Steingrímsson). For which generalized patterns β is avoiding β equivalent to avoiding the underlying classical pattern? For which pairs of sets B_1, B_2 of generalized patterns is avoiding B_1 and B_2 equivalent?

See Hardarson [15] for some recent progress on these questions.

7 Permutations of special form

Let D_n denote the set of permutation matrices of dimension $2n-1 \times 2n-1$ where ones can appear either on or below the main diagonal or on or below the main diagonal of the $n \times n$ submatrix in the upper right-hand corner. For example, the cells where ones are allowed for $n = 5$ are denoted by $*$ in the matrix below.

$$\begin{pmatrix} * & 0 & 0 & 0 & 0 & * & 0 & 0 & 0 \\ * & * & 0 & 0 & 0 & * & * & 0 & 0 \\ * & * & * & 0 & 0 & * & * & * & 0 \\ * & * & * & * & 0 & * & * & * & * \\ * & * & * & * & * & 0 & 0 & 0 & 0 \\ * & * & * & * & * & * & 0 & 0 & 0 \\ * & * & * & * & * & * & * & 0 & 0 \\ * & * & * & * & * & * & * & * & 0 \\ * & * & * & * & * & * & * & * & * \end{pmatrix}$$

Burstein and Stromquist [13] have proved that $|D_n|$ is the nth Genocchi number.

Problem 7.1 (contributed by Alex Burstein). Give a bijective proof that D_n is enumerated by the Genocchi numbers.

Finally, we return to Problem 2.1. The proof in the $k = 3$ case relies on the following notion: the permutation π is said to be k-rigid if every entry of π participates in some copy of $k \cdots 21$. Asymptotically, 4/9th of the permutations in Av(321) are 2-rigid. However, it is not known if this behavior continues:

Problem 7.2 (contributed by Mike Atkinson). Prove that a positive fraction of the permutations in Av$((k + 1) \cdots 21)$ are k-rigid.

Acknowledgments. In addition to the contributors of the problems, I would like to thank Anders Claesson, Sergey Kitaev, and Nik Ruškuc for their help in assembling this list.

References

[1] M. H. Albert, R. E. L. Aldred, M. D. Atkinson, H. P. van Ditmarsch, B. D. Handley, C. C. Handley, and J. Opatrny. Longest subsequences in permutations. *Australas. J. Combin.*, 28:225–238, 2003.

[2] M. H. Albert, M. D. Atkinson, R. Brignall, N. Ruškuc, R. Smith, and J. West. Growth rates for subclasses of Av(321). arXiv:0903.1999 [math.CO].

[3] M. H. Albert, M. Elder, A. Rechnitzer, P. Westcott, and M. Zabrocki. On the Wilf-Stanley limit of 4231-avoiding permutations and a conjecture of Arratia. *Adv. in Appl. Math.*, 36(2):95–105, 2006.

[4] M. H. Albert and S. Linton. Growing at a perfect speed. *Combin. Probab. Comput.*, 18:301–308, 2009.

[5] R. Arratia. On the Stanley-Wilf conjecture for the number of permutations avoiding a given pattern. *Electron. J. Combin.*, 6:Note, N1, 4 pp., 1999.

[6] M. D. Atkinson, M. M. Murphy, and N. Ruškuc. Sorting with two ordered stacks in series. *Theoret. Comput. Sci.*, 289(1):205–223, 2002.

[7] E. Babson and E. Steingrímsson. Generalized permutation patterns and a classification of the Mahonian statistics. *Sém. Lothar. Combin.*, 44:Article B44b, 18 pp., 2000.

[8] E. Babson and J. West. The permutations $123p_4 \cdots p_m$ and $321p_4 \cdots p_m$ are Wilf-equivalent. *Graphs Combin.*, 16(4):373–380, 2000.

[9] J. Backelin, J. West, and G. Xin. Wilf-equivalence for singleton classes. *Adv. in Appl. Math.*, 38(2):133–148, 2007.

[10] M. Bóna. Exact enumeration of 1342-avoiding permutations: a close link with labeled trees and planar maps. *J. Combin. Theory Ser. A*, 80(2):257–272, 1997.

[11] M. Bóna. A survey of stack-sorting disciplines. *Electron. J. Combin.*, 9(2):Article 1, 16 pp., 2003.

[12] M. Bóna. The limit of a Stanley-Wilf sequence is not always rational, and layered patterns beat monotone patterns. *J. Combin. Theory Ser. A*, 110(2):223–235, 2005.

[13] A. Burstein and W. Stromquist. Dumont permutations of the third kind. Extended abstract, FPSAC 2007.

[14] M. Elder and V. Vatter. Problems and conjectures presented at the Third International Conference on Permutation Patterns, University of Florida, March 7–11, 2005. arXiv:0505504 [math.CO].

[15] M. T. Hardarson. Avoidance of partially ordered generalized patterns of the form k-σ-k. arXiv:0805.1872v1 [math.CO].

[16] A. Marcus and G. Tardos. Excluded permutation matrices and the Stanley-Wilf conjecture. *J. Combin. Theory Ser. A*, 107(1):153–160, 2004.

[17] V. R. Pratt. Computing permutations with double-ended queues, parallel stacks and parallel queues. In *STOC '73: Proceedings of the fifth annual ACM symposium on Theory of computing*, pages 268–277, New York, NY, USA, 1973.

[18] N. Ray and J. West. Posets of matrices and permutations with forbidden subsequences. *Ann. Comb.*, 7(1):55–88, 2003.

[19] A. Regev. Asymptotic values for degrees associated with strips of Young diagrams. *Adv. in Math.*, 41(2):115–136, 1981.

[20] C. Schensted. Longest increasing and decreasing subsequences. *Canad. J. Math.*, 13:179–191, 1961.

[21] Z. Stankova and J. West. A new class of Wilf-equivalent permutations. *J. Algebraic Combin.*, 15(3):271–290, 2002.

[22] Z. E. Stankova. Forbidden subsequences. *Discrete Math.*, 132(1-3):291–316, 1994.

[23] V. Vatter. Permutation classes of every growth rate above 2.48188. *Mathematika*, 56:182–192, 2010.

[24] J. West. *Permutations with forbidden subsequences and stack-sortable permutations*. PhD thesis, M.I.T., 1990.

[25] J. West. Sorting twice through a stack. *Theoret. Comput. Sci.*, 117(1-2):303–313, 1993.

Printed in the United States
by Baker & Taylor Publisher Services